Electroanalysis in Biomedical and Pharmaceutical Sciences

Monographs in Electrochemistry

Series Editor: Fritz Scholz, University of Greifswald, Germany

Surprisingly, a large number of important topics in electrochemistry is not covered by up-to-date monographs and series on the market, some topics are even not covered at all. The series Monographs in Electrochemistry fills this gap by publishing indepth monographs written by experienced and distinguished electrochemists, covering both theory and applications. The focus is set on existing as well as emerging methods for researchers, engineers, and practitioners active in the many and often interdisciplinary fields, where electrochemistry plays a key role. These fields will range – among others – from analytical and environmental sciences to sensors, materials sciences and biochemical research.

More information about this series at
http://www.springer.com/series/7386

Sibel A. Ozkan • Jean-Michel Kauffmann •
Petr Zuman

in collaboration with
Ana Maria Oliveira Brett, Christopher Brett,
Bengi Uslu, Philippe Hubert, Eric Rozet,
Cobra Parsajoo, Stéphanie Patris, Ahmad Sarakbi

Electroanalysis in Biomedical and Pharmaceutical Sciences

Voltammetry, Amperometry, Biosensors, Applications

Springer

Sibel A. Ozkan
Faculty of Pharmacy
Department of Analytical Chemistry
Ankara University
Tandogan, Ankara
Ankara
Turkey

Jean-Michel Kauffmann
Faculty of Pharmacy
Université Libre de Bruxelles
Brussels, Belgium

Petr Zuman
Department of Chemistry and Biomolecular Science
Clarkson University
Potsdam, New York
USA

ISSN 1865-1836 ISSN 1865-1844 (electronic)
Monographs in Electrochemistry
ISBN 978-3-662-47137-1 ISBN 978-3-662-47138-8 (eBook)
DOI 10.1007/978-3-662-47138-8

Library of Congress Control Number: 2015944207

Springer Heidelberg New York Dordrecht London

Printed on acid-free paper

Springer-Verlag GmbH Berlin Heidelberg is part of Springer Science+Business Media (www.springer.com)

Preface

In the modern pharmaceutical industry, selective, sensitive, and fully validated analytical methods for drug analysis are major tools applied at all stages of drug discovery, development, and production. Fast and effective development of rugged analytical methods is efficiently undertaken with a thorough understanding of their principles, theory, and instrumentation.

The main focus of this book is devoted to electroanalytical methods applied to drug compound analysis. New electrode materials, sensors, biosensors, separation methods with EC detection, their theory, advantages, and limitations have been reported.

The unique performance, small size, and low cost of electrochemical devices have led to many useful measuring systems. Given the impressive progress in electroanalytical chemistry and its growing impact on drug analysis, this book offers also an up-to-date, easy-to-read presentations of numerous recent applications in drug compound analysis.

In preparing this volume, we have tried to find an appropriate balance between theory and practice, between "classical" and "modern" methods of electroanalysis, and between electroanalytical methods. Only brief consideration, however, has been given on sampling, method optimization, data analysis, and statistics.

This textbook contains nine contemporary chapters, sharing one goal in electroanalysis of pharmaceuticals. In the first chapter, an overview of drug analysis is presented to situate electrochemical methods in their analytical context. Chapter 2 contains polarography and its application on pharmaceuticals. The following chapter includes modern voltammetric methods which are used most frequently in drug analysis (Chap. 3). In Chap. 4, solid electrodes in detail for drug analysis are described. Screen printed electrodes (SPE) are presented for the determination of drug active compounds in Chap. 5, and the next chapter describes electrochemical biosensors used for drug analysis (Chap. 6). Recent developments in the field of separation methods, specifically liquid chromatography (LC) and flow injection analysis (FIA) with electrochemical detection and the hyphenated techniques related with it, are summarized in Chap. 7. Validation of electroanalytical methods in pharmaceutical analysis, necessary parameters, and their calculations are presented in Chap. 8. The last chapter includes examples of electrochemical application on drug assays.

Readers from various fields will likely find new ideas and approaches to solve typical analytical problems in this volume.

The completion of this book could not have been possible without the help, inspiration, and encouragement from many people. Last but not least, we warmly acknowledge the gracious support of our families, and we are very grateful to them for their understanding and support throughout the entire process of writing and editing.

We would like to thank colleagues (Mehmet Gumustas, Sevinc Kurbanoglu, Burcin Palabiyik, Nurgul Karadas, and Burcu D. Topal) of Sibel A Ozkan for their valuable help. We would like to special thanks to Dr. Heike Kahlert who made useful and valuable comments based on a detailed addition to Chap. 8.

Finally, we would like to express our sincere gratitude to a unique set of internationally leading authors, who accepted our invitation to join us and committed their valuable time and efforts to guarantee the success of this book.

We would like to express our appreciation and special thanks to Fritz Scholz for his accurate and inspired revision and continuous support and enthusiasm in editing this textbook.

Ankara, Turkey Sibel A. Ozkan
Brussels, Belgium Jean-Michel Kauffmann
Potsdam, NY Petr Zuman

Authors of Chapters

The chapters have been written by the following authors:

Chapter 1

Sibel A. Ozkan Faculty of Pharmacy, Department of Analytical Chemistry, Ankara University, Ankara, Turkey

Petr Zuman Department of Chemistry, Clarkson University, Potsdam, NY, USA

Chapter 2

Sibel A. Ozkan Faculty of Pharmacy, Department of Analytical Chemistry, Ankara University, Ankara, Turkey

Petr Zuman Department of Chemistry, Clarkson University, Potsdam, NY, USA

Chapter 3

Sibel A. Ozkan Faculty of Pharmacy, Department of Analytical Chemistry, Ankara University, Ankara, Turkey

Ana Maria Oliveira-Brett Faculty of Science and Technology, Department of Chemistry, Coimbra University, Coimbra, Portugal

Christopher M.A. Brett Faculty of Science and Technology, Department of Chemistry, Coimbra University, Coimbra, Portugal

Chapter 4

Sibel A. Ozkan Faculty of Pharmacy, Department of Analytical Chemistry, Ankara University, Ankara, Turkey

Christopher M.A. Brett Faculty of Science and Technology, Department of Chemistry, Coimbra University, Coimbra, Portugal

Ana Maria Oliveira-Brett Faculty of Science and Technology, Department of Chemistry, Coimbra University, Coimbra, Portugal

Chapter 5

Stéphanie Patris Faculty of Pharmacy, Lab. Instrumental Analysis and Bioelectrochemistry, Université Libre de Bruxelles, Bruxelles, Belgium

Jean-Michel Kauffmann Faculty of Pharmacy, Lab. Instrumental Analysis and Bioelectrochemistry, Université Libre de Bruxelles, Bruxelles, Belgium

Chapter 6

Cobra Parsajoo Faculty of Pharmacy, Lab. Instrumental Analysis and Bioelectrochemistry, Université Libre de Bruxelles, Bruxelles, Belgium

Jean-Michel Kauffmann Faculty of Pharmacy, Lab. Instrumental Analysis and Bioelectrochemistry, Université Libre de Bruxelles, Bruxelles, Belgium

Chapter 7

Sibel A. Ozkan Faculty of Pharmacy, Department of Analytical Chemistry, Ankara University, Ankara, Turkey

Bengi Uslu Faculty of Pharmacy, Department of Analytical Chemistry, Ankara University, Ankara, Turkey

Ahmad Sarakbi Faculty of Pharmacy, Lab. Instrumental Analysis and Bioelectrochemistry, Universite Libre de Bruxelles, Bruxelles, Belgium

Jean-Michel Kauffmann Faculty of Pharmacy, Lab. Instrumental Analysis and Bioelectrochemistry, Universite Libre de Bruxelles, Bruxelles, Belgium

Chapter 8

Sibel A. Ozkan Faculty of Pharmacy, Department of Analytical Chemistry, Ankara University, Ankara, TURKEY

Jean-Michel Kauffmann Faculty of Pharmacy, Lab. Instrumental Analysis and Bioelectrochemistry, Universite Libre de Bruxelles, Bruxelles, Belgium

Philippe Hubert Department of Pharmacy, CIRM, Laboratory of Analytical Chemistry, Université de Liège (ULg), Liège, Belgium

Eric Rozet Department of Pharmacy, CIRM, Laboratory of Analytical Chemistry, Université de Liège (ULg), Liège, Belgium

Chapter 9

Bengi Uslu Faculty of Pharmacy, Department of Analytical Chemistry, Ankara University, Ankara, Turkey

Sibel A. Ozkan Faculty of Pharmacy, Department of Analytical Chemistry, Ankara University, Ankara, Turkey

Jean-Michel Kauffmann Faculty of Pharmacy, Lab. Instrumental Analysis and Bioelectrochemistry, Université Libre de Bruxelles, Bruxelles, Belgium

Contents

Chapter 1
Introduction

Contents

1.1 Introduction

In the process of pharmaceutical research, aiming at the drug discovery (that leads to invention of new lead compounds), in the drug development (characterization of the candidate compounds and their clinical tests), as well as in the control of the manufacture of drugs, reliable analytical procedures are needed. In some countries, the final products must be analyzed in pharmacies as well.

By far, the most frequently used analytical methods in pharmaceutical research, development, and quality control, are separation-based methods, in particular HPLC and its variants [1–3]. The electrochemical methods are currently in more frequently used, except in the Middle East and Asia, where the main driving force is the considerably lower cost of the electrochemical equipment. To be more widely used, the electroanalytical methods must offer for a given analytical problem some considerable advantages. Among those perhaps the dominant are situations, in which it is necessary to analyze very low concentrations (up to three orders of magnitude below millimolar) or analyses of very small amounts of the working compounds. Another advantage of electroanalytical methods is the high reproducibility of some electroanalytical procedures with results accurate to $\pm 3\ \%$ or if needed even better. Therefore, electroanalytical methods can offer reliable results for comparison to those obtained in different laboratories [4, 5].

Electroanalytical methods are, of course, restricted to determination of compounds, which can undergo electrochemical oxidation or reduction in the available potential range, which depends on the structure of the analysed species [4–6]. For organic compounds, it usually means the presence of conjugation of single bonds,

S.A. Ozkan et al., *Electroanalysis in Biomedical and Pharmaceutical Sciences*,
Monographs in Electrochemistry, DOI 10.1007/978-3-662-47138-8_1

like that of the C-halogen. For inorganic species, the oxidation or reduction usually depends on the presence of at least two different stable oxidation states. Special situation exists for application of analyses of special (inorganic or organic) which form slightly soluble or complex compounds with mercury ions, such as halide ions and S^{2-} ions, and organic compounds, such as barbiturates, thiobarbiturates, and some other heterocyclic compounds [6–8].

For more basic information about applications of polarography consult [7–30] and for that about the theory of polarography [4, 31–33].

Currently, the most frequently used types of indicator or working electrodes in pharmaceutical analyses are, nevertheless, solid electrodes, which are in the shape of a disc, surrounded and shrouded by a chemically inert material, such as Teflon, glass or epoxy. The electrodes can be made from noble metals (like Au, Pt, Ru, etc.) or from various forms of carbon, such as carbon paste, glassy carbon, microfiber carbon, or simply solid carbon [34–38]. All these electrodes enable reaching positive potentials, and their main areas of application are oxidations, mostly of organic compounds.

Among the techniques used, predominantly are various forms of voltammetry: in practice it is predominantly (CV) or linear sweep voltammetry (LSV). Sometimes are also used various pulse and square wave methods [39–42]. Details about physical and chemical principles involved in applications of such techniques are found elsewhere [43–48]; here the stress is on the description of a typical analytical procedure and on the evaluation of electrochemical data. Attention is further paid to the size, form, and surface of the working electrode, the structure and state of the surface, and its cleaning process. It should always be kept in mind, that for a reliable analysis and its results, an understanding of the electrochemical and chemical processes involved is essential [41]. It is important to understand what takes place at the electrode surface and its immediate vicinity. It is essential to understand the sequence of electron transfers and the role of chemical processes, such as proton transfers as well as the role of the working electrode material and surface; all play a role in the development of reliable analytical procedures. They all are reflected in advantages and limitations of described analytical procedures.

Relatively new trends in electroanalytical methods, used in pharmaceutical analyses of drugs, are applications of screen-printed electrodes and of biosensors [49–55]. The screen-printed electrodes may be reused or may serve for a single use, so that contamination and memory effects are avoided. They may be applied for the analysis of volumes as small as a few microliters. SPEs are mass produced with a reproducibility of the working electrode area of less than 5 %. The advantage of the screen-printed electrodes, as working ones, is an easy modification of its surface, which opens many possibilities for extention of applications. They are most popular, but would need modification of the working electrode surface for selective and accurate determination of the studied compound. It is anticipated, that robust modified screen-printed electrodes, when applied with a proper validation procedure, could find a successful application in many areas, where rapid drug control is required [54–57].

Biosensors, thanks to their hybrid structures, with a biological element placed in close proximity of a physical transducer, can advantageously be considered in drug assays since they may offer high selectivity and sensitivity due to signal amplification [58]. They are well suited for studying the interaction of an immobilized biological element with the target analyte or a family of related substrates [59]. The results can be exploited for the quantification of a given analyte in a complex matrix provided that the nature of the chemical reactions involved is well understood. Thanks to their unique configuration, they are particularly suitable for investigations devoted to the study of drug–biocomponent interaction. Incredibly small microelectrodes [54, 56–59] are being used to probe biochemical events on the cellular level. Voltammetric methods, which amazingly low detection limits, are being used to monitor lead levels in the bloodstream.

Electrochemical methods of detection and determination of an electroactive component in the effluent from a chromatographic column offer in some cases advantage over other analytical methods, because of their sensitivity, relative rapidity, and specificity [1–3, 60–64]. This will be particularly of importance in recently developed variant UPLC using a shorter column and higher pressure, requiring shorter analysis time than HPLC. As numerous modern pharmaceuticals are electroactive, such analyses are possible. Electroanalysis of separated species can be detected and determined also in flow injection analyses (FIA) [1–3, 61, 65–70].

In this book the applications of various electrochemical modes, electrodes, modifications and hyphenated techniques, etc. in the analysis of drugs and pharmaceutical active compounds are presented in this book. Also, this book highlights the role of various electroanalytical techniques and their corresponding electroanalytical techniques in the analysis of pharmaceuticals and their validation procedures. Examples of different types of application have been reported and as with all other aforementioned techniques. Moreover, it should be kept in mind, that the electrochemical techniques have not only advantages but also limitations. Some of both types of such factors are pointed out in the following chapters.

References

1. Ahuja S, Scypinski S (eds) (2001) Handbook of modern pharmaceutical analysis, vol 3. Academic, London
2. Adamovics JA (ed) (1997) Chromatographic analysis of pharmaceuticals, 2nd edn. New York, Marcel Dekker
3. Synder LR, Kirkland JJ, Glajch JL (eds) (1997) Practical HPLC method development, 2nd edn. New York, Wiley
4. Scholz F (ed) (2010) Electroanalytical methods, guide to experiments and applications, 2 revised and extended ednth edn. Springer, Berlin
5. Smyth MR, Vos JG (eds) (1992) Analytical voltammetry, vol XXVII. Elsevier, Amsterdam
6. Patriarche GJ, Chareau-Gosselin M, Vandenbalck JL, Zuman P (1979) Polarography and related electroanalytical techniques in pharmacy and pharmacology. In: Bard AJ (ed) Electroanalytical chemistry, vol 11. Marcel Dekker, New York, pp 141–289

7. Zuman P (1967) Substituent effects in organic polarography. Plenum, New York
8. Heyrovsky J, Zuman P (1968) Introduction to practical polarography. Academic, London, UK
9. Dogan B, Uslu B, Ozkan SA, Zuman P (2008) Anal Chem 80:209–216
10. Abdine H, Belal L (2002) Talanta 56:97–104
11. Debnath C, Haslinger E, Ortner A (2006) Nat Prod Commun 1:487–494
12. Korany MA, Hewala II, Abdel-Hay KM (2008) J AOAC Int 91:1051–1058
13. El-Enany N, Al-Ghannam SM, Belal F (2003) Pharm Ind 65:76–81
14. Samota S, Garch A, Pandey R (2010) Port Electrochim Acta 28:87–94
15. Ramadan AA, Mandil H, Genco T (2009) Asian J Chem 21:7387–7397
16. Jain R, Mishra R, Dwivedi A (2009) J Sci Ind Res 68:945–950
17. El-Sayed GO, Yasin SA, Badawy AA (2009) J Chem Pharm Res 1:225–232
18. Ibrahim F, El-Enany N (2003) Farmaco 58:1313–1318
19. Ramadan AA, Mandil H, Hafez B (2011) Asian J Chem 23:403–406
20. Knoth H, Petry T, Gärtner P (2012) Pharmazie 67:987–990
21. Reddy GVS, Reddy CLN, Myreddy VN, Reddy SJ (2011) Clin Med Res 3:35
22. Hoda M (2003) Bull Fac Pharm (Cairo Univ) 41:299
23. Ghoneim MM, Tawfik A (2003) Can J Chem 81:889–896
24. El-Enany N, Belal F, Rizk M (2008) J Biochem Bioph Methods 70:889–896
25. Yardimci C, Ozaltin N (2001) Analyst 126:361–366
26. Singh M, Kumar Y, Meena O, Garg A, Pandey R (2010) Int J ChemTech Res 2:1738–1744
27. Song JF, Shao Y, Guo W (2001) Anal Sci 17:1145–1148
28. Alvarez-Lueje A, López C, Núñez-Vergara LJ, Squella JA (2001) J AOAC Int 84:649–658
29. Dumitrescu V, Dumitrescu N, Balu AM (2005) Rev Chim 56:345–348
30. Subba Reddy GV, Jayarama Reddy SJ (2001) Indian J Environ Prot 21:816–819
31. Garg A, Meena OP, Kumar Y, Pandey R (2011) Int J ChemTech Res 3:33–38
32. Brezina M, Zuman P (1958) Polarography in medicine, biochemistry and pharmacy. Interscience, New York
33. Heyrovsky J, Kuta J (1966) Principles of polarography. Academic, London, UK
34. Kissinger PT, Heineman WR (eds) (1996) Laboratory techniques in electroanalytical chemistry, 2nd edn. New York, Marcel Dekker
35. Trasatti S (ed) (1980) Electrodes of conductive metallic oxides, Part A. Elsevier, Amsterdam
36. Pierson HO (ed) (1993) Handbook of carbon, graphite, diamond and fullerenes. Properties, processing and applications. Noyes, New Jersey
37. Uslu B, Ozkan SA (2007) Anal Lett 40:817–853
38. Uslu B, Ozkan SA (2007) Comb Chem High Throughput Screen 10:495–513
39. Besenhard JO, Schultze A, Schur K, Jannakoudakis PD (1991) Preparation of voltammetric and potentiometric carbon fibre microelectrodes. In: Montenegro MI, Queirós MA, Daschbach JL (eds) Microelectrodes: Theory and applications, NATO ASI Series E. Springer, Netherlands
40. Bard AJ, Faulkner LR (2001) Electrochemical methods, fundamentals and applications, 2nd edn. Wiley, New York
41. Brett CMA, Oliveira-Brett AM (1993) Electrochemistry: Principles methods and applications. Oxford University Press, Oxford
42. Brett CMA, Oliveira-Brett AM (2003) Step and pulse techniques. In: Unwin P (ed) Encyclopaedia of electrochemistry. Wiley, Weinheim
43. Ozkan SA, Uslu B, Senturk Z (2004) Electroanalysis 16:231–237
44. Uslu B, Ozkan SA (2004) Electrochim Acta 49:4321–4329
45. Suryanarayanan V, Zhang Y, Yoshihara S, Shirakashi T (2005) Electroanalysis 17:925–932
46. Dogan B, Uslu B, Suzen S, Ozkan SA (2005) Electroanalysis 17:1886–1894
47. Jain R, Jadon N, Radhapyari K (2006) Talanta 70:383–386
48. Demircan S, Kir S, Ozkan SA (2007) Anal Lett 40:1177–1195
49. Dominguez Renedo O, Alonso-Lomillo MA, Arcos Martinez MJ (2007) Talanta 73:202–219
50. Branina KZ, Kositzina AN, Ivanova A (2007) Electrochemical sensor analysis. In: Alegret S, Merkoci A, Barcelo D (eds) Comprehensive analytical chemistry, 1st edn. Elsevier, Amsterdam

51. Sljukic B, Malakhova NA, Branina KZ, Banks CE, Compton RG (2006) Electroanalysis 18: 928–930
52. Hart JP, Crew A, Crouch E, Honeychurch KC, Pemberton RM (2004) Anal Lett 37:789–830
53. Hart JP, Wring SA (1997) Trac-Trends Anal Chem 16:89–103
54. Ozkan SA (ed) (2012) Electroanalytical methods in pharmaceutical analysis and their validation, 1st edn. HNB Pub, New York
55. Kauffmann JM, El Kaoutit M, Parsajoo C (2012) Biosensors for drug testing and discovery. In: Higson S (ed) Biosensors for medical applications. Woodhead, Cambridge
56. Li M, Li YT, Li DW, Long YT (2012) Anal Chim Acta 13:31–44
57. Jubete E, Loaiza OA, Ochoteco E, Pomposo JA, Grande H, Rodriguez J (2009) J Sens 2009: 1–13
58. Hodes G (ed) (2001) Electrochemistry of nanomaterials. Weinheim, Wiley
59. Avramescu A, Andreescu S, Noguer T, Bala C, Andreescu D, Marty J-L (2002) Anal Bioanal Chem 374:25–32
60. Snyder LR, Kirkland JJ, Dolan JW (2010) Introduction to modern liquid chromatography, 3rd edn. Wiley, Chichester
61. Štulík K, Pacáková V (1987) Electroanalytical measurements in flowing liquids. Ellis Horwood, Chichester
62. Yu D, Blankert B, Viré JC, Kauffmann JM (2005) Anal Lett 38:1687–1701
63. Weber SG (1986) In: Yeung ES (ed) Detectors for liquid chromatography. New York, Wiley
64. Shoup RE (1986) High Performance Liquid Chromatography 4:91
65. Ruzicka J, Hansen EH (1988) Flow injection analysis, 2nd edn. Wiley, New York
66. Ruzicka J, Hansen EH (1975) Anal Chim Acta 78:145–157
67. Ruzicka J, Stewart JWB (1975) Anal Chim Acta 79:79–91
68. Ruzicka J, Stewart JWB, Zagatto EA (1976) Anal Chim Acta 81:387–396
69. Ruzicka J, Hansen EH (1998) TrAC Trends Anal Chem 17:69–73
70. Karlberg B, Pacey GE (1989) Flow injection analysis. A practical guide. Elsevier, The Netherlands

Chapter 2
Polarography in Studies of Pharmaceuticals

Contents

2.1 Principles of Polarography

Polarography is an electroanalytical technique based on recording of current–voltage (i–E) curves using electrolysis, with a dropping mercury electrode (DME) as an indicator electrode. This technique has been invented in 1922 by Heyrovsky [1]. For his invention, he was awarded in 1959 the Nobel Prize. According to the Analytical Chemistry, polarography has been over the period from 1950 to 1970 one of the five most frequently used analytical techniques. The scope of its worldwide applications suffered a decline more recently, because of the wide range of separation techniques in organic analysis. Similarly, the use of atomic absorption (AA) and related techniques became predominant in the determination of metal ions. As reflected by the number of references in the Tables, polarography is currently used in those laboratories, where limited budgets do not allow acquiring of more expensive instrumentation, as needed in AA and modern separation techniques.

S.A. Ozkan et al., *Electroanalysis in Biomedical and Pharmaceutical Sciences*, Monographs in Electrochemistry, DOI 10.1007/978-3-662-47138-8_2

The fundamentals of polarographic theory have been published in several basic books [2–5], but the most important features for applications will be discussed here to allow the reader a simple understanding of the contents of this monograph. In this contribution the pre-1960 convention is used. This means that the reduction currents are recorded as positive, the oxidation currents (anodic) as negative. The recording of each curves starts from left, at the least negative (or most positive) potential and increasing voltage from left to right.

In the simple direct current polarography (DCP), a dropping mercury electrode (DME) is immersed into the investigated solution placed in an electrolytic cell together with a reference electrode. Linearly increasing voltage is applied to the two electrodes, and the current, passing between the two electrodes, is recorded as a function of the applied voltage. Such simple arrangements are sufficient for investigations of aqueous solutions, commonly used in drug analysis, discussed in this chapter. In the studies in nonaqueous media, a third electrode is used to eliminate the role of resistance of the solution between the electrodes. In most electrode processes, taking place at the surface of the DME, either reduction or oxidation of the investigated species takes place (Table 2.1). In reductions, electrons are transferred from the surface of the electrode to the investigated species. In electro-oxidations, electrons are transferred in the opposite direction, from the investigated species to the electrode (DME). In polarographic convention, the reduction currents are recorded above the current zero line. Such currents are denoted cathodic currents and are recorded from positive voltages on the left to negative voltages

Table 2.1 Examples of polarographically active groupings

Bond types	Bond	Examples
Single Bonds	C–X	C–F, C–Cl, C–Br, C–I
	C–O	CO–CH–OH, 4-Py CH_2OH
	C–N	CO–CH–NH_3^+, 4-Py CH_2 NH_3^+
	O–O	R–O–O–R, R–O–O–H
	S–S	R–S–S–R, R–S–S–H, R–S–S–Ar
Double Bonds	C=O	R_1R_2C=O, Ar_1COAR_2, BzCOBz, BzCOR, BzCOH, Bz(1-CHO), Bz(1,4-CHO), Bz(1,2-CHO), Bz(1-3-CHO), Py-4-CHO, Py-3-CHO, corresponding CORCO in alicyclic and heterocyclic rings
	C=C	Ar–CH=CH–R, R–CH=CH–CO–, R–CH=CH–CN, R–CH=CH–C=NH_2^+, R–CH=CH–COOH, R–CH=CH–COOR
	C=N	R^1–C=N- R^2, Ar–C=N–R, Ar–C=N–Ar, R–CO–CH=NR, RN=C–C=NR, Ar–C=NR, R–CH=NOH, R–CH=N–NH_2
	N=O	R–NO, CH=CH–NO, Ar–NO, Het–NO
Triple Bonds	C≡C	R–C≡C–R, R–C≡C–COR, R–C≡C–Ar, R–C≡C–Het,
	C≡N	R–C≡N, Ar–C≡N, Het–C≡N
Other Types of Bonds	NO_2	R–NO_2, Ar–NO_2, Het–NO_2,
	Rings	Benzenoid aromatic, Heterocyclic aromatic, Others heterocyclic
	COOH	Acids—oxalic, phthalic, terephthalic

Abbreviations used: *R* aliphatic chain, *Ar* Aromatic ring in general, *Bz* Benzenoid aromatic ring, *Het* heterocyclic ring in general, *Py* Pyridine ring, substituent in any position, *X* Halogen

on the right. The oxidation currents are recorded below the current zero line and starting at negative to positive potentials (Fig. 2.1). Such currents are called anodic.

Most recently IUPAC, based on conventions in other areas of electrochemistry, proposed another convention, in which anodic currents are positive and recorded starting from positive voltages on the right. Cathodic currents are then considered to be negative and recorded starting at a negative voltage on the left. Nevertheless, it is strongly recommended to check in each publication, which kind of directions of axes in used.

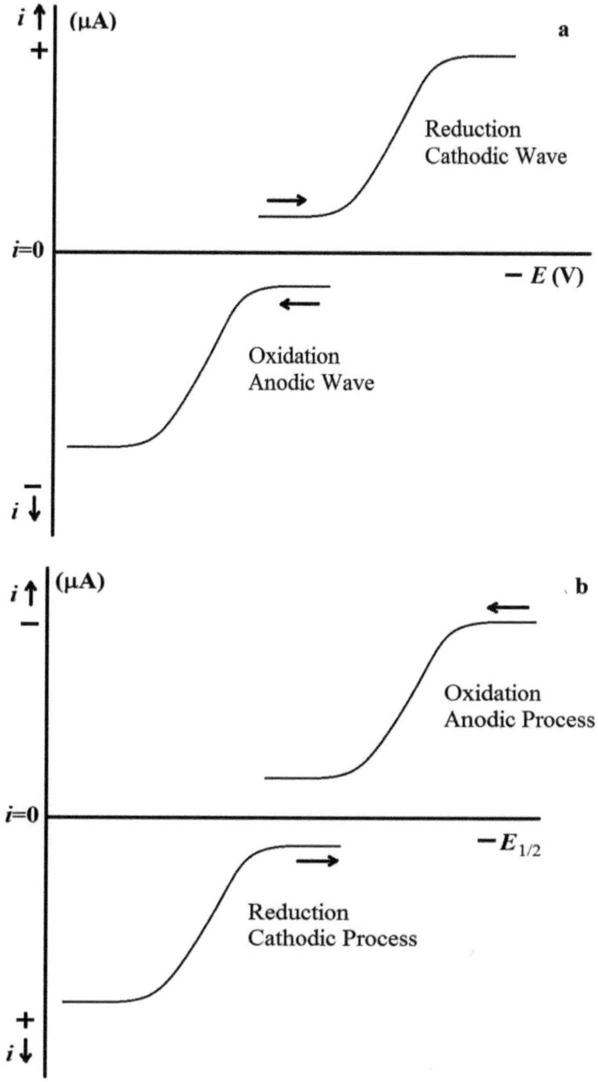

Fig. 2.1 Two conventions used for reporting the current–voltage curves: (**a**) Used in polarography; (**b**) recommended by IUPAC

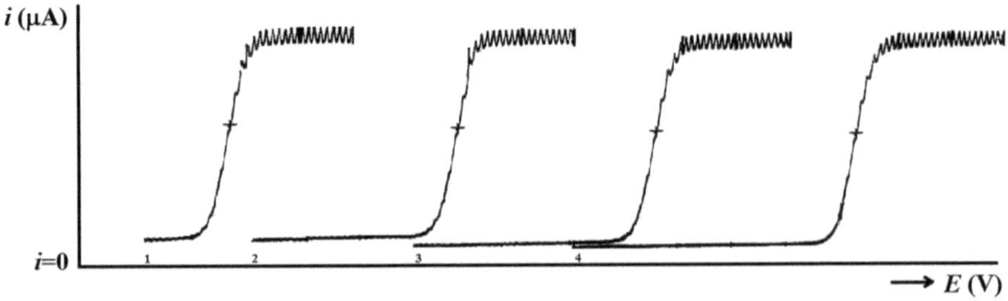

Fig. 2.2 Shapes of polarographic *i–E* curves, indicating formation of the limiting current. Polarographic waves of Pb^{2+} ions in solutions of sodium hydroxide. The concentration of OH^- ions increases from *left* to *right* and shifts the signal due to the complex formation between OH^- and lead ions. Observe the reproducibility of limiting currents

In the presence of a reduced or oxidized species, the current depends on the applied voltage, as shown in Fig. 2.2 (Tables 2.2 and 2.3). The current remains low, before a potential is reached when electrolysis takes place. The current increases in a shape, which is denoted as "polarographic wave." The part of the recorded curve, where the increased current reaches a practically constant value, is called "limiting current" (i_L). Another important factor is the potential at this point of the recorded *i–E* curve where the current reaches one half of the limiting value. This potential is called "the half-wave potential ($E_{1/2}$)." The value of the limiting current offers information about the concentration of the species undergoing electrolysis (that is quantitative information) whereas the $E_{1/2}$ offers information about the nature of the electrolyzed species (that is qualitative information).

The DMEs used in polarography are of two different types; both consist in a small drop of mercury, formed at an orifice of a glass capillary of inner diameter between 0.03 and 0.1 mm. The capillary (usually 5–10 cm long) is attached to a reservoir of mercury. In the older type of this electrode, it is connected to the mercury reservoir by a rubber or plastic tubing. This allows changing the mercury pressure by varying the height of the mercury reservoir. The height of mercury reservoir (h) is useful for identification of the kind of the transport of the electroactive species from the bulk of the solution to the electrode surface.

This type of electrodes was used, when $i-E$ curves were recorded using older, analog instruments (polarographs). More modern instruments (polarographs) are digital. They must be used in connection with an artificially controlled drop time. This is advantageous for analytical applications. The problems with using this type of electrode for basic research are that all of the theory of polarographic $i-E$ curves has been developed for transport to a naturally growing mercury drop. Dependence of the limiting current on the drop time t_1 is less sensitive to the controlled drop time than those obtained using electrodes with natural drop time for the dependence on mercury pressure (h). The drop time (t_1), that is given by the time interval between dropping off of the two consecutive drops, is usually between 2 and 5 seconds. The measured limiting current depends also on the flow velocity of the out flowing mercury (m), usually about 2 mg L s^{-1}.

Table 2.2 Names and structures of pharmaceuticals, the polarographic analyses of which was reported between 1990 and 2011

Compound	Structure	Class (group; sub-group)	Pharmacological Effect	References
Abacavir	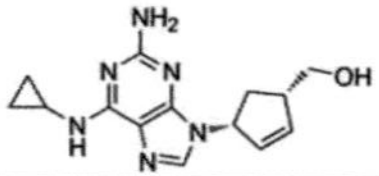	Nucleoside-reverse transcriptase inhibitor	Anti-HIV; Anti-viral	[6]
Acrivastine	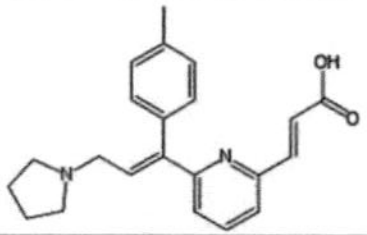	H_1-receptor antagonist	Antihistaminic	[7]
Amphetamines; 2-Phenyl Ethylamines	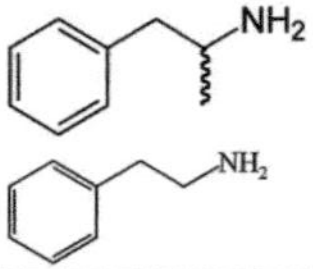	Psychoanaleptic	Central effect sympathomimetic	[8]
Artemisinin	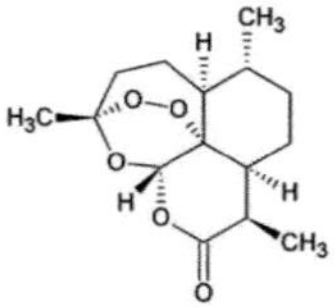	In plant	Anti-viral Anticancer	[9]
Atorvastatin	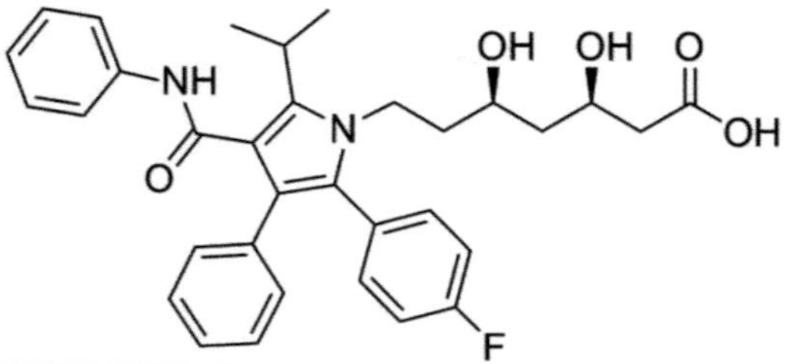	HMG-CoA reductase inhibitory	Antihyperlipidemic	[10]

(continued)

Table 2.2 (continued)

Compound	Structure	Class (group; sub-group)	Pharmacological Effect	References
Etofibrate	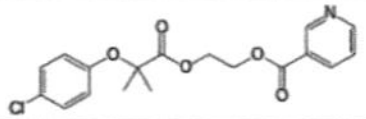	HMG-CoA reductase inhibitory	Antihyperlipidemic	[10]
Fenofibrate	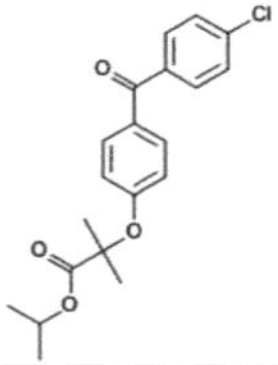	HMG-CoA reductase inhibitory	Antihyperlipidemic	[10]
8-Azaguanine	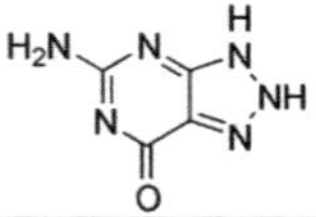	Purin analog	Anti-cancer	[11]
Azintamide	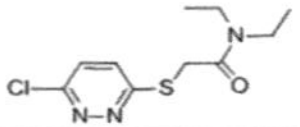	Choleretic activity	Abdominal distension	[12]
Dibenzazepines	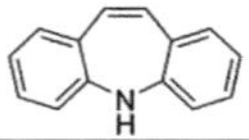	Non selective monoamine reuptake inhibitory	Antidepressant	[13]
Aztreonam	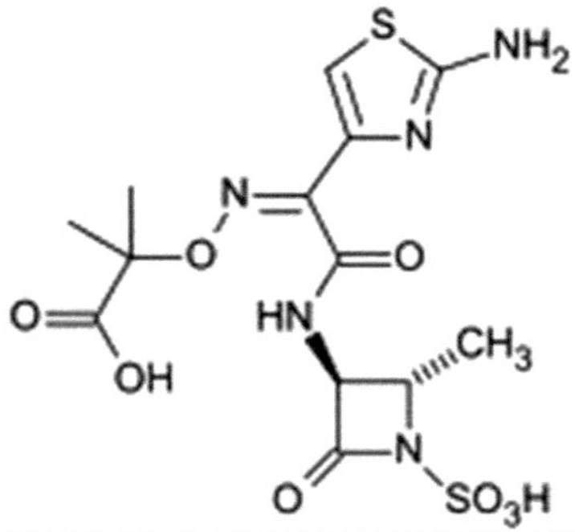	Monobactam group antibiotic	Antibacterial	[14]

Benzydamine	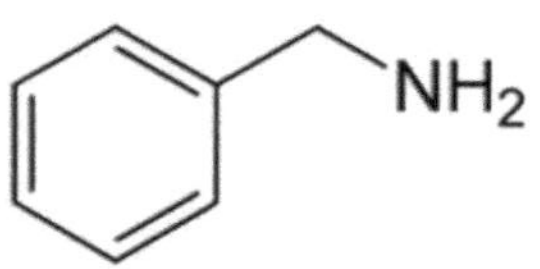	Topical nonsteroidal anti-inflammatory agent	Anti-inflammatory	[15]
Bupropion	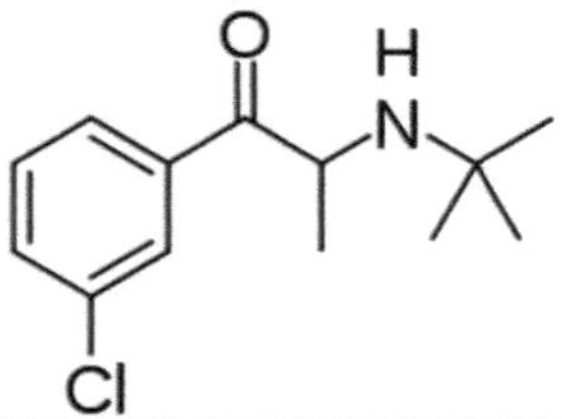	Chlorpropio-phenone group	Antidepressant	[16]
Carbinoxamine maleate	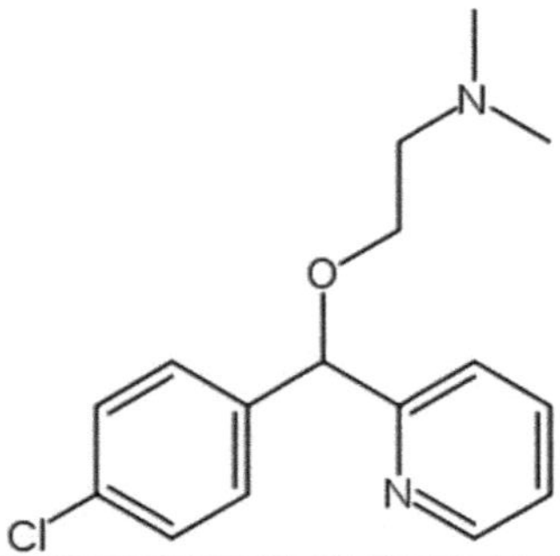	Amino-alkylether	Antihistaminic	[17]
Cefpodoxime	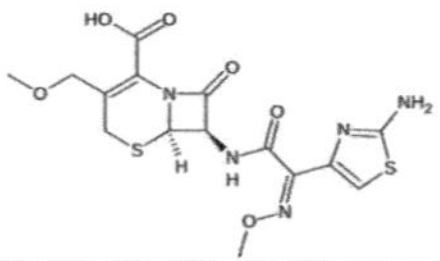	Third generation cephalosporins	Antibacterial; antibiotic	[18]

(continued)

Table 2.2 (continued)

Compound	Structure	Class (group; sub-group)	Pharmacological Effect	References
Cephalosporine (Ceftadizine)	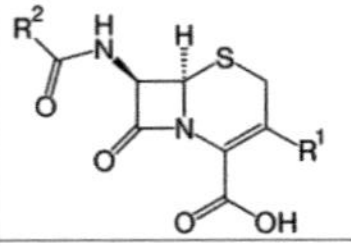	3rd generation cephalosporin	Biliary-tract infections; antibacterial	[19]
Ceftriaxone	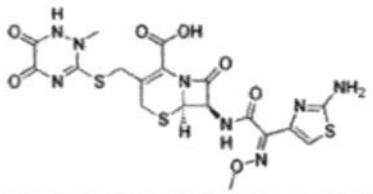	3rd generation cephalosporin	Biliary-tract infections; antibacterial	[20]
Chlordiazepoxide	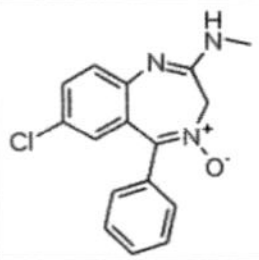	Benzodiazepine derivates	sedative/hypnotic	[21]
8-Chlorotheophylline	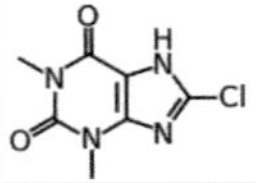	Xanthine derivates	Bronchial smooth muscle; bronchodilatator	[22]
Chloramphenicol	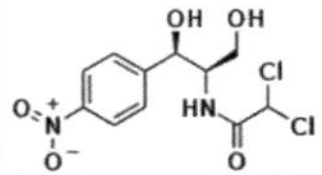	Topical anti-infective; Antiparasitic	Antibacterial Anti-protozoal	[23]
Nitrofurantoin	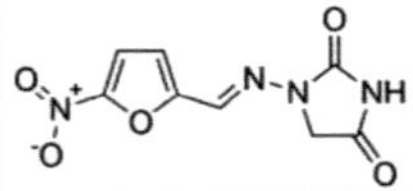	Topical anti-infective; Antiparasitic	Antibacterial Anti-protozoal	[23]
Nitrofurazone	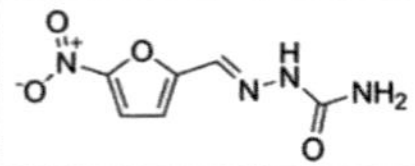	Topical anti-infective; Antiparasitic	Antibacterial Anti-protozoal	[23]
Furazolidone	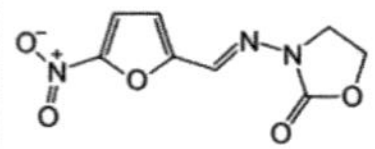	Topical anti-infective; Antiparasitic	Antibacterial Anti-protozoal	[23]

Chloramphenicol α-Sodium Succinate	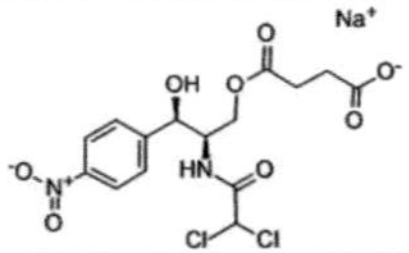	Topical anti-infective	Antibacterial	[24]
Ciclopiroxolamine	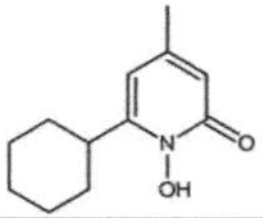	Topical antifungal	Antimycotic; Antifungal	[25]
Cinolazepam	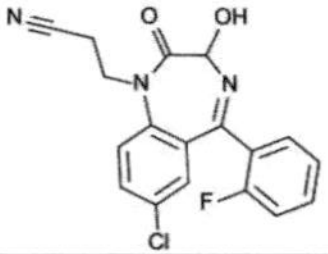	Benzodiazepine group	Hypnotic, sedative	[26]
Cytarabine	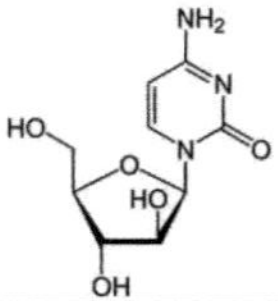	Pyrimidine group	Anti-neoplastic	[27]
Furaltadone	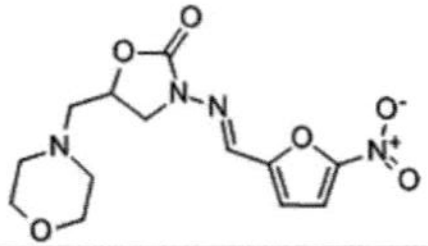	Nitrofuran derivative,	Antimycobacterial, bactericidal activity	[28, 29]
Dipyrone	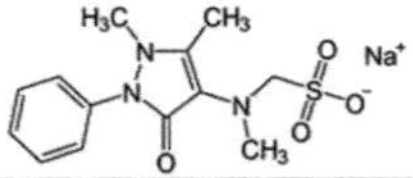	Nonselective COX inhibitory	Anti-inflammatory, rheumatoid arthritis	[28, 30, 31]
Enoxacin	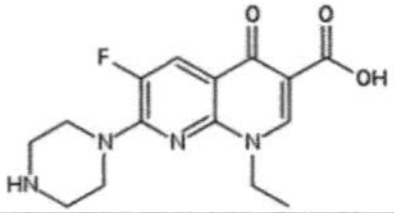	Fluoroquinolone	Antibacterial	[32]

(continued)

Table 2.2 (continued)

Compound	Structure	Class (group; sub-group)	Pharmacological Effect	References
Erythromycin	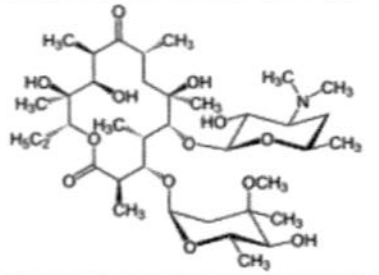	Macrolide Antibiotic	Antibacterial	[33]
Flunidazole	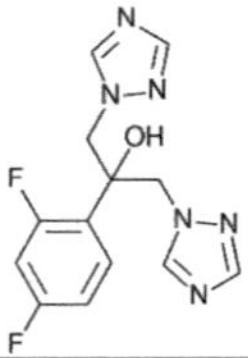	Imidazole and Triazole Class Antifungal	Antifungal	[34]
Flutamide	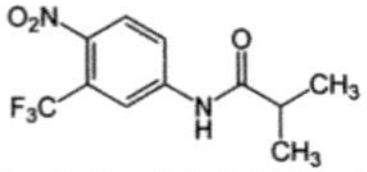	Gonadal hormones, estrogens	Antiestrogen; treatment of breast cancer; prevention of Osteoporosis	[35]
Flunitrazepam	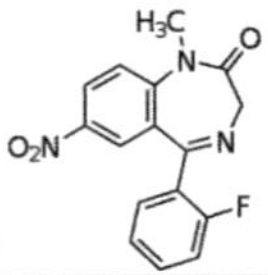	Diazepam; benzodiazepine group	Hypnotic, sedative	[36]
Fosinopril	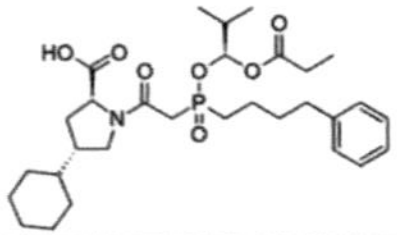	Angiotensine converting enzyme (ace) inhibitory	Antihypertensive	[37]
Ketanserin	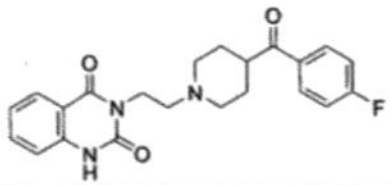	Seratonin antagonist	Antihypertensive	[38]

Ketoprofen	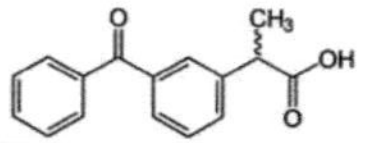	Nonsteroidal anti-inflammatory; propionic acid	Anti-inflammatory; Analgesic	[39, 40]
Omeprazole	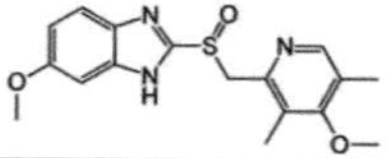	Proton pump inhibitory	Anti-ulcer	[41]
Lansoprazole	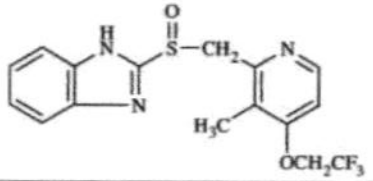	Proton pump inhibitory	Anti-ulcer	[41, 42]
Lamivudine	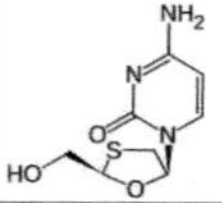	Nucleoside-reverse transcriptase inhibitory	Anti-viral; Anti-HIV	[43]
Lomefloxacin	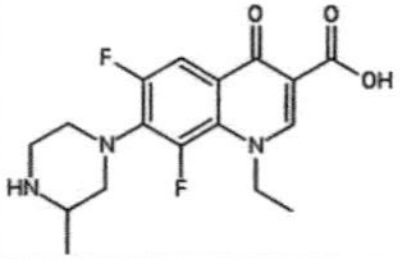	Quinolone group	Antibacterial	[44, 45]
Lopirazepam	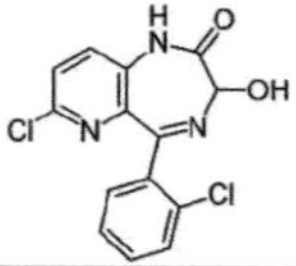	Benzodiazepine analog, the pyridodiazepine type antipsychotic	Antixiolytic, hypnotic	[46]
Loratadine	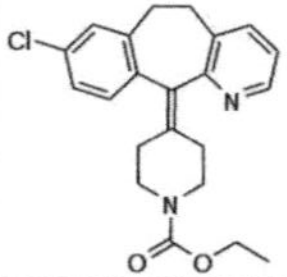	H_1-receptor antagonist	Antihistaminic	[47]
Metronidazole	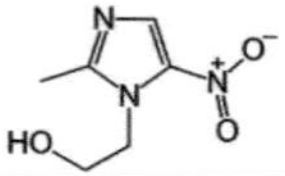	Nitro-imidazol group	Anti-protozoal	[48]

(continued)

Table 2.2 (continued)

Compound	Structure	Class (group; sub-group)	Pharmacological Effect	References
Methotrexate	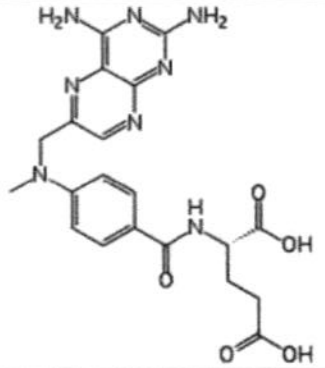	Immunosuppressant	Anti-neoplastic	[49]
Nifedipine	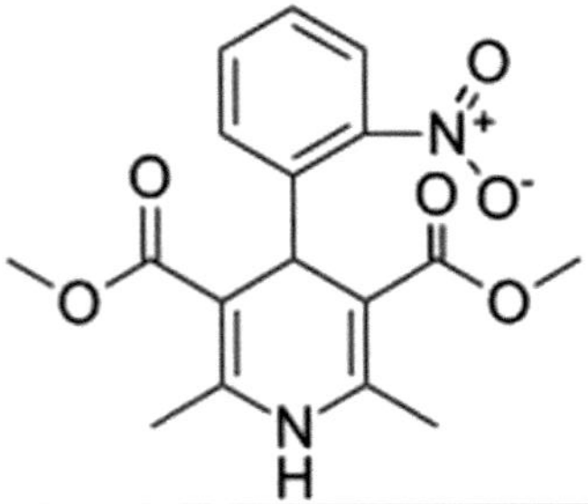	Dihydropyride group Ca-channel blocker	Antihypertensive	[50, 51]
Nitrendipine	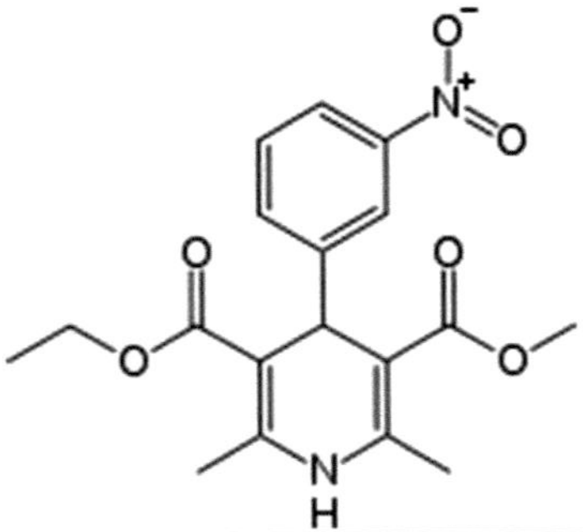	Dihydropyride group Ca-channel blocker	Antihypertension	[52]

Nitroxazepine	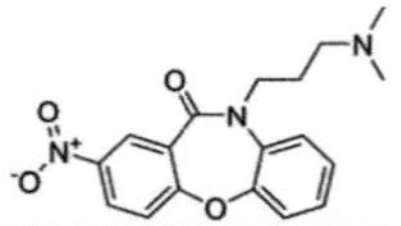	Benzodiazepine group	Antitranquilizer	[53]
Nitroxynil	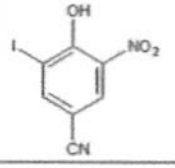	Veterinary drugs	Anti-helmintic	[54]
Ornidazole	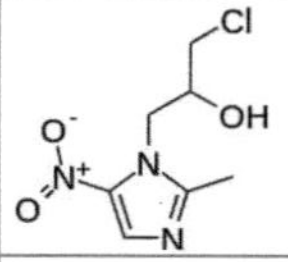	Nitro imidazol group	Anti-protozoal	[55]
Pentamidine isethionate		Aromatic diamidine	Anti-protozoal	[56]
Piperacillin	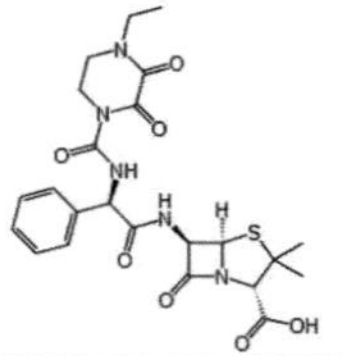	Penicillin	Antibacterial	[57]
Ranitidine	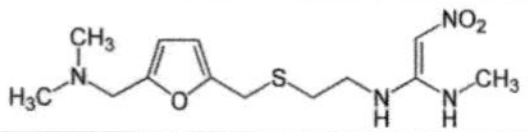	H_2 receptor antagonist	Anti-ulcer	[58]
Rofecoxib	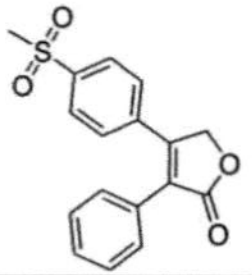	Selective COX-2 inhibitory	Anti-inflammatory, rheuma-toid arthritis	[59]
Salazosulfapyridine	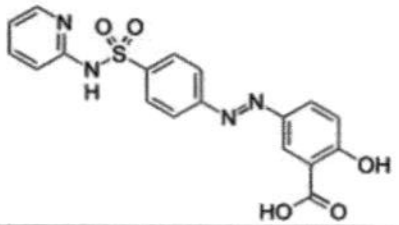	Sulfonamides	Antibacterial; Anti-infective	[60]

(continued)

Table 2.2 (continued)

Compound	Structure	Class (group; sub-group)	Pharmacological Effect	References
Sulfamethoxazole	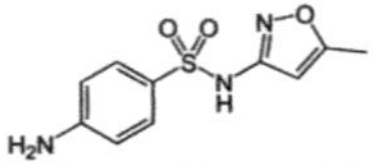	Sulfonamides	Antibacterial; Anti-infective	[60]
Sulfonamides	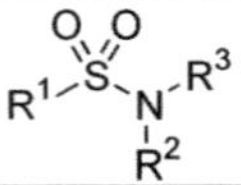	Sulfonamides	Antibacterial; Anti-infective	[60]
Sparfloxacin	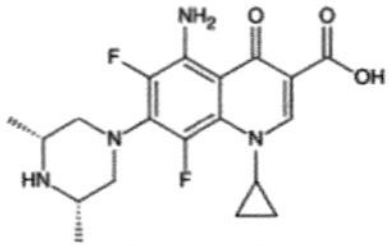	Quinolone group	Antibacterial	[61]
Tenoxicam	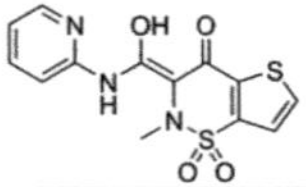	Nonselective COX inhibitory	Anti-inflammatory; Rheuma-toid arthritis.	[62]
Tetracycline	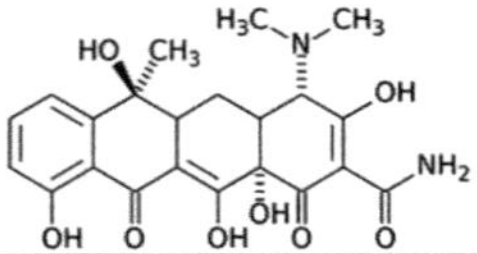	Tetracycline group antibiotic	Antibacterial	[63]
Timelotem	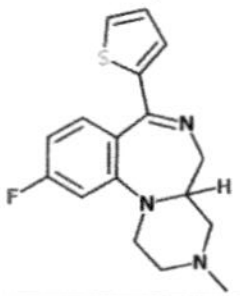	A typical antipsychotic	Antipsychotic	[64]
Timolol Maleate	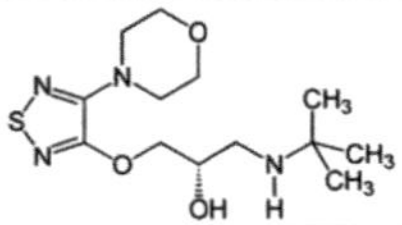	ß-Blocker	Anti-glaucoma, myotic	[65]

Tolmetin	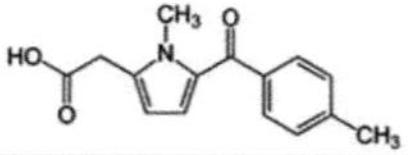	Nonselective COX-2 inhibitory	Anti-inflammatory, rheumatoid arthritis	[66]
Topiramate	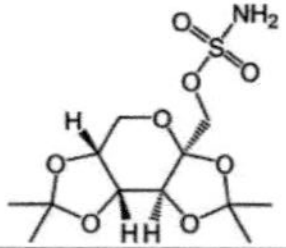	Sulfamate-substituted monosaccharide	Anti-convulsant, Anti-epileptic	[67]
Zileuton	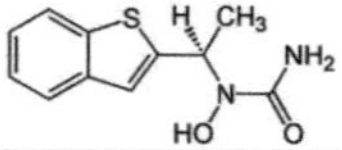	Lypooxygenase inhibitory	Anti-inflammatory, rheumatoid arthritis	[68]
Zopiclone	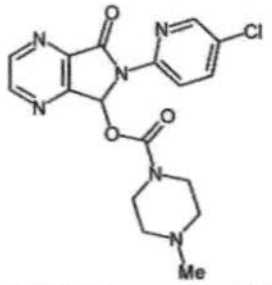	Non-benzodiazepine	Psycholeptic, sedatif, hypnotic	[69]

Table 2.3 Pharmaceuticals arrangement according to physiological activity

Pharmacological effect	References
Analgesic	[39, 40]
Anthelmintic	[54]
Anti-estrogen	[32]
Antibacterial; Antibiotic	[14, 18, 19, 24, 29, 32, 33, 44, 48, 57, 60, 61, 63]
Anti-cancer; Anti-neoplastic	[9, 27, 49]
Anti-convulsant, Anti-epileptic	[67]
Anti-depressant	[13, 16, 22]
Anti-fungal	[25]
Anti-glaucoma	[65]
Anti-histaminic	[7, 17, 47]
Anti-HIV; Anti-viral	[6, 9, 43]
Anti-hyperlipidemic	[10]
Antihypertensive	[37, 38, 40, 41]
Anti-inflammatory	[15, 28, 30, 39, 40, 59, 62, 66, 68]
Anti-protozoal	[23, 48, 55, 56]
Anti-psychotic	[64]
Anti-tranquilizer	[53]
Anti-ulcer	[41, 42, 58]
Anxiolytic	[69]
Bronchodilator	[22]
Choleretic activity	[12]
Hypnotic	[26, 35, 36, 44, 46, 69]
Psycholeptic, sedative	[26, 33, 69]

Provided that the values of m and t_1 are known as well as the solution composition, when the same supporting electrolyte and the same electroactive species of known concentration are used, the i–E curve can be reproduced, if the starting and final potentials and the rate of voltage scanning are known. Such curves can be reproduced within ±2–3 % from laboratory to laboratory as well as from country to country, so no validation is needed.

2.1.1 Comparison with Voltammetric Methods

Current–voltage curves can be recorded using various electrodes, e.g., mercury, gold, platinum, silver, and carbon. For historical reasons, the technique using dropping mercury electrodes is called polarography, whereas techniques employing any other electrode are summarized under the term voltammetry. These techniques and their applications are discussed in another chapter in this volume [70].

The limits of applications for the most frequently used voltammetry with some form of a carbon electrode and those of DME can be compared as follows: DME

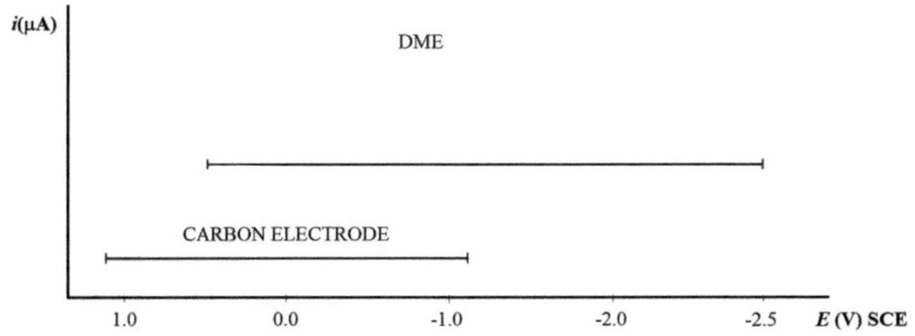

Fig. 2.3 Potential ranges available when dropping mercury electrode (DME) and a carbon electrode are used in recording of current–voltage curves

can be used within potential window from +0.4 V to −2.5 V, whereas a typical carbon electrode can be used from +1.1 V to −1.2 V (Fig. 2.3). This limits the types of electroactive substances that can be investigated and the electrochemical processes involved. Numerous oxidations of organic components take place at potentials more positive than +0.4 V and can be studied using a carbon electrode. On the other hand, numerous reductions occur at potentials more negative than about −0.5 V. Thus, the potential windows at, e.g., carbon electrodes enable following voltammetric oxidations of hundreds of organic compounds. On the other hand, the much wider potential window of DME and much wider use of this electrode for the recording of i–E curves resulted in observations of several thousands of reductions of organic compound (Table 2.4).

There is one more difference between the use of carbon electrodes and DME: The difference in the scan rate: in DC polarography (DCP) the typical scan rate is between 100 mV/min and 300 mV/min, whereas in cyclic voltammetry it is between few millivolts to several V/s. In these cases, when a chemical equilibrium is established before the first electron uptake, it is possible to use such fast scan rate to neglect the role of the re-establishment of the antecedent equilibria. This has been demonstrated for the hydration–dehydration equilibrium of formaldehyde [110] and involving 1,2-alicyclic diketones [111]. The effect of the scan rate on the rate of acid–base equilibria, preceding electron transfer, has not been systematically investigated by voltammetry. By controlling the pH of the investigated solution it is, on the other hand, possible to obtain information about the rate of the re-establishment of the acid–base equilibria in the vicinity of the electrode, using the pH dependence of the limiting currents, obtained using DME.

Table 2.4 Other pharmaceutically important species and reactions

Properties and reactions	References
Alkaloids	
Nicotine	[71]
Blood components	
Hemin, protoporphyrin	[72]
Hemoglobin	[73]
Heparin	[74]
Enzymes	
Liposomes	[75]
Lysozyme	[76]
Enzyme inhibitor	[77]
Materials in plants and herbs	
Artemisin in artemisia	[9]
Balcain	[78]
Cimophilin	[79]
5-Hyroxynapthoquinones	[80]
Indigo	[78]
Paeonol	[78]
Piperin	[78]
Thymoquinone	[81]
Steroids	
Progesterone	[82]
Vitamins	
Nicotinamide	[83]
K_1	[84]
Reactions with metal ions	
Me (II) with furazolidone	[85]
Me (II) with tetracyclines	[86]
W (VI) with cephalosporins	[87]
Cd (II) with antibiotics	[88]
Cu (II) with isoniazid	[89]
Mn (II)	
with amino acids and vitamin B7 (biotin)	[90]
with antibiotics and vitamin B6	[91]
with antibiotics and cefotaxin	[92]
with antibiotics and cephradine	[93]
with sulfodiazine and cefazolin	[94]
with sulfonamides and cephalothin	[95]
Ni (II)	
with curcumin	[96]
with amoxycilin and ampicillin	[97]
with antibiotics and cephradine	[98]
with penicillamine	[99]

(continued)

Table 2.4 (continued)

Properties and reactions	References
Zn (II)	
With antibiotics	[100]
With amino acids and biotin	[101]
With amino acids and nicotinamide	[102]
With tetracyclines and antibiotics	[103]
Investigations of other reactions	
Nitazole, cleavage of	[104]
Amino compounds, with 3,5-dinitrobenzoic acid	[105]
Heparin, with Malachite Green	[106]
Lysozyme, with Alizarin S	[107]
Salicylamide, with HNO_2	[108]
NO generating	[109]

2.2 Variants of Polarography

In addition to DC polarography (DCP) (Fig. 2.4a) lately numerous modifications of the way in which $i-E$ curves are recorded have been proposed. Among them, the most frequently reported are: the differential pulse polarography (DPP) (Fig. 2.4b), the square wave polarography (SWP) (Fig. 2.4c), and alternating current polarography (ACP) (Fig. 2.4d). These techniques differ principally in the ways of the time dependence of the applied voltage. In DPP the linearly increasing voltage is superimposed during the lifetime of a single drop by an increase of voltage of small amplitude. This increase is applied towards the end of the drop-time (Fig. 2.4). The difference of the current before and during the voltage step is measured and plotted as a function of the applied voltage. The recorded current–voltage curve is a derivative of the $i-E$ curve. The maximum current (a peak) is usually a linear function of concentration.

In the SWP, the voltage is increased by steps (Fig. 2.4c), and the current is often measured during a selected time interval, usually during the last 20 % of the voltage step. Finally in ACP (Fig. 2.4d), the linearly increasing voltage sweep is superimposed by an alternating voltage of small amplitude. The $i-E$ curve (as a function of the increasing voltage) is recorded.

The DPP and SWP were developed to increase the sensitivity of the measured current, but this applies only to reversible systems, where the equilibrium between the oxidized and reduced forms at the electrode surface is rapidly established. For such system, the low limit of concentration, in which the current can be accurately measured, is decreased from about 5.10^{-5} M in DCP to about 5.10^{-7} M in DPP. Such increase in sensitivity is rarely needed in analyses of pharmaceuticals.

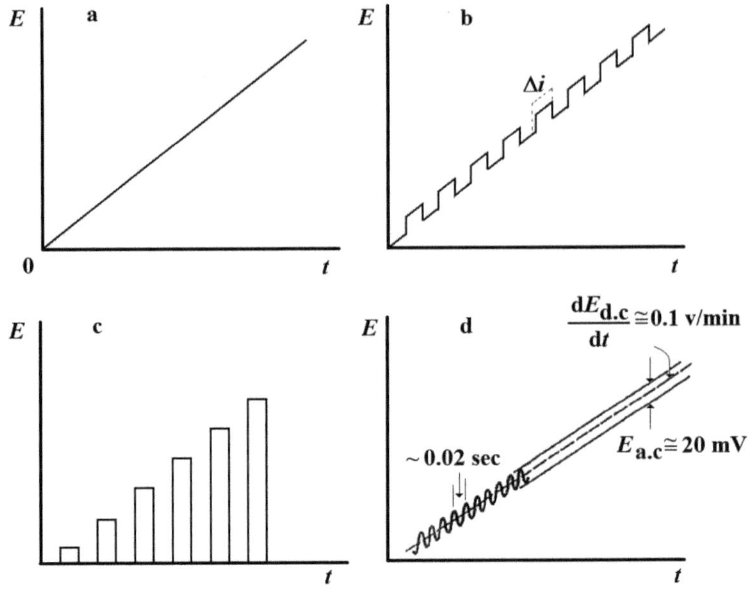

Fig. 2.4 Types of voltage ramps used in various polarographic techniques: (**a**) DC polarography (DCP); (**b**) differential pulse polarography (DPP); (**c**) square wave polarography (SWP) [112]; (**d**) AC polarography (ACP). Plotted: Applied voltage as a function of time

2.3 Record Keeping

For keeping records, describing the obtained polarographic i–E curves, the following information should be kept: (1) Composition of the studied solution, consisting of information about the kind of the supporting electrolyte (such as pH, nature of the buffer used, including concentration and kind of all components. The presence of salts necessary for controlling the ionic strength—where needed, presence of surfactants—when needed, and kind and concentration of all electroactive components should be stated. (2) Temperature of the analyzed solution should be given, if different from 25 °C. (3) Nitrogen or argon gases which are needed for removal of oxygen should be described together with the way of their introduction. (4) The kind of the reference electrode used and the way of its connection with the investigated solution should be mentioned. (5) The scales of axes on the polarogram should be given. For example: On axis x: How many mV correspond to 1 cm on this axis. And for the y axis: How many microamperes correspond to 1 cm on this axis. (6) Importantly, the characteristics of the glass capillary used should be stated: The drop time (t_1) in seconds, the outflow velocity of mercury from the capillary (m) in mg/cm, as well as the pressure of the column of mercury (h) given by the distance in cm between the level of mercury in the reservoir and the orifice of the capillary used. (7) The starting and the final potentials for the recorded curve are reported.

(8) All this should be recorded in a hard cover booklet. All of this will be important in the case of some legal or authorship problems.

If the above information is available, polarographic i–E curves can be reproduced in laboratories around the world.

2.4 Supporting Electrolytes

In order to use successfully the polarographic method of analysis, it is necessary to record polarographic curves with well-developed limiting currents. To obtain such curves, a proper supporting electrolyte must be used. In cases of analyses, in which the sample contains two or more electroactive species, the individual waves of each kind of species present must be well separated, that means that they must occur at sufficiently different potentials. To achieve such a goal is possible by a proper choice of the supporting electrolytes.

The choice of supporting electrolyte depends on (a) the potential range available in the given supporting electrolyte; (b) on the type and degree of interaction, either with the investigated species or on its electrolysis product.

In investigation of reductions of metal ions, the complex formation of the cation with the compound of supporting electrolyte is of importance. If the investigated species can exist in two oxidation states (like Fe^{3+} and Fe^{2+}), the complex formation with both states can play a role. For reductions and oxidations of organic species, most commonly the acidity or basicity of the supporting electrolyte, particularly the pH value plays an important role. In some instances, for example, when borate buffers are used, complex formation with the oxidized (e.g., 1,2-diols) or reduced forms of the investigated compound can play a role.

As in analyses of pharmaceuticals, the establishment of the content of the organic components usually plays a more important role than that of the inorganics; the further discussion in this chapter will be restricted predominantly to investigation of the reduction and oxidation of organic species. Exceptions are anodic waves, observed in solution of species, which can form slightly soluble compounds or complexes with mercury ions. In such processes, it is mercury—the material of the indicator electrodes—that undergoes anodic oxidation, and the generated mercury ions [both of Hg(I) and Hg(II)] react in the vicinity of the electrode surface with studied species. Examples of such inorganic species, which can be determined, using anodic waves of mercury compounds, are anions like Cl^-, Br^-, or OH^-. An example of complex forming species is CN^- ions. Examples of organic species are represented by anions like RS^- and some heterocyclic compounds, like barbiturates, thiobarbiturates, uracils, etc.

The first consideration for the choice of the supporting electrolyte is the range of available potentials. The range of potential limits, between which polarographic i–E curves can be recorded, depends on two types of processes, which have to be considered: At positive potentials, the limit is given by the potential, at which in a given supporting electrolyte, the oxidation of mercury takes place. Commonly used anions of the supporting electrolyte are sulfates and perchlorates, which allow recording up to about +0.4 V and chlorides which allow recording of i–E curves up to +0.1 or 0.0 V (exact values depend on concentration of the anion). At negative potentials, the available voltage ranges, to be reached, are limited by the kind of cations of the supporting electrolyte which are reduced. Up to pH at about 10, it is the potentials of reductions of hydrogen ions. This naturally depends on pH, becoming more negative with increasing pH. In unbuffered solutions and pH > 10, the most negative potential depends on the nature of the cation of the supporting electrolyte. The nature of this cation controls the potential at which the cation of the supporting electrolyte is reduced. In more frequently used buffered solutions at pH > 10, the available range at negative potentials depends on the nature of the cation of the electrolyte. Thus, if Na^+ or K^+ are the cations of the supporting electrolyte, the available limit at negative potentials is about −1.9 V. For solutions containing Li^+, this value is about −2.0 V. In the presence of tetraalkylammonium ions, even −2.5 V can be reached.

Thus, from the practical point of view, the DME can be used between potentials of +0.4 V and −2.5 V. Numerous organic compounds are reduced between about +0.2 and −2.5 V, but only those, which are most easily oxidized (like ascorbic acid and hydroquinone), form slightly soluble species with mercury ions or complexes, with mercury ions can be successfully (directly) determined by DCP. Generally, DME is an excellent electrode for reductions, but is also suitable for following the most easily oxidized species. It is also useful for following anodic processes involving formation of mercury ions and their reactions, yielding formation of slightly soluble or complex species. For other oxidations, solid metal electrode or a carbon based electrode are electrodes of choice.

The situation is reflected by the fact that reductions of more than 20,000 organic compounds have been studied by DCP. Dropping mercury electrodes are thus mostly used in studies of reductions, in which electrons are transferred from the DME, functioning as a cathode, to the reduced species.

The choice of the supporting electrolyte is very important and must be paid attention to. If no information is found in the literature for the given electroactive compound or related compounds, a number of various supporting electrolytes must be used, to find one in which the best developed waves are obtained, that are easy to measure. It must be taken into consideration that the supporting electrolyte can have several functions, which can be summarized as follows:

1. To keep low the resistance of the solution.
2. To minimize the migration of ionized species between the cathode and the anode.
3. To convert the investigated species into a certain form: a complex for inorganic, a protonated, dissociated, or uncharged form for organics.
4. To establish conditions, when best developed, polarographic waves are obtained and their limiting currents are most accurately measurable.
5. To allow establishment of optimum conditions when the acid, or the base, or the hydrated, or unhydrated form predominates.

6. If the investigated solution contains more than one electroactive species to be determined, to establish conditions (by applying a complexing agent or a buffer of suitable pH or a strongly acidic or strongly alkaline solution), where the waves of two or more electroactive species yield well resolved, separated waves.

As reductions of organic compounds are in most instances in aqueous solutions accompanied by a proton transfer (either before or after the electron transfer), electroreductions of organic compounds depend frequently on the composition of the supporting electrolyte used. Role of pH reflects the general rule: "The conjugate acid, the protonated form of the investigated species, is reduced more easily—which means at more positive potentials—than the corresponding conjugate base." When oxidations are followed, it is the conjugate base, which is oxidized more easily, that is at more negative potentials than the corresponding acid.

2.5 The Half-Wave Potentials and Limiting Currents

When the examined solution contains only supporting electrolyte and the applied potential is gradually increased from positive to negative values, a very small current increases (almost linearly). This increase continues until a potential is reached, where hydrogen ions or the cation of the supporting electrolyte undergoes reduction. In this potential range, the current continuously increases (Fig. 2.2); the increase is very small and at the sensitivity used hardly noticeable. This small current is known as "charging current," as it is due to the charging of the double layer of ions at the surface of the electrode. When DC polarography is used, the analyzed solution usually contains between 1 mM and 0.01 mM concentration of solution of the species to be determined. Under such conditions, the charging current remains much smaller than the current due to the reduction. For the measurement of the reduction current, it is usually sufficient to linearly extrapolate the charging current before the rise of the wave to the potential range, where the electroreduction takes place. In less than 10^{-5} M solutions of the electroactive species, the role of the charging current must be taken into account. This can be achieved by passing a linearly increasing current of opposite direction or using some other advanced techniques, like the derivative polarography, the square wave polarography (SWP), the differential pulse polarography (DPP), or AC polarography (ACP). In pharmaceutical analysis, such problems are not frequently encountered, as usually there is enough material available to prepare analyzed solutions containing more than 10^{-5} M electroactive compound.

In the presence of a reducible species, starting at a positive potential and applying gradually more negative potential, the current first increases slightly due to the charging current in the supporting electrolyte. At a certain potential, the value of which depends on the nature of the electroactive (reducible) species present, the current gradually increases. After a couple of hundred mV, the current reaches a constant (practically) limiting value, which remains practically constant for at least a few couple of hundred mV. The increase of the current above the charging current is called "limiting current" (Fig. 2.2). Often the shape of the i–E curve does not depend on the direction of the applied voltage. That means that if the voltage is applied from positive to the negative values or vice versa from the negative to positive potentials, practically, the same curve is obtained. Only if the anodic wave is due to a formation of slightly soluble or complex mercury compounds, the potential must be applied from the negative to the positive potentials. Mostly, the limiting current is a function of concentration of the investigated species, and often the dependence of the limiting current (i_L) on concentration is linear. The role of concentration on the shape of the dependence of the limiting current on the concentration of the electroactive species for various kinds of limiting currents is discussed below. For anodic waves, corresponding to oxidation, the dependences are a mirror image of that discussed above. Starting at negative potentials, a gradually more positive voltage is applied. At a potential, characteristic for oxidation of the given species in the supporting electrolyte, the anodic current starts to increase, until it reaches a limiting value. Formation of an anodic wave is observed. A potential, corresponding to one-half of the limiting current, and at the inflection point of the i–E wave, is the half-wave potential of the anodic wave. This half-wave potential of the anodic wave depends on the nature of the oxidized species and on the supporting electrolyte used.

In a given supporting electrolyte, the half-wave potential depends on the nature of the reduced or oxidized species. $E_{1/2}$ is a function of the nature and structure of the organic compound in the given supporting electrolyte, for example, of pH.

Hence, polarographic i–E curves offer:

(a) Both qualitative ($E_{1/2}$) and quantitative (i_L) information about the nature and concentration of the reduced or the oxidized species.
(b) The nature of the process involving the reduction or oxidation of organic compounds.
(c) Kind of limiting currents.
(d) Nature of the transport of the reduced or oxidized species to the electrode surface.

When a potential sufficiently more negative is applied to DME in a solution of a reducible organic compound, electrons are transferred from the mercury electrode to the reduced species. This results in a decrease of concentration in the immediate vicinity of the electrode surface. To re-establish the equilibrium condition in the vicinity of the electrode, where electron transfers can occur, more of the electroactive species is transported from the bulk of the solution towards the electrode surface. This transport is caused by a process called diffusion. The area around the electrode, from which the electroactive species can be transported to the electrode, is called "diffusion layer." Limiting currents which are controlled by the rate of diffusion are called "diffusion currents" (i_d).

Some organic compounds are present in aqueous solutions in equilibria, which are relatively rapidly established—for example between some acids [113] and conjugate bases or hydrated and non-hydrated forms [111, 114]. If one of the forms is more easily reduced than the other, the limiting current of the more easily reduced acid form undergoes reduction. The reduction of the more easily reduced form results in perturbation of the conditions in the vicinity of the electrode. To re-establish the perturbed equilibrium, some amount of the more easily reduced acid form is generated by conversion of some of the electro-inactive form into the more easily reducible species. In such situations, the limiting current of the more easily reducible species is converted by the rate of the conversion—for example, of an acid from a base or of an unhydrated form from the hydrated one [115]. The limiting currents in such situations depend on the rate of conversion of the less reactive into more reactive (the acid or the unhydrated forms). Such currents, controlled by the rate of a chemical reaction, are called "kinetic currents," (i_k).

In above discussed types of currents—i_d and i_k—it is assumed that the processes occur in homogeneous media. In some cases, adsorption either of the starting material or of the product of electrolysis is taking place. In such cases, the dependences of limiting currents on concentration are nonlinear [116]. Distinction between two types of adsorption is discussed below, but all have in common a nonlinear dependence of the limiting current on concentration.

There is one more type of processes, which yields a nonlinear dependence on concentration. These are the catalytic currents. Most common type of such currents involved an oxidation–reduction couple, like Fe(II)/Fe(III) and a component (the catalyst) of the supporting electrolyte which is able to reconvert Fe(II), formed by electroreduction, back into reducible Fe(III). Another example of catalytic currents involves the reduction of oxygen. This reduction occurs on DME in two steps: At rather positive potentials O_2 is reduced to H_2O_2 which at more negative potentials is reduced to water. In the presence of a catalyst, H_2O_2 is converted back to O_2 and the limiting current of O_2 is increased. In all these cases, the limiting current is a nonlinear function of the concentration of the catalyst.

A completely different type of current, which is also denoted as "catalytic," are processes which result in a shift of the reduction of H^+ ions to move positive potentials (so-called catalytic hydrogen waves [117]). The current at this potential depends on the structure, on concentration of the catalyst, and on pH. To this group belong organic substances bearing –SH and $-NH_3^+$ groups, such as amino acids, peptides, and some proteins.

2.6 Identification of the Type of the Limiting Current

The nature of the electrode process involved in the electrolysis can be distinguished, based on the dependence of the limiting current involved on concentration (C), mercury pressure controlled by the position of the reservoir of mercury (h), and the role of composition of the supporting electrolyte, in particular of pH.

The most frequently encountered type of limiting currents are diffusion currents. This type of current is controlled by the rate of the transport of the electroactive species from the bulk of the solution to the electrode surface. As this transport is

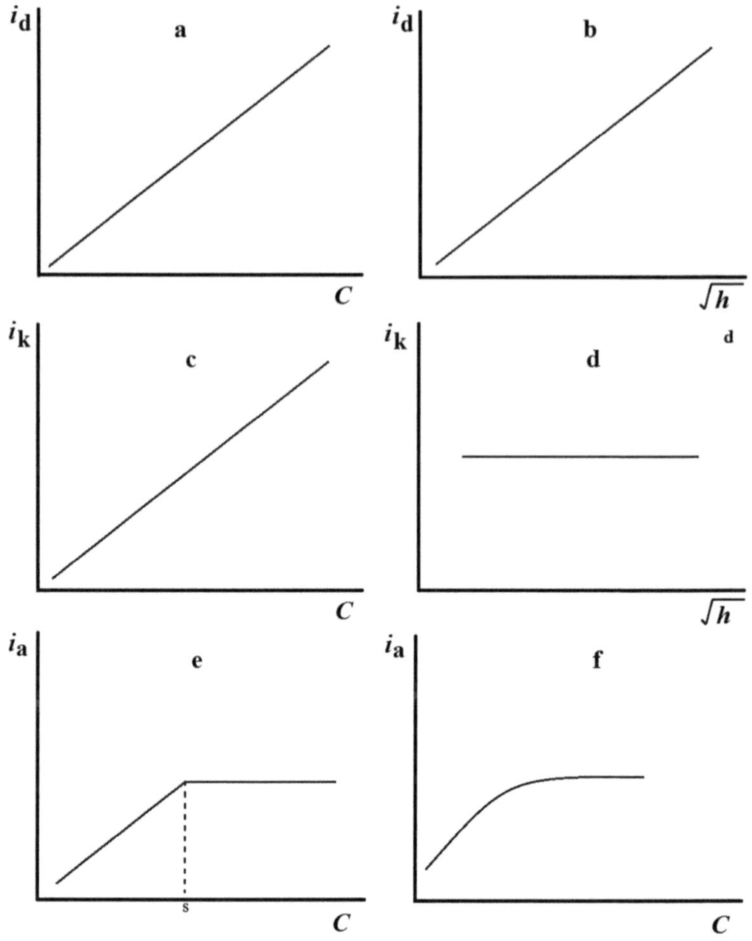

Fig. 2.5 Characterization of the types of currents, most frequently encountered in DCP: (**a**) Dependence of diffusion currents on concentration of the studied species; (**b**) dependence of diffusion limiting current on the square root of mercury pressure (**h**); (**c**) dependence of the limiting kinetic current on concentration of the studied species; (**d**) dependence of the independence of the limiting kinetic current on mercury pressure (**h**); dependence of a limiting current controlled by a rapidly established adsorption–desorption equilibrium on concentration of the studied compound (**e**); dependence of the limiting current controlled by a relatively slowly established adsorption–desorption process (**f**)

carried out by diffusion, this type of currents is called "diffusion currents" (i_d). These currents increase linearly with increasing concentration. The plot of $i_d = f(C)$ is linear and passes through origin (Fig. 2.5a). The diffusion current is also a linear function of the square root of the height of the mercury column above the orifice of the capillary electrode (Fig. 2.5b).

In another type of current, the electroactive form is generated from an excess of electro-inactive form present in excess in equilibrium. This chemical reaction takes place in the vicinity of the electrode within a reaction layer. The resulting current is called "kinetic current" (i_k). The kinetic current is also a linear function of concentration (Fig. 2.5c), but it is independent of mercury pressure (h) (Fig. 2.5d). These currents often increase much more (5–10 %) with increasing temperature, whereas the increase of i_d is just 1.84 %, which is simply because the rate of chemical reactions depends much more on temperature than the rate of diffusion.

Finally, there are adsorption currents (i_a). These currents are observed, when the oxidized, the reduced form, or both are adsorbed at the electrode surface. There are two types of adsorption currents, depending on the rate of adsorption, which can be distinguished based on the shape of the dependence of the limiting current on concentration of the adsorbed species. If the formation of the adsorbed layer is fast relative to the drop time, the reduction current first increases linearly with concentration of the reduced species and then remains independent of concentration. The concentration at the observed intersection of the two linear segments (Fig. 2.5e) corresponds to a complete coverage of the electrode surface by the adsorbed species, and the dependence of the limiting current as a function of concentration shows a nonlinear increase (Fig. 2.5f). When the formation of the adsorbate is slow, the current changes gradually with concentration of the adsorbed species.

As adsorption currents may represent an unwanted complication, solubility might be increased by 5−25 % of an organic co-solvent miscible with water, such as acetonitrile. At higher concentrations of the co-solvent, the solvation of the organic compound and its chemistry may be changed compared to its in water.

There are some other, less frequently encountered types of currents that may be important in analytical applications that will not be discussed here. Examples are catalytic currents. There are of two types: In one, in a reversible couple like Fe(III) $+e^- \leftrightarrows$ Fe(II) or $O_2 + 2e^- \leftrightarrows H_2O_2$, the reduction product (Fe(II) or H_2O_2) is converted by a catalyst (that may be an enzyme) back to the oxidized form and again reduced. In the second type, the catalyst causes a decrease of hydrogen overvoltage, and the current of the reduction of H^+ is shifted to more positive potentials. Such catalytic currents are used in investigations of proteins [118].

2.7 Role of pH

Most reductions as well as of oxidation of organic compounds, the transfer of one or more electrons is accompanied by transfer of one or more protons. This is manifested on polarographic i–E curves by changes in pH. Such changes can involve the values of half-wave potentials ($E_{1/2}$), or changes in the limiting current (i_L), or in both. Such changes represent an important tool in the elucidation of mechanisms of electrode process involved. The proton transfers can occur either before or after the electron transfer.

This situation involves the reduction processes will be discussed first. Simplest situation will be encountered in the pH range, where the limiting current (i_L) and the half-wave potential remain pH independent. The plot of $E_{1/2} = f$ (pH) (Fig. 2.6a) in this pH range corresponds to a linear segment at pH < pK_a, parallel with the pH axis. This reflects the situation, in which the species of the reduced compound is in the same form at the electrode surface and in the bulk of the solution. At pH > pK_a, the value of $E_{1/2}$ is shifted to more negative potentials. The plot of $E_{1/2} = f$ (pH) in this potential range is linear, with slope different from zero. This slope depends on the number of transferred electrons, on the number of involved protons, and the rate of protonation. The limiting current remains pH independent as long as the conjugate base is completely protonated within the pH range studied. In this pH range, the rate of protonation is sufficient to convert in the vicinity of the electrode all the predominant base form into the more easily reducible acid form. From the pH at the intersection of the linear segments, it is possible to determine the value of pK_a of the investigated compound. As within the investigated pH range, the rate of protonation is sufficiently rapid to convert all of the base form into the more easily reducible acid form; the limiting current in this pH range remains pH independent (Fig. 2.7). In the case of some reducible organic compounds, it is possible to extent the pH range higher than in the previous case. At a sufficiently high pH, the rate of the protonation of the predominant base form becomes insufficient to convert in the vicinity of the electrode surface all of the predominant base form into the more easily reducible conjugate acid form. As the rate of protonation decreases with increasing pH, the resulting i_L decreases with increasing pH. In this pH range, both the limiting current and $E_{1/2}$ change with changes in pH. The plot of $i_L = f$ (pH) has a shape of decreasing dissociation curve. The pH at the inflection point of such a curve is denoted pK' (Fig. 2.6b). At pH = pK' also the $E_{1/2} = f$ (pH) plot becomes pH independent.

When values of pK' are obtained from polarographic data (from $i_L = f$ (pH) or/and $E_{1/2} = f$ (pH) and the value of pK_a is either available from the literature or estimated from $E_{1/2} = f$ (pH) plots, it is possible to calculate the rate constant of the protonation. These are fast reactions with rate constant of the order between 10^7 and 10^{11} L mol^{-1} s^{-1}.

Situation dealing with evaluating dependences of $E_{1/2}$ and i_L on pH for oxidative processes is different and resembles mirror images of those discussed for reductions alone. This reflects the fact that a base form is more easily oxidized than the corresponding conjugate acid. When the investigation is limited to the pH range in the alkaline region, where the $E_{1/2}$ is constant and independent of pH and so is the limiting current, the plot of $E_{1/2} = f$ (pH) corresponds to a line parallel with the pH axis. At pH < pK_a, the $E_{1/2}$ is shifted gradually to more positive values. The slope of the $E_{1/2} = f$ (pH) linear plot depends on number of electrons in the potential determining step, number of H^+ (or OH^-) transferred before the potential determining step, and the rate of this transfer (Fig. 2.6b). The value of pH at the intersection of pH-independent line and of the line corresponding to potential shift due to the formation of the electroactive anionic form corresponds to a pK_a value of the acid–base reaction (Fig. 2.6c).

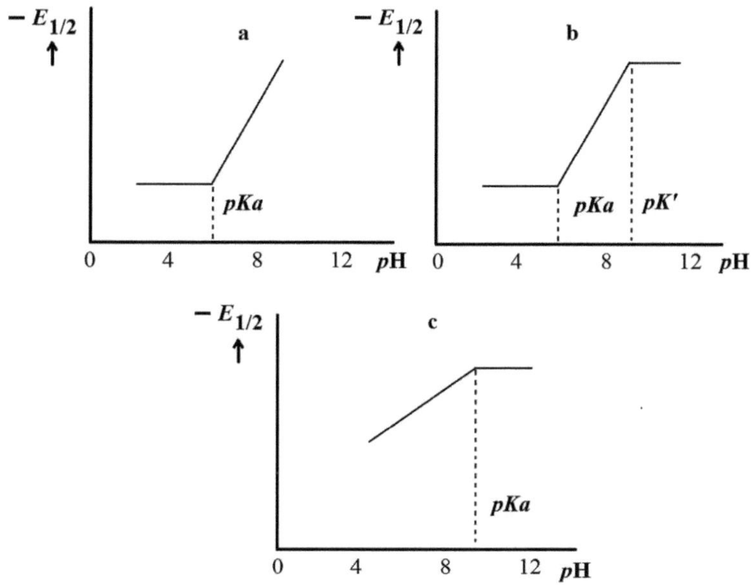

Fig. 2.6 Most frequently observed types of dependences of half-wave potentials ($E_{1/2}$) on pH: (**a**) reduction process, in which only the rapidly established acid–base equilibrium is accessible. From intersection of the two linear segments, the approximate value of pK_a can be found; (**b**) similar type of dependence, when investigation is possible also to higher pH values. The intersection at lower pH corresponds to pK_a, that at higher pH to pK'. That is a pH, where the rate of pre-protonation in the vicinity of the DME surface is rapid enough to convert 50 % of the available conjugate base back into free acid; (**c**) pH dependence of half-wave potentials of anodic waves, corresponding to the oxidation of the conjugate base. The pH at the intersection of the linear segments corresponds to the pK_a of a reaction in which one half of the base is converted into conjugate acid

Finally, in more acidic media, the $E_{1/2}$ becomes again pH independent. The case, where the limiting current decreases with decreasing pH and half-wave potentials became pH independent, has been observed for oxidation of substituted benzaldehydes [119], where the acid–base reaction is the addition of OH^- ions to the CH=O group. With accessible pK_a and pK' values, it was possible to calculate the rate constants of the addition of OH^- with substituted benzaldehydes, in good agreement with values obtained by other techniques. Slowly established acid–base equilibria that are observed for some C-acids involving cleavage of the activated C–H bond result in pH-dependent waves, the i_L of which is diffusion controlled (Fig. 2.7).

Alternatively, the pH can affect i–E curves, where the establishment of an equilibrium involving an electroactive species is pH dependent. This is the case of hydration–dehydration equilibria, where the rate of the dehydration (Fig. 2.8), yielding the reducible unhydrated form, is acid or base catalyzed. This results in a diffusion controlled current in the medium pH range, with increasing current in acidic solution controlled by the acid catalyzed dehydration. Similarly at pH higher

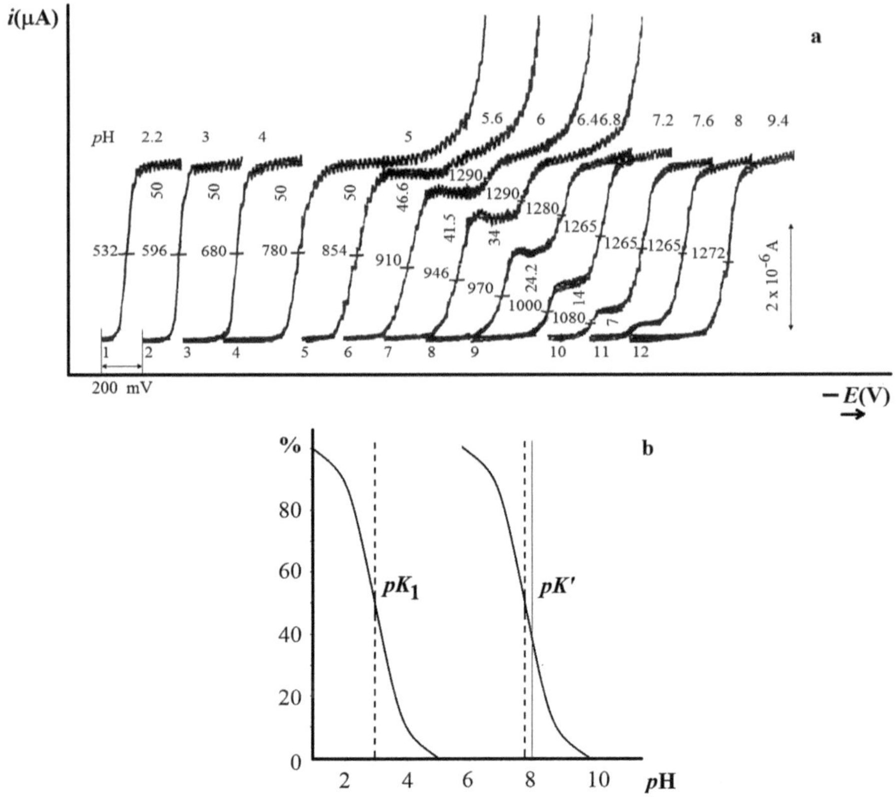

Fig. 2.7 Example of the role of pH on polarographic i–E curves involved in the reduction of an acid. (**a**) Dependence of DC polarographic waves obtained in solutions containing 0.2 mM solution of phenylglyoxylic acid in buffers of varying pH value. Limiting currents of the more positive reduction wave of the acid form decreases with increasing pH; decrease is accompanied by an increase of the wave of the conjugate base, the anion, at more negative potentials. (**b**) Plot of the bulk concentration of the phenylglyoxylic acid (obtained by titration or spectrophotometry) on pH. The inflection point of this curve (on the *left hand side*) corresponds to the pK_a of phenylglyoxilic acid. The plot at the *right* hand side corresponds to the dependence of the limiting current of the more positive wave of phenylglyoxylic acid on pH. The inflection point of this curve, where the limiting currents of the acid and the anion are equal, corresponds to pK'. The difference between pK' and pK_a is the greater the faster the pre-protonation in the vicinity of the electrode takes place

than about 7 or 8, the current increases due to a base catalyzed dehydration (Fig. 2.8).

So the role of pH should be investigated for all studied reductions or oxidations of organic compounds.

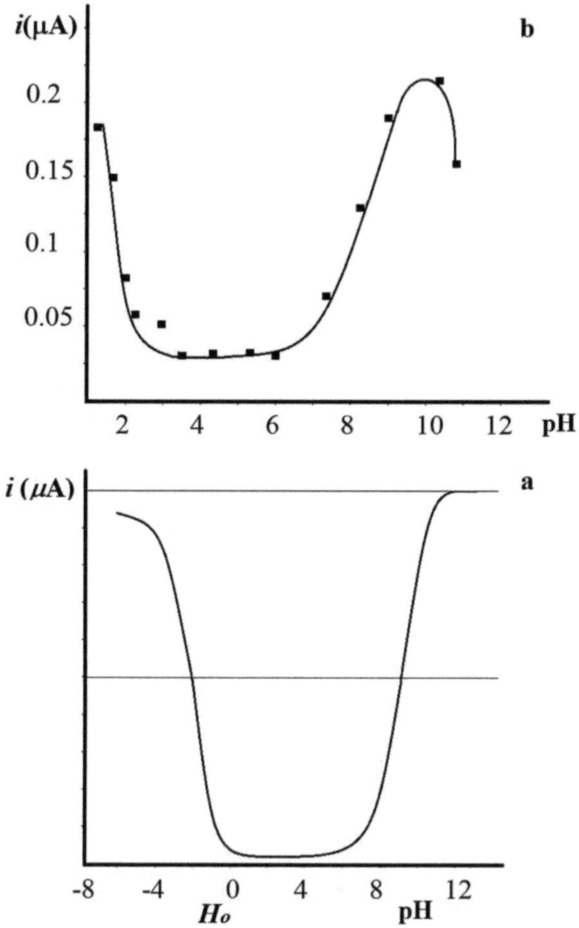

Fig. 2.8 Examples of the dependences of limiting currents of waves of compounds, involved in hydration-dehydration equilibria on pH. (**a**) Reduction of solutions of pyridoxal, (**b**) reduction waves of solutions of orthophthalaldehyde. Increases in limiting currents with increasing acidity correspond to the increase of the rate of acid catalyzed dehydration, increases above pH about 7 correspond to base catalyzed dehydrations. The decrease of the current of orthophthalaldehyde at pH values higher than about 12 are due to the dissociation of one OH group of the hydrated form

2.8 Investigation of a New Compound

A "new compound" is meant here a compound, the behavior of which has not been studied, or a compound, related compounds of which have not been critically studied. So the first action should be a literature search for data [120] dealing with electrochemical information for this or related species. If the search is negative, the following approach proved to be useful for development of a polarographic method of analysis.

The solubility of the investigated compound is tested first to prove if it is possible to prepare a 0.01 M stock solution of this compound in water. If the solubility is too low, it is possible to add an electroinactive co-solvent. The best is to add 5–25 % of organic co-solvent, such as acetonitrile, DMF, or THF.

The concentration of co-solvent in the investigated solution should be kept below about 25 %. At higher concentration of the co-solvent, changes in solvation of the electroactive species as well as changes of the mechanism of the electrode process might occur. Ethanol may be a suitable co-solvent, provided that it does not change the solution chemistry of the electroactive species. It is also possible to add 0.01 M HCl or 0.01 M NaOH to increase the solubility of the investigated compound, provided that no additional chemical reactions are involved.

For the initial investigation, a 0.01 M stock solution of the studied compound is prepared in water, and if necessary, in the presence of an organic co-solvent. A stream of nitrogen or argon gas is then passed through the investigated solution to remove soluble, reducible oxygen. Usually 1–2 min of a vigorous stream of nitrogen is sufficient to remove all of the interfering oxygen. While still introducing nitrogen stream, an aliquot of the stock solution is added to a chosen volume of the supporting electrolyte. It proved best to prepare the solution of the species to be studied in 0.2 mM final solution for recording of the $i–E$ curves. If necessary, 0.02 mM solutions can also be investigated. This situation occurs when the studied compound is slightly soluble or if a limited amount of the investigated compound is available. At final concentrations lower than 0.01 mM, the role of the charging current has to be considered. At concentrations higher than about 0.3 or 0.5 mM, adsorption phenomena can play a role. Next the effect of pH on the behavior of studied compound is followed. To obtain initial information, 0.2 mM solution is prepared by addition of chosen volume of the stock solution of the studied compound to obtain a 0.2 mM final concentration in the same volume of the following solutions:

1. 0.1 M HCl (pH about 1.0)
2. Acetate buffer pH 4.7
3. Phosphate buffer pH 6.8
4. Borate buffer pH 9.3
5. 0.1 M NaOH (pH about 13).

From resulting $i–E$ curves, the following information is sought: (a) In which pH range are polarographic waves present? (b) Is there a single wave or more waves observed? (c) How does the wave height (i_L) and the half-wave potential ($E_{1/2}$) change with pH? (d) Between which pH values are the changes in i_L and $E_{1/2}$ largest?

It is also recommended to record $i–E$ curves immediately after the removal of oxygen and compared them with $i–E$ curves obtained in solutions kept under nitrogen for 15–30 min. Changes in wave heights or numbers with time indicate presence of chemical reactions, occurring in the bulk of the solution, which can be acid or base catalyzed. If such reactions take place, extrapolation of i_L to $t = 0$ is necessary.

Comparison of changes of i_L and $E_{1/2}$ on curves obtained in the five supporting electrolytes mentioned above enables distinguishing the range (or ranges) of pH in which the values of i_L and $E_{1/2}$ show largest variations. This information is used when making the choice of the supporting electrolyte to obtain enough data in those pH region(s), where they are most informative and their variations are most marked.

Such well-planned experiments offer sufficient information about the role of pH within the electrode process. In the pH range, where the change of obtained data is most marked, the buffer used should differ by less than about 0.3–0.5 pH-units. On the other hand, in the range where i_L and $E_{1/2}$ show little variations, a difference between two successive curves can be of the order of a unit of pH.

For preliminary, less detailed, information, the use of a general buffer (such as the Britton–Robinson) is sufficient. This buffer contains boric acid, which can interact with numerous organic compounds. Therefore for more reliable information, the use of the following supporting electrolytes is recommended: 0.1 M $HClO_4$ or 0.1 M HCl (if the presence of chloride ions does not interfere) for pH about 1.0; 0.01 M $HClO_4$ with 0.1 M $NaClO_4$ (or 0.01 M HCl with 0.1 M KCl) for pH about 2.0; phosphate buffers pH 2.3–3.5; acetate buffers pH 3.7–5.7; phosphate buffers pH 5.8–7.8; borate buffers pH 8.3–10.3; phosphate buffers pH 10.5–11.7; 0.01 M NaOH with 0.1 M $NaClO_4$ (or NaCl) for pH about 12.0; and 0.1 M NaOH for pH about 13. If the i_L and $E_{1/2}$ obtained in borate buffers do not fit the plots obtained in phosphate buffers, an interaction of the studied compound with boric acid is indicated. Then for the range between pH 8.3 and 10.3, different buffers have to be used. If the investigated compound does not contain a C=O or N=O group, buffer pH 8.3–10.3 containing 0.1 M NH_3 and varying concentration of NH_4^+ ions (e.g., from 0.01 M to 0.1 M NH_4Cl) can be used. Alternatives can be veronal or p-phenol sulfonate buffers.

From recording of *i–E* curves over the investigated pH range, plots of $i_L = f$ (pH) and $E_{1/2} = f$ (pH) are constructed. Plots of $i_L = f$ (pH) and $E_{1/2} = f$ (pH) may already offer information about processes involved in the course of electroreduction. The distinguishing of characters of individual observed waves is often helpful. This can be achieved based on dependences $i_L = f(C)$, $i_L = f$ (pH), and $E_{1/2} = f$ (pH). Such information may not only help to understand the principles of the electrochemical processes involved, but also to allow finding optimum conditions for analysis, where the *i–E* recorded curves are best measurable and time independent.

2.9 Structure of Organic Compounds and Polarographic Activity

When polarographic analysis of a pharmaceutical is considered, it is the structure of the physiologically active component that must be first taken into consideration. A search of literature is carried out to learn, if such or a similar compound has been investigated and the course of reduction or oxidation has been reported. If the result of such search is positive and sufficient information about the role of the supporting electrolyte is available, conditions for polarographic analysis can be chosen. When this is not the case, it is checked, if the studied compound bears one of the groupings listed in Table 2.1.

In some cases, the presence of a single such grouping is sufficient for polarographic activity. In such situation, cathodic or anodic waves are observed on polarographic $i{-}E$ curves, the limiting current of which is a linear function of concentration of the studied species, in a suitable supporting electrolyte. If the investigated compound contains one of the reducible groups depicted in Table 2.1 and no polarographic wave is observed on the $i{-}E$ curve, in order to obtain a well measurable polarographic wave within the available potential range, the studied compound must contain an activating group.

In aliphatic and alicyclic compounds, the role of the activating group depends strongly on the distance between the electroactive center and the activating grouping. The facilitation of the electrode process depends on the polar effect of the activating grouping. When the activator is present on an aromatic ring, the activation depends on the mutual positions of the electroactive center and the activator, for example, in meta- or para-position. In such cases, both polar and resonance effects of the activator are of importance. Some effects of activating groupings can be treated quantitatively [121]. In ortho-position also steric effects of the activator play a role.

Among activating effects perhaps the most important one is the effect of conjugation, which plays an important role both in aliphatic chains and in alicyclic or aromatic rings. Such effects play an important role in heterocyclic rings, especially those containing –N=N– and >C=N– groupings.

Two rules have a general validity: In compounds, which can exist as an acid or conjugate base, the acid form is always reduced at more positive potentials than the conjugate base. Thus, the reduction of the conjugate acid takes place at more positive potentials than that of the corresponding base. On the other hand, the conjugate base form is more easily oxidized (that means at more negative potentials) than the corresponding acid. In polarography, the normally used rate of potential scan (100–200 mV/min) allows some acid–base reactions to occur in the vicinity of the electrode surface, in the so-called reaction layers. Depending on the rate of the establishment of the acid–base equilibria, this can be manifested either by a shift of the half-wave potentials (for very rapidly established acid–base equilibria) or by a change of limiting currents with pH.

Hence for investigation of any organic compound, that can be protonated in the reaction layer, the half-wave potential, or wave height of the limiting currents, the pH of the studied solution must be controlled. The same applies to anodic oxidations, facilitated with increasing pH. Thus, the role of pH on reductions or oxidations on DME cannot be overstressed.

In some cases, when the organic compound to be studied does not bear any of the groupings given in Table 2.3 or when the potentially electroactive group is not supported by an activating grouping, it is possible to introduce into the organic molecule by a chemical reaction a grouping that is reducible. An example of this type is the determination of morphine, which is converted into its nitro derivative by a reaction with potassium nitrite in acidic media (Fig. 2.9).

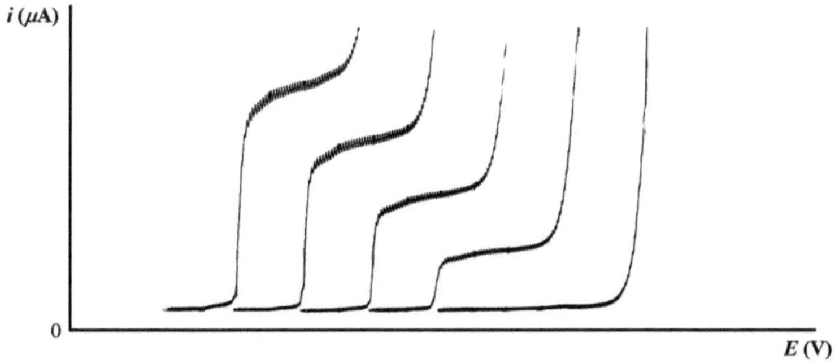

Fig. 2.9 Example of waves of a compound formed in a chemical bulk reaction taking place before recording of the DCP curve. For this determination a sample of morphine was converted into nitromorphine in a reaction with potassium nitrite in acidic media. Curve 1—supporting electrolyte, curves 2–5 obtained at varying concentrations of morphine

2.10 Conclusion

General principles of polarography, experimental conditions, and mathematical treatment of the theory involved can be found in monographs and original papers [1–5, 70, 110–131]. Analytical applications reported between 1990 and 2012 are summarized in Tables 2.2, 2.3, and 2.4 [6–69, 71–109]. In Table 2.2, the information is arranged alphabetically according to the name of the drug. In Table 2.3 the information is arranged alphabetically according to the physiological effect and kind of application of the drug. Finally, in Table 2.4 are listed some applications of polarography in the course of reaction of medications with heavy metal ions and references of some other reactions of pharmaceuticals are given.

As HPLC and AA and related techniques are currently predominant and LSV an CV among electrochemical methods, polarography is currently used in countries where research cannot afford more expensive analytical techniques. Use of LSV and CV, in which solid electrodes (most often carbon electrodes) are used for oxidations, polarography offers the possibility to analyze compounds that contain reducible groupings. In analyses of pharmaceuticals, DCP is usually sufficient, as analyses of trace amounts are not usually requested in analyses of pharmaceuticals. The high reproducibility of polarographic *i–E* curves may allow avoiding validation procedures. Hence, the use of polarography in routine analyses is rather limited, but variations of polarographic limiting currents with time offer the use of polarography in following chemical reactions in following the again of pharmaceuticals, following kinetics of their degradations with time or solution components in both chemical and enzymatic reactions as well as equilibria by following changes of reacting species is possible from following such reactions based on the variations of limiting currents with time makes polarography complementary to spectrophotometric techniques.

References

1. Heyrovsky J (1922) Chem Listy 16:256–264, Philosophical Magazine 45:303–315
2. Heyrovsky J, Kuta J (1966) Principles of polarography. Academic, London, UK
3. Bard AJ, Faulkner LR (2001) Electrochemical methods: fundamentals and applications, 2nd edn. Wiley, New York, NY
4. Galus Z (1994) Fundamentals of electrochemical analysis, 2nd edn. Ellis Horwood/Polish Scientific Publishers PWN, New York/Warsaw
5. Scholz F (2010) Electroanalytical chemistry, 2nd edn. Springer, Berlin
6. Dogan B, Uslu B, Ozkan SA, Zuman P (2008) Anal Chem 80:209–216
7. Abdine H, Belal L (2002) Talanta 56:97–104
8. Keller T, Mutz M, Aderjan R, Latscha MP (1999) Fresenius J Anal Chem 363:270–276
9. Debnath C, Haslinger E, Ortner A (2006) Nat Prod Commun 1:487–494
10. Korany MA, Hewala II, Abdel-Hay KM (2008) J AOAC Int 91:1051–1058
11. Zhang Y, Fan H, Kong X, Liu Y, Jinghao P (1998) Fenxi Huaxue 26:729–732
12. Belal F (1992) Microchim Acta 108:329–334
13. Weber JM, Volke J (1991) Pharmazie 46:853–857
14. El-Maali NA (1998) Bioelectrochem Bioenergy 45:281–286
15. El-Enany N, Al-Ghannam SM, Belal F (2003) Pharm Ind 65:76–81
16. Samota S, Garch A, Pandey R (2010) Port Electrochim Acta 28:87–94
17. Ramadan AA, Mandil H, Genco T (2009) Asian J Chem 21:7387–7397
18. Jain R, Mishra R, Dwivedi A (2009) J Sci Ind Res 68:945–950
19. Tanase IG, Matachescu C, David I, Ioneci I (1993) Rev Roum Chim 38:1133–1141
20. Ramadan R, Hasna M, Barra H (1996) Analusis 24:154
21. El-Sayed GO, Yasin SA, Badawy AA (2009) J Chem Pharm Res 1:225–232
22. Pinila GE, Calvo BL, Garcia-Monco Carra RM, Sanchez AM (1993) Electroanalysis 5: 343–347
23. Tanase I, Stangioiu I (1993) Rev Roum Chim 38:495–504
24. Sreedhar NY, Reddy JS (1992) Bull Electrochem 8:341
25. Ibrahim F, El-Enany N (2003) Farmaco 58:1313–1318
26. Oelschlaeger H, Volke J, Belal F (1992) Arch Pharm 325:65–68
27. Dogrukol D, Tuncel M (1991) Port Electrochim Acta 9:331–338
28. Reddy YVR, Reddy PRK, Mohan SK, Reddy SJ (1996) Bull Electrochem 12:446–448
29. Reddy YVR, Reddy PRK, Mohan SK, Reddy SJ (1996) Analusis 24:154–156
30. Ramadan AA, Mandil H, Hafez B (2011) Asian J Chem 23:403–406
31. Altinoz S, Temizer A (1990) J Pharm Sci 79:351–353
32. Squella JA, Alvarez-Lueje A, Sturm JC, Nunez-Vergara LJ (1993) Anal Lett 26:1943–1957
33. Guo R, Ding A, Zhong Y, Zhu C, Li J (1996) Zhongguo Kangshengsu Zazhi 21:406–407
34. Knoth H, Petry T, Gärtner P (2012) Pharmazie 67:987–990
35. Reddy GVS, Reddy CLN, Myreddy VN, Reddy SJ (2011) Clin Med Res 3:35
36. Sreedhar NY, Reddy PKR, Reddy SJ (1997) J Indian Chem Soc 74:477–479
37. Hoda M (2003) Bull Fac Pharm (Cairo Univ) 41:299
38. Squella JA, Lemus I, Nunez-Vergara LJ (1995) Anal Lett 28:467–484
39. Ghoneim MM, Tawfik A (2003) Can J Chem 81:889–896
40. Emara KM, Ali AMM, Abo-El Maali N (1994) Talanta 41:639–645
41. El-Enany N, Belal F, Rizk M (2008) J Biochem Bioph Methods 70:889–896
42. Yardimci C, Ozaltin N (2001) Analyst 126:361–366
43. Singh M, Kumar Y, Meena O, Garg A, Pandey R (2010) Int J ChemTech Res 2:1738–1744
44. Song JF, Shao Y, Guo W (2001) Anal Sci 17:1145–1148
45. Alvarez-Lueje A, López C, Núñez-Vergara LJ, Squella JA (2001) J AOAC Int 84:649–658
46. Abu-Khurmah M, Oelschlager H, Engel J (1997) Pharmazie 52:775–780
47. Squella JA, Sturm JC, Diaz MA, Pessoa H, Nunez-Vergara LJ (1996) Talanta 43:2029–2035
48. Dumitrescu V, Dumitrescu N, Balu AM (2005) Rev Chim 56:345–348

49. Ordieres AJM, Garcia AC, Alvarez JMF, Jose M, Blanco PT (1990) Anal Chim Acta 233: 281–287
50. Tompe P, Halbauer NA (1990) Acta Pharm Hung 60:130–142
51. Richter P, Toral MI, Quiroz G, Jaque P (1997) Lab Rob Autom 9:255–262
52. Subba Reddy GV, Jayarama Reddy SJ (2001) Indian J Environ Prot 21:816–819
53. Reddy CS, Sridevi C, Reddy SJ (1995) Bull Electrochem 11:591–593
54. Alvarez-Lueje AP, Bastias M, Bollo S, Nunez-Vergara L, Squella JA (1995) J AOAC Int 78: 637
55. Garg A, Meena OP, Kumar Y, Pandey R (2011) Int J ChemTech Res 3:33–38
56. Valnice M, Zanoni B, Fogg AG (1993) Analyst 118:1163–1166
57. El-Maali NA, Ghandour NA, Khodari M (1993) Talanta 40:1833–1838
58. Richter P, Toral MI, Munoz-Vargaz F (1994) Analyst 119:1371–1374
59. Galal SM, Abdel-Hay MH, Ragab MA, Alexanderia (2008) J Pharm Sci 22:38
60. Kotoucek M, Skopalova J, Michalkova D (1997) Anal Chim Acta 353:61–69
61. Jain S, Jain NK, Pitre KS (2002) J Pharm Biomed Anal 29:795–801
62. Atkosar Z, Tuncel M (1996) Anal Lett 29:2383–2397
63. Thanase IG, David IG, Radu GL, Iorgulescu EE, Magearu V (1996) Analusis 24:281–284
64. Abu-Khurmah M, Oelschlaeger H (1998) Pharmazie 53:32–38
65. Alvarez-Lueje A, Nunez-Vergara LJ, Squella JA (1991) Farmaco 46:593–600
66. Al-Khamees HA, Al-Obaid AM, Al-Rashood KA, Bayomi MS, Mohamed EM (1990) J Pharm Biomed Anal 8:225–228
67. Mandil H, Sakur AA, Alulu S (2010) Asian J Chem 22:2129–2135
68. Sreedhar NY, Sankara NM, Srinivasa KP, Prasad PR, Nageswar RC (2010) E-J Chem 7:166–170
69. Squella JA, Sturm JC, Alvarez-Lueje A, Nunez-Vergara LJ (1994) J AOAC Int 77:768–773
70. Ozkan SA (ed) (2012) Electroanalytical methods in pharmaceutical analysis and their validation, 1st edn. HNB Pub, New York
71. Hannisdal A, Mikkelsen O, Schroeder KH (2007) Collect Czech Chem Commun 72: 1207–1213
72. Xu D, Luo D (2007) Anal Lett 40:3060–3069
73. Shi J, Luo D (2005) Instrum Sci Technol 33:533–540
74. Sun W, Ding Y, Jiao K (2006) J Anal Chem 61:359–364
75. DeNardis NI, Zutic V, Svetlicic V, Frkanec R, Tomasic J (2007) Electroanalysis 19: 2444–2450
76. Sun W, Zhao N, Niu X, Wang Y, Jiao K (2009) J Chem Sci 121:217–223
77. Al-Shabrawy Y, Rizk M, Belal F, İbrahim F, Mesbah Alex AO (2004) J Pharm Sci 18:21–25
78. Liang Y (1997) J Chinese Pharm Soc 6:154
79. Michelitsch A, Zellnig K, Baumeister A, Schubert-Zsilavecz M, Likussar W (1997) Pharmazie 52:451–453
80. Michelitsch A, Wurglics M, Schubert-Zsilavecz M, Likussar W (1999) Phytochem Anal 10: 64–68
81. Michelitsch A, Rittmannsberger A (2003) Phytochem Anal 14:224–227
82. Hu S, Yan Y, Zhao Z (1991) Anal Chim Acta 248:103–108
83. Gonzaelez HM, Cosya CJ, Alpizar J, Font G (1999) Sci Pharm 3:152–156
84. Cheng Y, Wang F, Wang S, Li X (2005) Fenxi Kexue Xuebao 21:396
85. Fotouhi L, Kohestanian E, Heravi MM (2006) Electrochem Commun 8:565–570
86. Ghandour MA, Azab HA, Hassan A, Ali AM (1992) Monatsh Chem 123:853–864
87. Guru Prasad AR, Rao VS (2010) Res Pharm Sci 5:57–63
88. Dhingra P, Sharma S, Pandey RS (2006) J Ultra Chem 2:219–223
89. Garg A, Samota S, Singh M, Pandey R (2011) Int J ChemTech Res 3:70–74
90. Khanam A, Khan F (2009) J Indian Chem Soc 86:711–716
91. Chadar SN, Khan F (2006) J Indian Chem Soc 83:1242–1247
92. Khan F, Agrawal R (2009) Oxid Commun 32:216–225

93. Khan F (2009) Oxid Commun 32:463–472
94. Parihar MS, Khan F (2008) J Indian Chem Soc 85:765–768
95. Parihar MS, Khan F (2008) Port Electrochim Acta 26:377–384
96. Kumar Y, Garg A, Pandey R (2011) Int J Anal Bioanal Chem 1:110–115
97. Lyle SJ, Yassin SS (1993) Anal Chim Acta 274:225–230
98. Khan F (2008) Oxid Commun 31:380
99. Banica FG, Ion A (1998) Collect Czech Chem Commun 63:995–1006
100. Chadar SN, Khan F (2009) Khimiya 18:73–81
101. Khan F, Khanam A (2009) Chemija 20:154–161
102. Khan F, Khanam A (2009) Port Electrochim Acta 27:87–98
103. Khan F (2008) Oxid Commun 31:685–693
104. Levina VI, Azizov OV, Pyatakova NV, Severina IS, Arzamastsev AT, Grigorev NB, Granik VG (2002) Vopr Biol Med Farm Khim 4:6
105. Kohler W, Oelschlaeger H (2002) Pharmazie 57:41–44
106. Niu X, Zhang W, Zhao N, Sun W (2008) Bull Chem Soc Ethiop 22:162–172
107. Sun W, Jiang H, Jiao K (2005) J Chem Sci 117:317–322
108. Walash MI, El-Brashy AM, Sultan MA (1994) Microchim Acta 113:113–124
109. Levina VI, Khmelnitskaya EY, Chechekin GV, Danilov AV, Grigorev NB (2000) Zavod Lab 66:771–775
110. Valenta P (1960) Collect Czech Chem Commun 25:853–861
111. Segretario J, Sleszynski N, Partch R, Zuman P, Horak VJ (1986) Org Chem 51:5393–5396
112. Mirčeski V, Komorsky-Lovrič Š, Lovrič M (2007) Square-wave voltammetry, theory and application. In: Scholz F (ed) Monographs in electrochemistry. Springer, Berlin
113. Brdicka R, Wiesner K (1947) Collect Czech Chem Commun 12:138–150
114. Vesely K, Brdicka R (1947) Collect Czech Chem Commun 12:313–332
115. Brdicka R (1942) Z Elektrochem 48:278–288
116. Brdicka R, Wiesner K, Schaeferna K (1943) Naturwiss 31:390–391
117. Mairanovskii SG (1968) Catalytic and kinetic waves in polarography. Plenum, New York
118. Heyrovsky J, Babicka J (1930) Collect Czech Chem Commun 2:370–378
119. Bover WJ, Zuman P (1975) J Electrochem Soc 122:368–377
120. Meites L, Zuman P (eds) (1977–1983) Handbook series in organic electrochemistry, vols I–VI. CRC Press, Boca Raton, FL
121. Zuman P (1967) Substituent effects in organic polarography. Plenum, New York
122. Palecek E, Bartosik M, Ostatna V, Trefulka M (2012) Chem Record 12:27–45
123. Brdicka R (1933) Collect Czech Chem Commun 5:148–164
124. Koutecky J, Brdicka R (1947) Collect Czech Chem Commun 12:337–355
125. Koutecky J (1953) Collect Czech Chem Commun 18:597–610
126. Brezina M, Zuman P (1958) Polarography in medicine, biochemistry and pharmacy. Interscience, New York
127. Patriarche GJ, Chareau-Gosselin M, Vandenbalck JL, Zuman P (1979) Polarography and related electroanalytical techniques in pharmacy and pharmacology. In: Bard AJ (ed) Electroanalytical chemistry, vol 11. Marcel Dekker, New York, pp 141–289
128. Zuman P (2011) J Solid State Electrochem 15:1753–1759
129. Patriarche GJ, Vire JC (1987) Anal Chim Acta 196:193–204
130. Heyrovsky J, Zuman P (1968) Introduction to practical polarography. Academic, London, UK
131. Meites L (1965) Polarographic techniques, 2nd edn. J Wiley Interscience, New York

Chapter 3
Electroanalytical Techniques Most Frequently Used in Drug Analysis

Contents

3.1 Introduction

Amongst the variety of electrochemical methods and techniques available for drug analysis, voltammetry, i.e. the recording of the current, as a function of applied potential, has become the most important and widespread.

Sweep voltammetry in unstirred solution, based on continuously varying the potential that is applied across the electrode–solution interface and measuring the resulting current, where the predominant mode of mass transport is diffusion, is one of the most important and useful techniques for the study of electrochemical reactions and their mechanisms [1–10]. The form of the current–potential curves at stationary electrodes, in unstirred solution, depends on the rate of change of applied potential. Analysing these curves can give information about the kinetics and mechanism of reactions associated with electron transfer at the electrode

S.A. Ozkan et al., *Electroanalysis in Biomedical and Pharmaceutical Sciences*, Monographs in Electrochemistry, DOI 10.1007/978-3-662-47138-8_3

surface, since the faradaic current that flows at any time is proportional to the rate of the electrochemical reaction taking place at the electrode.

The current observed in a voltammetric experiment is dependent upon:

1. The rate at which material arrives from the bulk of solution to the electrode, mass transport
2. The rate of electron transfer across the electrode–solution interface, the charge transfer kinetics.

When the electron transfer rate is sufficiently fast, the electrochemical reaction is limited by the rate of mass transport of ions or molecules from bulk solution to or from the electrode surface.

There are three types of mass transport:

Convection
Migration
Diffusion

When any one of these types of movement is insufficient to transport the reactant to, or the product from, the electrode surface concentration polarisation is observed [1–10].

The simplest way of transferring reactants to the vicinity of the electrode surface is by mechanical means such as agitation or stirring, i.e. forced convection. Forced convection decreases concentration polarisation and is often not important on a short time scale. In an unstirred solution, there may be effects of natural convection, due to vibrations and temperature and density gradients. There is a group of techniques, hydrodynamic techniques, based on forced convection of the solution, in which the mass transport is defined by a combination of convection and diffusion. An example is the rotating disc electrode.

Migration is the movement of charged particles in response to a local electric field gradient. In bulk solution, ion movement is caused primarily by the influence of the electric field. The influence of the migration of electroactive species is generally made negligible by the addition of a fully dissociated inert electrolyte of sufficiently high ionic strength, which acts as the principal ionic charge carrier.

Diffusion is the movement of molecules or ions from a region of higher concentration to one of lower concentration. During an electrode reaction, the concentration of electroactive species at the electrode surface decreases, and a concentration gradient develops between the interfacial region and the bulk solution, which causes diffusion. Diffusion is common to all voltammetric experiments; however, the current response depends on the flux at the electrode surface, which in turn depends on the electrode size and geometry [1–10].

All electrochemical reactions are governed, at least in part, by the Nernst equation [1–8]. This equation specifies the relationship between the potential of an electrode and the concentrations of both oxidised (C_{Ox}) and reduced (C_{Red}) species involved in the redox reaction at that electrode at equilibrium:

$$\text{Ox} + \text{ne}^- \leftrightarrows \text{Red}$$

$$E = E^{0'} + \frac{RT}{nF} \ln \frac{[C_{\text{Ox}}]}{[C_{\text{Red}}]}$$

In this equation, $E^{0'}$ is the formal potential of the couple involving Ox and Red, C_{Ox} is the concentration of the oxidised species and C_{Red} is the concentration of the reduced species. If an electrode reaction occurs at a particular applied potential, then C_{Ox} and C_{Red} at the electrode surface do not have the same value as in bulk solution. Hence, there is a driving force for transport by diffusion of analyte to or from the electrode surface. The transports of the analyte to the electrode surface, and the rate of oxidation or reduction of analyte at the electrode surface, together determine the magnitude of the current. If electron transfer is sufficiently fast, i.e. reversible electrode reactions, then the current is determined only by the rate of mass transfer of analyte to the electrode surface and the concentration difference between the electrode/solution interface and the bulk at the value of the applied potential used, according to the Nernst equation.

The rate of diffusion can be expressed mathematically by Fick's laws of diffusion. Fick's first law states that material will diffuse from a region of high concentration to low concentration, in our case the electrode surface. This mass flux, J, in one dimension is:

$$J = -D\left(\frac{\partial C}{\partial x}\right)$$

where D is the diffusion coefficient of the species being transported to the electrode surface, and $(\partial C/\partial x)$ is the partial derivative of concentration with respect to distance from the electrode surface.

The change of concentration with time and space is described by Fick's second law:

$$\frac{\mathrm{d}C}{\mathrm{d}t} = D\left(\frac{\partial^2 C}{\partial x^2}\right)$$

Fick's equations in three dimensions allow mathematical solution of all diffusion problems, which can be used for system characterisation.

The flux of species, J, is converted into a current by Faraday's law, which states that the current is equal to the flux multiplied by the number of moles of the species undergoing oxidation or reduction, the number of electrons transferred and the Faraday constant.

Thus, at a uniformly accessible planar electrode, the current is proportional to the concentration gradient:

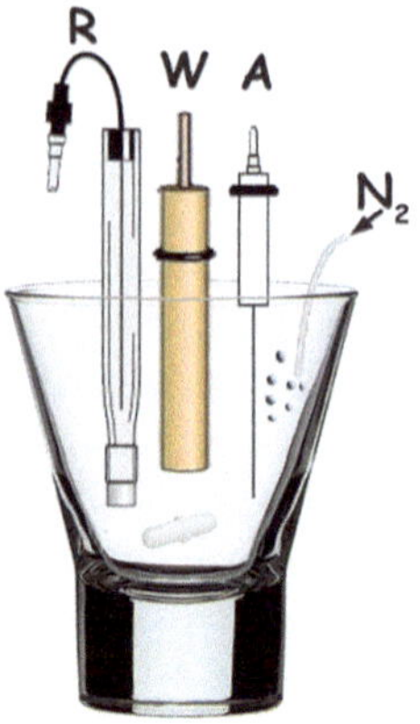

Fig. 3.1 Electrochemical cell with three electrodes: (W) working, (A) auxiliary and (R) reference; and (N_2) a nitrogen gas line

$$I = \frac{dQ}{dt} = -nFAJ = -nFAD\left(\frac{\partial C}{\partial x}\right)_{x=0}$$

This is the fundamental relationship that is used to calculate the current in any electrochemical experiment. Generally, the term $(\partial C/\partial x)_{x=0}$ is dependent on time, and on how the applied potential is varied with time and thus can be difficult to evaluate, since the potential determines the ratio (C_{Ox}/C_{Red}) at the electrode surface. Diffusion always occurs and is thus often the most significant transport process for many electrochemical reactions.

Generally, the potential of the working electrode is controlled versus a reference electrode possessing a constant and known potential and the electrical circuit is completed with an auxiliary electrode, which should be inert, e.g. platinum, and has a large surface area compared to the working electrode. This three-electrode arrangement is schematised in Fig. 3.1. A potentiostat that controls the potential of the working electrode vs. the reference in such a way that the current flows almost exclusively between the working electrode and the auxiliary electrode, while a very small, practically negligible current is passing through the reference electrode (usually Ag/AgCl or calomel (Hg/Hg_2Cl_2) electrodes are used). The essential elements needed for such an electrochemical measurement are:

- The working electrode: usually made of an inert metal or carbon based material (such as gold, platinum, glassy carbon, carbon paste and boron-doped diamond)
- The solvent: this usually has a high dielectric constant (e.g. water or acetonitrile) to enable the electrolyte to dissolve and thus aid the passage of current.
- A background electrolyte: this is an electrochemically inert salt (e.g. NaCl in water or tetrabutyl ammonium perchlorate (TBAP) in organic solvents) or buffers (e.g. phosphate, acetate, borate and Britton-Robinson in water) and is usually added in high concentration (0.1 M) to allow the current to pass and suppress migration effects.
- The reactant: in low concentration typically between 10^{-6} and 10^{-3} M.

Voltammetry is employed to investigate electrode reaction mechanisms and for electroanalysis. A plot of potential vs. current is called a voltammogram. Normally, the applied potential is controlled and the current response recorded but the opposite, i.e. applying a current and record the change in potential, can also be useful as will be seen in Sect. 3.5.3. Analysis of voltammograms can yield analytical, kinetic and thermodynamic information as well as information on other phenomena such as adsorption on the electrode surface.

There are many forms of voltammetry, the most important of which will be addressed in the following sections. Some selected applications in pharmaceutical analysis will be discussed in other chapters.

3.2 Linear Sweep and Cyclic Voltammetry

Potential sweep methods, i.e. linear sweep voltammetry (LSV) and cyclic voltammetry (CV), probably represent the most widely used diagnostic tool for studying electrode processes [6–11]. The difference between LSV and CV is that the former is just one single sweep whilst in CV the potential sweep is inverted to form a cycle or multiple cycles. LSV and CV techniques were first reported in 1938 and described theoretically in 1948 by Randles and Sevcik. CV has become a very popular technique for initial electrochemical studies of new systems and has also proven very useful in obtaining information about fairly complicated electrode reactions particularly in the field of organic chemistry and of pharmaceutical compounds. It is often the first experiment performed in an electrochemical study. It is the most widely used technique for acquiring qualitative information about electrochemical reactions and mechanism and is also a powerful tool for the rapid determination of formal potentials, detection of chemical reactions that precede or follow electron transfer or evaluation of electron transfer kinetics. It enables the electrode potential to be rapidly scanned to search for redox couples and makes possible the elucidation of the kinetics of electrochemical reactions taking place at electrode surfaces. One can simultaneously activate molecules by electron transfer and probe subsequent chemical reactions.

In conventional LSV and CV experiments, electrochemical measurements are performed in a quiescent solution of an electroactive species in a chosen supporting (inert) electrolyte. The concentration of the electroactive species is typically between 10 μM and 1 mM while the supporting electrolyte typically has a concentration of 0.10 M. The resulting solution may need to be deoxygenated, if reactions in the negative potential range are to be studied, to avoid contributions from oxygen reduction and possible secondary reactions with dissolved oxygen. Purging the solution with nitrogen or argon for 5–15 min is done for this purpose.

The LSV/CV technique consists in the application of a potential, which increases (or decreases) linearly as a function of time according to:

$$E = E_{\mathrm{i}} + vt$$

where E_{i} is the initial potential and v is the scan rate (expressed in mV s^{-1} or V s^{-1}). The starting potential value, E_{i}, is usually selected so that the chemical species under investigation is not initially oxidised or reduced. The potential is then swept to a voltage E_{max} or E_{min} (positive or negative sweep, respectively), at which point the scan direction is reversed back to the original, or to a different, potential value, E_f. The potential limits, E_{max} and E_{min}, are selected so that the potential interval scanned includes the oxidation or reduction process of interest. The current vs. applied potential curve is recorded over one sweep (linear sweep), one cycle or multiple cycles (Fig. 3.2).

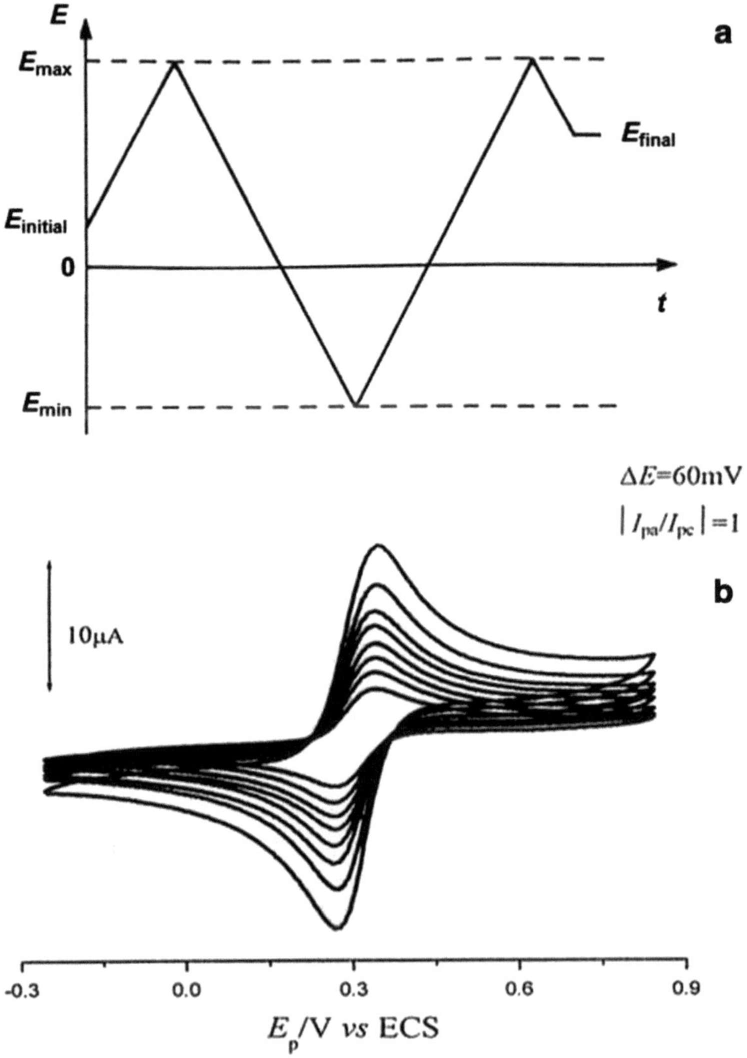

Fig. 3.2 Cyclic voltammetry (CV): (**a**) typical potential triangular waveform, (**b**) current–potential curve for a reversible system at different scan rates

If, in a CV experiment, only the reduced (Red) form is initially present in solution, a positive-going potential scan is chosen for the first half-cycle, starting from a value where no oxidation occurs. As the applied potential approaches the characteristic formal potential ($E^{0'}$) for the redox process, an anodic current appears and begins to increase, reaching a maximum peak value, after which the current decreases owing to depletion of electroactive species at the electrode surface. After scan inversion at E_{max}, Ox species in the vicinity of the electrode surface can be reduced back to Red, resulting in a cathodic peak (Fig. 3.2b). Thus, CV is capable of generating a species during the forward scan and then probing its fate on the reverse scan.

The important parameters in CV are the magnitudes of the peak currents, $I_{p,a}$ and $I_{p,c}$, and the potentials at which peaks occur, $E_{p,a}$ and $E_{p,c}$. Measuring accurate values of peak currents is perhaps the biggest difficulty of analysing cyclic voltammograms.

A redox couple in which both species rapidly exchange electrons with the electrode, i.e. the potential–current profile is governed solely by diffusion and by the Nernst equation, is termed an electrochemically reversible couple.

However, in most cases, the CV profile is more complicated. The main reasons for deviations from reversible behaviour include:

- Slow kinetics of interfacial electron transfer
- The presence of parallel chemical reactions in the solution phase
- The occurrence of surface effects such as gas evolution and/or adsorption/desorption and/or formation/dissolution of solid deposits

Furthermore, CV curves can be distorted by uncompensated ohmic drop and capacitive effects in the cell [1–5, 11–16].

For a reversible system, the anodic peak current, $I_{p,a}$, is given by the Randles–Sevcik equation ($T = 25$ °C).

$$I_{p,a} = 2.69 \times 10^5 n^{3/2} A\, D^{1/2} v^{1/2} C$$

where $I_{p,a}$ is the anodic peak current (in amperes), v is the scan rate (V s^{-1}), n is the number of electrons transferred per species, A is the electrode area (cm^2), D is the diffusion coefficient of the electroactive species (cm^2 s^{-1}) and C is the concentration of analyte in bulk solution (mol cm^{-3}). For a cathodic peak, $I_{p,c}$ is given by the same expression but with a negative sign. Beyond the current peak, the current falls due to mass transport limitations to the electrode surface.

The value of E_p is useful in characterising compounds and can be used for identification. The midpoint potential of the anodic and cathodic peaks in the CV is given by:

$$E_{\mathrm{mid}} = \frac{E_{\mathrm{p,a}} + E_{\mathrm{p,c}}}{2} = E^{0'} + \frac{RT}{nF} \ln\left(\frac{\sqrt{D_{\mathrm{Red}}}}{\sqrt{D_{\mathrm{Ox}}}}\right)$$

where $E^{0'}$ is the formal potential, and D_{Ox} and D_{Red} are the diffusion coefficients of the oxidised and reduced species. It can usually be assumed that D_{Ox} and D_{Red} are equal and in such cases the midpoint potential is nearly equal to the formal potential.

The separation between anodic and cathodic peaks of the CV is given by:

$$\Delta E_{\mathrm{p}} = E_{\mathrm{p,a}} - E_{\mathrm{p,c}} = \frac{57}{n} \quad \mathrm{mV} \quad (\text{at } 25\ ^{\circ}\mathrm{C})$$

All the information concerning LSV and CV for a reversible reaction controlled by analyte diffusion can be summarised [1–5, 8–16].

For LSV

- I_{p} is proportional to $\sqrt{v}$
- E_{p} is independent of scan rate, v
- $E_{\mathrm{p}} - E_{\mathrm{p/2}} = 56.6/n$ mV
- I_{p} increases linearly with concentration

and for CV

- $\Delta E_{\mathrm{p}} = E_{\mathrm{p,a}} - E_{\mathrm{p,c}} = 57.0/n$ mV
- $|I_{\mathrm{p,a}}/I_{\mathrm{p,c}}| = 1$

Hence, in CV the peak separation can be used to determine the number of electrons per molecule of analyte oxidised or reduced, and as a criterion for Nernstian behaviour. In particular, if ΔE_{p} is greater than $57/n$ mV then the reaction is not fully reversible.

Larger differences or nonsymmetric reduction and oxidation peaks are an indication of a quasi-reversible or irreversible reaction. For a quasi-reversible system, the electron transfer reaction does not respond instantaneously to a change in potential. The current is controlled by both charge transfer and mass transport. Compared with a reversible voltammogram, the anodic peak is shifted to more positive potentials and the cathodic peak to more negative potentials. The quasi-reversible CV parameters can be summarised as [1–5, 8–16]:

- E_{p} is dependent on v.
- ΔE_{p} depends on v and increases with v.
- I_{p} depends on $\sqrt{v}$ in a non-linear way, due to the transition from a reversible to an irreversible system.
- I_{p} is linearly dependent on concentration.

Moving towards irreversible CV systems, when the kinetics has a successively greater effect in slowing the electron transfer, the current peaks become reduced in height and broader, and anodic and cathodic peak potentials become more widely separated. In the irreversible limit, no reverse reaction is observed, as happens with the oxidation of most pharmaceutical compounds. Totally irreversible electrochemical systems are characterised by a shift of the peak potential with the scan rate, according to:

$$E_{\mathrm{p}} = E^{0'} - \frac{RT}{\alpha n F}\left[0.78 - \ln\frac{k^0}{D^{1/2}} + \ln\left(\frac{\alpha n F v}{RT}\right)^{1/2}\right]$$

where α is the transfer coefficient, n is the number of electrons which are involved in the charge transfer step and k_0 is the standard rate constant of the electron transfer reaction. For an irreversible system, the peak current is:

$$I_{\mathrm{p}} = 2.99 \times 10^5 n(\alpha n_{\mathrm{a}})^{1/2} A D^{1/2} v^{1/2} C$$

where n_{a} is the number of electrons in the rate-determining step of the electrode process. In this irreversible case, LSV and CV are equivalent.

The parameters for irreversible reactions can be given as [1–5, 8–16]:

- E_{p} depends on v.
- I_{p} is proportional to $\sqrt{v}$,
- I_{p} is linearly proportional to concentration.

For irreversible systems, the peak potentials of anodic or cathodic processes are shifted toward more positive or negative potential values, respectively. In many cases, the value of $n_{\mathrm{a}} = 1$ but, in other cases where n appears to be 2, much care must be taken in using the above equation as it can lead to incorrect deductions for the value of α, for example, when $n = 2$ in a mechanism corresponding to a pre-equilibrium followed by a rate-determining step, each involving one electron. For further information on such cases, see [3].

In some cases, the electroactive species, or its oxidation or reduction products, adsorbs on the electrode surface and continues to give a voltammetric response (rather than just blocking the surface by adsorption). Alternatively, it may be confined close to the surface within an immobilisation matrix or be covalently linked to the surface. In the CV of such systems, the anodic and cathodic current peaks are closer together, since diffusion from bulk solution has no role. In the CV of reversible surface-confined species, the potentials of the anodic and cathodic peaks are exactly the same and the peak current is directly proportional to the scan rate (rather than square root of scan rate) [1–5, 8–16]. In this case the peak current, exemplified for an oxidation, is given by:

$$I_{\mathrm{p,a}} = \frac{n^2 F^2 v A \Gamma}{4RT}$$

in which Γ is the surface concentration (sometimes called surface excess) of electroactive species, expressed in mol cm^{-2}.

In modern digital potentiostats, it is not possible to directly apply a true linear waveform without making use of digital to analogue and analogue to digital converters, and a staircase waveform is used as an approximation. Hence, staircase voltammetry (SV) can be considered as the digital version of LSV and CV. However, there are some differences between LSV, CV and SV techniques. LSV or CV furnishes a continuous current vs. potential curve, while SV consists of many discrete current–potential points. Also, the peak currents obtained under conditions of identical scan rates in linear scan and SV techniques may differ considerably. Thus, care must be taken to ensure that the "SV" equivalent of CV is implemented in such a way as to avoid any differences. Some more details on SV will be given below in Sect. 3.3.

One of the most important applications of CV is for the qualitative diagnosis of coupled chemical reactions that precede or succeed the redox process of inorganic, organic or pharmaceutical compounds. It can also be used for clarifying and understanding the redox mechanisms of many biologically important molecules and pharmaceutically active compounds. The results of CV investigations into the redox mechanism of pharmaceutically active compounds and biomolecules may have profound implications for the understanding of in vivo redox processes, as in neuroscience, or pharmacological activity [1–7, 10–19]. For example, dopamine can easily be distinguished from serotonin by LSV or CV. In addition, by increasing the potential scan rate, the current response of ascorbic acid, which normally overlaps with that of the catecholamines, may be eliminated.

Most quantitative determinations are usually performed solely by LSV since they are based on measurement of the current peak in the first scan. However, some researchers prefer to use CV for the quantitative determination of pharmaceutical compounds. The detection limit (LOD) in both techniques is mainly governed by the ratio of the faradaic current to the background charging (capacitive) current. Typical scan rates are between 25 and 1,000 mV s^{-1} for analytical purposes. The detection limits of CV and LSV are usually about 10^{-6} M, so that these techniques are not sensitive enough to determine drugs in body fluids after therapeutic doses. However, they are well suited for the rapid, simple, sufficiently sensitive and accurate determination of drug-active compounds in raw materials or in their dosage forms. In routine analytical work, the main advantages of CV and LSV are:

1. 1.Specificity due to both a large available timescale and a good resolution of peak-shaped curves
2. Quickly recorded current–potential curves
3. Sufficient sensitivity
4. Using LSV electrochemical detection in flow analysis, such as coupled with HPLC, capillary electrophoresis and FIA

During recent years, there has been extraordinary progress in the discovery, synthesis, sensitive analysis and means of delivery of pharmaceutically active compounds used in the diagnosis, prevention and treatment of human diseases. CV is an appropriate analytical tool, in complex media where chemical reactions are coupled with electron transfer reactions, as is the case for many of these compounds. As a general rule, many drug-active compounds can be readily oxidised or reduced in contrast to the excipients of pharmaceutical dosage forms (tablets, sugar-coated tablets, capsules, syrup, etc.). Thus, sample preparation is simple and usually consists of dissolving the drug-active compound in a suitable solvent, filtered (or not) and directly analysed electrochemically without interferences from commonly encountered excipients. In the analysis of pharmaceutically active compounds in their dosage forms, electrochemical methods have quite often been shown to be superior to classical analysis techniques and to spectrophotometric methods.

Biological sample analyses cannot be carried out directly. First of all, the serum proteins and endogenous substances should be removed. Depending on the study, sometimes the biological samples can be analysed after precipitation of the endogenous substances but in other cases extraction procedures are needed [6–11].

Most publications are related to the application of LSV and CV techniques to drug dosage forms as seen in the numerous review articles and books published in the field [6–11, 18–26]. Also, the literature shows [27–50] that CV is a convenient technique for revealing the redox properties of drug-active compounds.

3.3 Step and Pulse Voltammetry

A variety of pulse electrochemical techniques has been developed which makes use of the data acquisition capabilities in modern instrumentation [1–7, 15]. The potential step is the basis of pulse voltammetry, which comprises the application of combinations of potential steps to the electrode.

The diffusion current resulting from a potential step (pulse) is given by the modified Cottrell equation:

$$I_{\mathrm{d}} = \frac{nFAD^{1/2}C}{\sqrt{\pi t_{\mathrm{s}}}}$$

where A is the electrode area, n is the number of electrons transferred, D is the diffusion coefficient of the electroactive species and t_{s} is the sampling time after application of the step (pulse). The increase in sensitivity, compared to sweep or similar voltammetric methods, is due to the short pulse duration, between 20 and 50 ms. The idea of imposing potential pulses and measuring the currents at the end of each pulse was proposed by Barker and Gardner in 1958 [51]. However, the first examples of affordable potentiostats which enabled pulse voltammetric techniques to be performed appeared on the market only in the 1970s, because of limitations of the electronics. In the 1990s, substantial progress in pulse voltammetric instruments occurred with the advent of digitally controlled potentiostats.

Pulse techniques were initially developed for the dropping mercury electrode (DME), the objective being to synchronise the pulses with drop growth and reduce the capacitive current contribution by current sampling at the end of drop life, as well as increasing sensitivity—after applying a potential pulse, the capacitive current dies away faster (exponentially) than the faradaic current, (dependence on $t^{-1/2}$) referred to as pulse polarography [1–7, 52–54]. Nowadays, the DME is little used, because of its large charging currents and difficult-to-control convection at the electrode surface. Instead of the DME, static mercury drop electrode (SMDE), hanging mercury drop electrode (HMDE), controlled growth mercury drop electrode (CGMDE) or mercury film electrode (MFE) can be used for reduction studies of pharmaceutically active compounds. At solid electrodes, where oxidations can also be investigated and monitored, there is the additional advantage of discrimination against blocking of the electrode reaction by adsorption, to which mercury is particularly prone.

Pulse voltammetric techniques enable lowering the detection limits of electroanalytical methods. By substantially increasing the ratio between the faradaic and non-faradaic currents compared to sweep techniques, and increasing sensitivity, pulse techniques may permit a limit of quantitation at about 10^{-8} M concentration level and, with correctly chosen parameters, the measured current consists almost solely of the faradaic current, Fig. 3.3. At solid electrodes there is the additional advantage of discrimination against blocking of the electrode reaction by adsorption.

Some important parameters in pulse techniques are [15]:

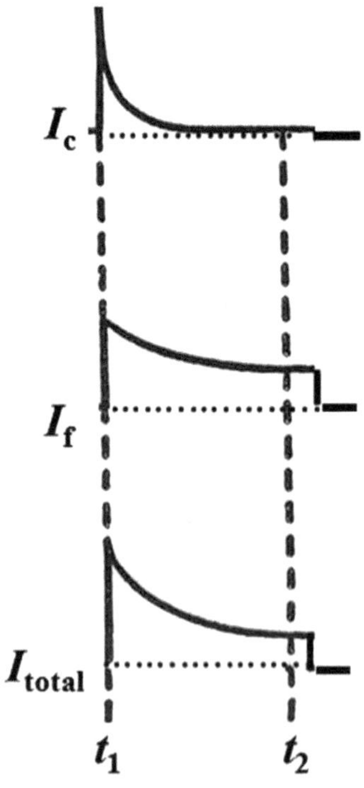

Fig. 3.3 Voltage pulse resulting current–time behaviour: I_c—charging current, I_f—Faradaic current, I_{total}—total current, t_1—pulse applied and t_2—time current measuring starts

- Pulse amplitude ($\Delta E \sim 5$–200 mV)
- Pulse width ($t \sim 5$–50 ms)
- Sampling period (4–15 ms)
- Pulse period ($\tau \sim 0.5$–5 s)

Pulse amplitude is the height of the potential pulse, and pulse width is its duration ($t \sim 5$–50 ms). Sampling period is the period of time near the end of the pulse during which the current is measured (4–15 ms). For some pulse techniques, the pulse period must also be specified, which is the time required for one full potential cycle ($\tau \sim 0.5$–5 s).

In order to increase speed and sensitivity, many forms of potential modulation have been evaluated over the years. The most common are staircase (SV), normal pulse (NPV), differential pulse (DPV), square wave (SWV) voltammetric techniques, the most widely used being DPV and SWV.

3.3.1 Staircase Voltammetry

Staircase voltammetry (SV) has similarities with LSV and is the simplest step pulse voltammetric technique. The theory of SV is more complex than LSV or CV because more parameters must be taken into account in the derivation of the theoretical equations. The "staircase sweep" is a series of equal potential steps and was initially proposed as a useful tool for rejecting the background charging current. Successive potential step height and duration are typically <10 mV and 50 ms, respectively. The current is measured at the end of each step, where the charging current has decayed to a negligible value. Hence, this waveform couples discrimination against charging current with the experimental speed of LSV/CV. It had been estimated that the SV sampling time, expressed as a fraction of step width (Δt), should be equal to either 0.25 or 0.5, depending on the type of working electrode. Figure 3.4 shows a typical staircase waveform.

The current response is similar to that in LSV/CV. Indeed, as the steps became smaller, the equations for the SV response converge with those of LSV [1–7]. Step heights between 1 and 2 mV are sufficiently small to yield identical peaks for identical scan rates in LSV/CV and SV [1–7]. The potential scan can be reversed in SV, as in CV to produce a cyclic staircase voltammogram. Whilst LSV or CV gives rise to a continuous current vs. potential curve, each staircase voltammogram consists of a number of current–potential points.

SV was first suggested by Barker [1–7, 74]. It is faster than DPV and NPV techniques since there is no delay time. The practical conditions have been outlined where CV and SV give identical results. This technique can be used in LC detection [75–77] since it has greater flexibility than other waveforms in flowing streams.

However, the SV technique implemented as SV, rather than assuming it behaves as CV, is little used in pharmaceutical analysis.

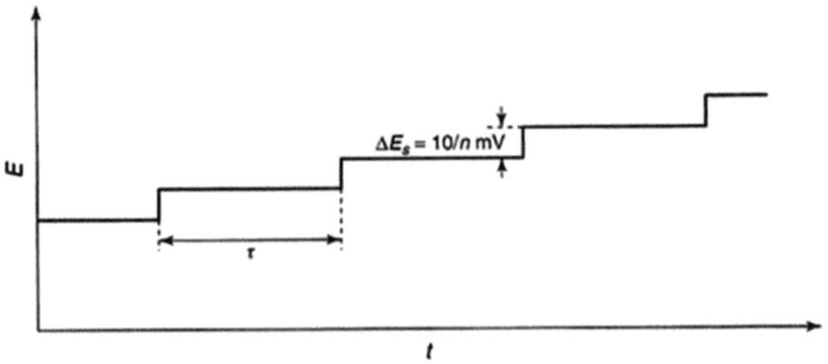

Fig. 3.4 Staircase voltammetry wave form. [Reprinted with permission from [15]]

3.3.2 Normal Pulse Voltammetry

The normal pulse technique was designed for use with the DME and was invented in 1958 by Barker [1–7, 53, 54]. It was introduced to minimise the effect of the DME charging current and hence to increase the sensitivity, consisting in the application of pulses of increasing amplitude at each successive drop, near the end of drop life.

At solid electrodes it is important to allow sufficient time between consecutive pulses to restore the initial conditions at the electrode surface before application of the next pulse, Fig. 3.5. Usually, the electrode is held for about 0.1–5 s between pulses at a constant initial (base) potential, E_{in}, where no reaction of the analyte occurs. The duration of each pulse, i.e. pulse width, t, ($=\tau-\tau$' where τ' is the interval between pulses, see Fig. 3.5) is usually between 20 and 50 ms, and the increase in pulse height, ΔE, between each successive pulse is 2–20 mV.

The current, which is almost faradaic near the end of the pulse, is sampled for a few ms, averaged and then plotted versus pulse potential. The applied waveform and resulting sigmoid-shaped voltammogram are shown in Fig. 3.4.

NPV is about ten times more sensitive than linear sweep voltammetry. At the diffusion-limited maximum current, the NPV response does not depend on the electron transfer rate, so is appropriate for the determination of diffusion coefficients of electroactive species. With this technique 10^{-6} M analyte concentrations can easily be measured. Since the initial (base) potential is applied during most of the experiment, possible surface fouling problems, due to adsorbed reaction products, are lessened although an adsorption pre-wave may be observed. If the reactant is adsorbed, a peak-shaped voltammogram may be obtained instead of a sigmoidal shape [1–7, 11–14]. Stirring is recommended for NPV of irreversible systems in order to remove depletion effects [55].

NPV does not allow the direct investigation of the products of the electrode reaction. For carrying out this type of study, reverse pulse voltammetry (RPV) may be employed. RPV consists of a pulse sequence that is a mirror image of NPV [1–7, 11–14, 56–59]. In RPV the electrode is held at an initial potential corresponding to the NPV plateau before the application of a reverse pulse with current sampling. For

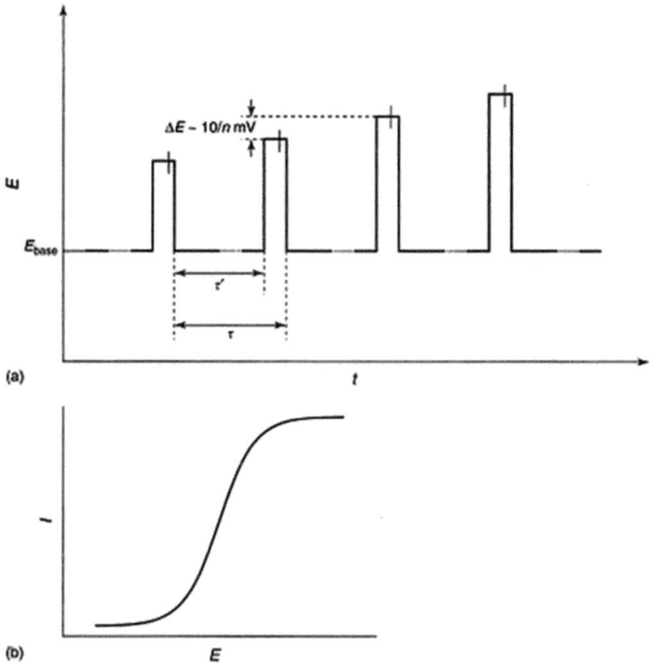

Fig. 3.5 Normal pulse voltammetry: (**a**) potential waveform and (**b**) Schematic voltammogram. [Reprinted with permission from [15]]

example, for a reversible redox couple one can easily record the oxidation current of a previously reduced species. The current is sampled at the end of the reverse pulse, and the RPV curve is a plot of the sampled current vs. the value of the pulse potential (Fig. 3.5). RPV can permit a greater control of the reaction conditions, minimisation of charging current effects, simplification of the mathematical treatment and may be more convenient than RPV for quantitative assays [56–59].

3.3.3 *Differential Pulse Voltammetry*

Differential pulse voltammetry (DPV) differs from NPV in that it uses a waveform that consists of pulses of constant amplitude superimposed on a staircase waveform (in older, analogue, potentiostats superimposed on a voltage ramp). The DPV technique was proposed by Barker and Gardner [54] for the DME in order to enable lower detection and quantification limits for electrochemically active compounds.

DPV is one of the most sensitive voltammetric techniques because the charging current is strongly discriminated against with respect to the faradaic current, and the ratio of faradaic to charging current is large. At solid electrodes, a better response is clearly seen in many cases, especially involving organic pharmaceutically active compounds, allowing the detection of analytes present in solution at a concentration as low as 10^{-8} M. The higher sensitivity of DPV than of NPV is from differential current sampling with constant height, small pulses.

In DPV potential, pulses of fixed, small amplitude (between 10 and 100 mV) are superimposed on a changing base potential. The minimisation of charging current is achieved by sampling the current twice, before pulse application and at the end of the pulse. In typical conditions, the potential is changed from an initial value in small steps (between 2 and 5 mV), and a voltage pulse (pulse width) of a short duration (50 ms) is superimposed at the end of a long step (between 0.5 and 5 s) [125].

As for NPV, the charging current is minimised by sampling the current near the end of each pulse when the faradaic contribution is still large, but DPV differs from NPV in that current is also sampled just before pulse application, Fig. 3.6a. By subtracting this current from that at the end of the pulse, and plotting the difference, ΔI, against the staircase potential, a voltammogram is obtained that is a measure of the change in current with potential.

A consequence of double sampling and representing ΔI against potential is that the DPV curves are peak-shaped, and the peak height is directly proportional to the concentration of electroactive species, Fig. 3.6. The response using DPV is, therefore, similar to the derivative of the NPV response. It differs from NPV in that each potential pulse is of fixed, small amplitude.

In a differential pulse voltammogram, Fig. 3.6b, the peak potential, E_p, can be related to $E_{1/2}$, and E_p can be used for the identification and for an indication of the reversibility or irreversibility of the system.

With increasing irreversibility, E_p moves away from the reversible value of $E_{1/2}$, at the same time as the peak width increases and peak height diminishes [1–7, 11–15]. The peak potential (E_p) for an oxidation is related to $E_{1/2}$ by the following equation:

$$E_p = E_{1/2} - \frac{\Delta E}{2}$$

where ΔE is the pulse amplitude, the negative sign being replaced by a positive sign for a reduction. Thus, for an oxidation process the peak is shifted in the negative direction as the pulse amplitude increases and for a reduction in the positive direction. As ΔE increases, some deviations from theoretical calculations occur. The faradaic current increases linearly with ΔE in the small-amplitude region, as does the double layer charging current. Increasing the amplitude of ΔE improves signal response and the signal-to-noise ratio whilst the faradaic to charging current ratio stays constant.

The peak current, I_p, is directly proportional to the concentration of electroactive species:

$$I_p = \frac{nFAD^{1/2}C}{\sqrt{\pi t_m}}\left(\frac{1-\sigma}{1+\sigma}\right)$$

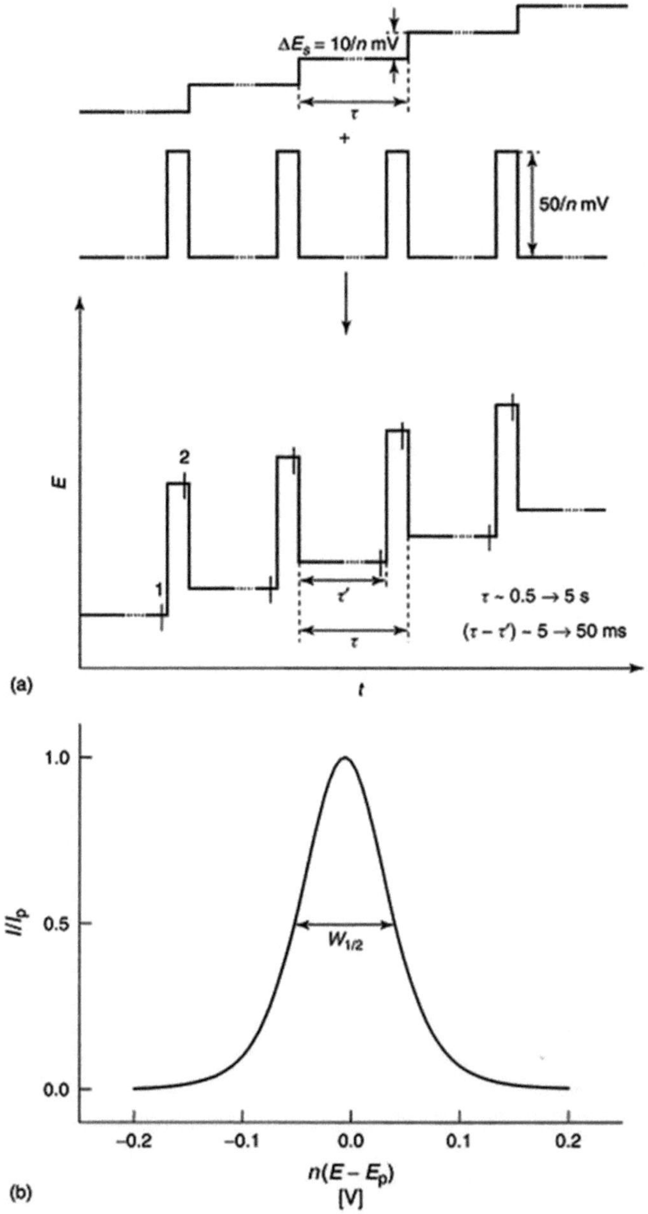

Fig. 3.6 Differential pulse voltammogram (DPV): (**a**) potential–time waveform: sum of staircase and synchronised pulses and (**b**) schematic voltamogram. [Reprinted with permission from [15]]

where $\sigma = \exp(nF\Delta E_p/2RT)$, and ΔE is the pulse amplitude. As ΔE increases, the quotient $\left(\frac{1-\sigma}{1+\sigma}\right)$ increases and finally reaches unity, but resolution may become worse.

The selection of pulse amplitude and potential scan rate usually requires a trade off between sensitivity, resolution and speed and should be done in agreement with the theory [125]. Larger pulse amplitudes can lead to broader peaks, the pulse width (t_m) is usually set at about 50 ms and pulse amplitudes of 25–50 mV coupled with a 5 mV s^{-1} scan rate are commonly employed in electroanalytical studies.

The DPV peak shape and current magnitude are strongly influenced by the chemical and electrochemical steps involved in the electrode process. This can have important implications in electroanalytical applications to pharmaceutically active compounds.

Irreversible systems have lower and broader peak currents with less sensitive and lower resolution than reversible systems [1–7, 11–15, 54]. Peak width at half height, $W_{1/2}$, can give an indication of the reversibility of the electrode reaction, which is

$$W_{1/2} = 90.4/n \text{ mV}$$

for a reversible reaction. The minimum peak width for a reversible one-electron reaction is 90.4 mV and for a reversible two-electron reaction it is 45.2 mV, and so a value of $W_{1/2}$ of less than 90 mV shows that at least two electrons are transferred.

DPV is an extremely useful technique for measuring trace levels of pharmaceutically active compounds. The DPV technique can be used for the simultaneous analysis of more than one electroactive species if the peak potentials are sufficiently far apart (normally at least 0.15 V). When the peak potentials differ by less than 0.15 V, the peaks overlap and a deconvolution treatment is necessary. Peaks separated by 50 mV can usually be distinguished in this way.

The greatest advantage of DPV is the reduction of charging current owing to calculating the difference in current at two potentials differing by less than 100 mV, while in NPV the charging current increases as E_p increases. The increase in faradaic to charging current ratio is manifested in a lower LOD value of about 10^{-8} M in a suitable supporting electrolyte. Hence, using DPV measurements of pharmaceutical compounds in biological samples following the administration of drugs on a routine basis can be more easily carried out. In addition to improvements in sensitivity and resolution, DPV can provide information about the chemical form of the analyte such as oxidation moiety and complexation. Due to its high sensitivity, DPV is particularly useful for trace analysis, for drug-active compounds and for forensic or environmental science. Because of these advantages and the availability of low-cost instruments, DPV is often the choice for the determination of drug-active compounds in their dosage forms and in body fluids.

3.3.4 Square Wave Voltammetry

Square wave voltammetry (SWV) is one of the major voltammetric techniques in current use for electroanalysis. In SWV, a square wave is superimposed on a staircase ramp, with current sampling just before the end of each forward and each backward pulse followed by subtraction of the currents; a plot is made of the forward (I_f), backward (I_b) and the total net ($I_t = I_f - I_b$) currents vs. the staircase potential. It has similarities with DPV, although it is faster, that can lead to drawbacks with kinetically slow processes, when DPV is to be preferred. It is a powerful electrochemical technique that can be applied in both analytical and kinetic measurements [1–7, 11–18, 60–65].

The application of SWV has become very popular in the last decade, it is highly sensitive to surface-confined electrode reactions and the theory is well-developed. Although it was invented by Barker et al [53] in 1952, it was little used at the time owing to difficulties with the controlling electronics. The theory and applications of SWV are described in books and reviews [61, 62, 66–70].

This technique can be divided three basic types, namely the Barker, Kalousek and Osteryoung types, the last of these being almost exclusively used in electroanalysis.

1. Barker type SWV is a direct analogue to sinusoidal ac voltammetry with a symmetric square wave of specified frequency and amplitude superimposed on either a ramp or slow staircase waveform. Barker employed a small-amplitude symmetrical square wave superimposed on a ramp and recorded the difference in currents measured at the ends of two successive half-cycles, with the objective of discriminating against the charging current.
2. Kalousek type SWV [64] is a lower frequency method, which measures the current only on the reverse half-cycle of the square wave. The potential is switched between a slowly changing ramp and a certain constant value to study the reversibility of the electrode reactions [71–73].
3. Osteryoung type SWV [65, 68] is today the most common form, Fig. 3.7. The Osteryoung waveform differs from the other SW waveforms in that the base staircase potential increases by a constant increment for each full cycle of the

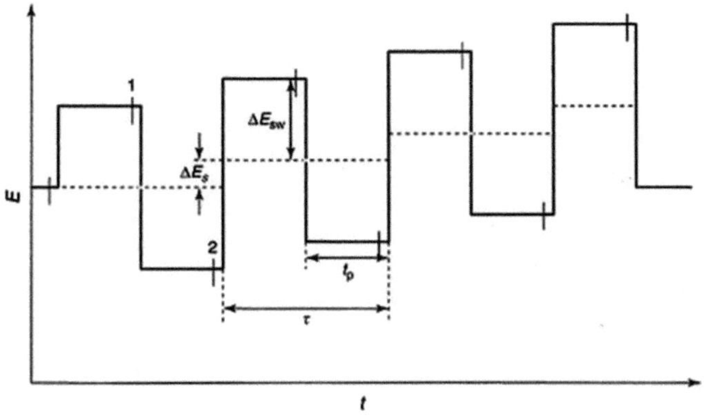

Fig. 3.7 Square wave voltammogram: potential–time waveform. [Adapted with permission from [15]]

square wave. The current is measured at the end of each half-cycle, Fig. 3.7. The difference between this and Barker SWV is the use of a staircase rather than a ramp.

As in DPV, the current is sampled at the end of the forward and reverse pulses, and the values are subtracted so that the plot of current difference vs. staircase potential yields a peak-shaped voltammogram. The main advantage of SWV is that a response can be found at high effective scan rates of up to 1 V s^{-1}, thus reducing the scan time, whereas DPV normally functions with effective sweep rates between 1 and 10 mV s^{-1}. Because of this, SWV is employed more often than DPV in pharmaceutical analysis. Besides greater speed, there are less problems with blocking of the electrode surface. Many slow reactions such as the irreversible reduction of dissolved oxygen produce very little faradaic response in a fast SWV scan. Thus, the removal of oxygen by nitrogen or argon bubbling is not usually necessary to obtain a good response for many species in the negative potential range.

The two components of the SWV response, the current response of the forward and reverse series of pulses, can be displayed as well as the net current. The theoretical forward, reverse and difference currents ($\Delta I = I_t = I_f - I_b$) are given in Fig. 3.8a for a reversible process. The sensitivity is usually increased due to the fact that the net total current (I_t) is larger than either the forward or reverse components. Hence, the sensitivity of SWV is often higher than that of NPV and DPV.

SWV experiments can be performed on solid electrodes or on static mercury drop electrodes. SWV has very low detection limits of about 10^{-8} M for fast electrode kinetics, and the theoretical sensitivity is 4 and 3.3 times higher than the DPV response, for reversible and irreversible systems, respectively.

Advantages of SWV to the electroanalyst are:

1. Detrimental effects of charging current are reduced, and so the scan rate can be increased dramatically. Typical SWV measurements take only 1–5 s whereas DPV may require much longer analysis times of 2–4 min. Frequencies of

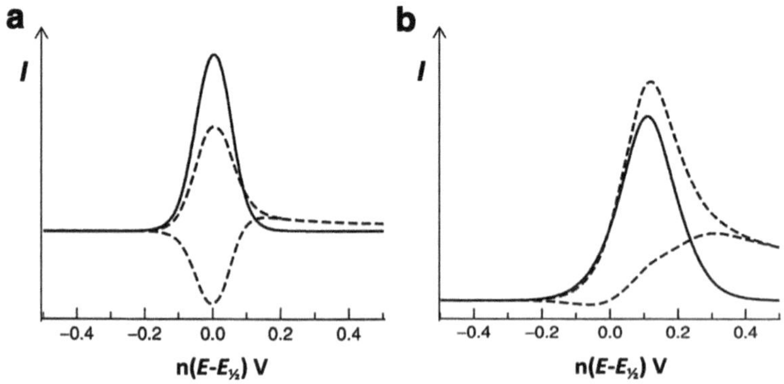

Fig. 3.8 Square wave voltammetry (SWV): (**a**) reversible and (**b**) irreversible reaction

1–100 Hz permit the use of extremely fast equivalent potential scan rates. This speed, coupled with computer control and signal averaging, allows experiments to be performed repetitively and increases the signal-to-noise ratio. However, detection limits for SWV are similar to the detection limits for DPV (10^{-7}–10^{-8} M).
2. Oxygen need not be excluded from the analyte solution, in cases where its reduction is not very slow, provided that the voltammetric peak is more negative than that for oxygen reduction. The reason for this is that since forward and reverse oxygen currents are equal at such potentials (diffusion-limited region), they cancel out and do not appear in the net total current.
3. The difference in currents, the net total current, is larger than either the forward or reverse current, so the peak height is usually quite easy to measure with greater accuracy than in DPV.

Because of its fast response, SWV can be used to record an entire voltammogram in a flow system such as FIA, LC or CE electrochemical detection. The inherent speed of SWV can greatly increase sample throughput in batch and flow analytical operations. Additionally, SWV detection can also be used to resolve co-eluting or co-migrating species in these methods.

3.4 Alternating Current Voltammetry

Alternating current (AC) voltammetry is a faradaic impedance technique that involves the application of a small-amplitude sinusoidal potential waveform, 5–50 mV, on top of a slowly changing linear potential ramp, Fig. 3.9a. Usually, the alternating potential has a frequency of 50–100 Hz, and the linear potential ramp has a scan rate between 2 and 5 mV s^{-1}. Measuring the current response at the same frequency as that of the applied sinusoidal signal, extracted from the total signal, gives a symmetrical peak shape in the potential range where an oxidation or reduction occurs, Fig. 3.9b. In such a voltammogram for reversible systems, the peak potential is the same as the half-wave potential, the potential where the sinusoidal perturbation has maximum impact on the value of the surface

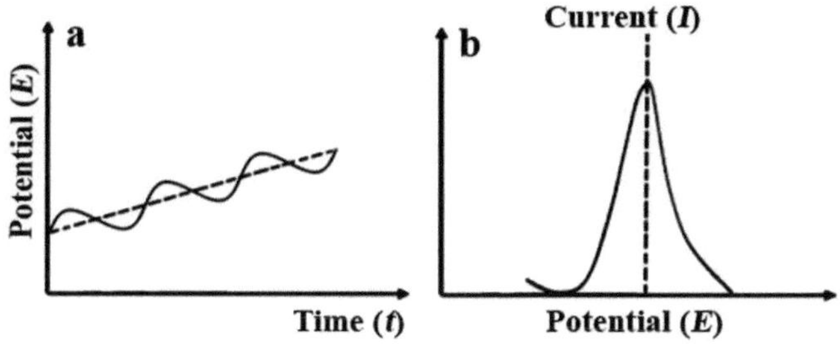

Fig. 3.9 AC voltammetry: (**a**) potential–time waveform and (**b**) current–potential curve

concentration. However, for irreversible electrode reactions the peak potential is shifted, as with pulse techniques. In addition to this, for the same number of electrons the faradaic current for an irreversible process is smaller than that obtained for a reversible process.

The usefulness of AC voltammetry derives from the fact that it is possible to use electronic techniques to discriminate between the faradaic and charging (capacitive) current [1–7, 11–18, 75]. The former shows a phase angle of 45° for a diffusion-controlled reaction while the latter shows a phase angle of 90° relative to the applied sinusoidal potential. This discrimination is possible because an AC current passing through a capacitor is 90° out of phase with the applied potential waveform, whereas faradaic contributions have both components that are in phase and 90° out of phase with the applied potential. The charging current can thus be rejected using a phase-sensitive lock-in amplifier or frequency response analyser. For convenience, AC voltammetric processes can be divided into:

1. Reversible AC electrochemical processes
2. Quasi-reversible AC electrochemical processes, of two types:
   (a) Reversible DC charge transfer
   (b) Quasi-reversible DC charge transfer
3. AC electrochemical processes with adsorption or coupled chemical reactions.

The quasi-reversible AC category reflects the dual time domain of an AC voltammogram. These processes are more common than complete reversibility [1–7, 11–18, 75, 76].

For a reversible system, the response is the derivative of the DC voltammetric response. The height of the AC voltammetric peak is proportional to the concentration of electroactive species, important for analytical applications, and to the square root of the frequency (ω) of the sinusoidal perturbation (in rad s^{-1}), according to:

$$I_{\mathrm{p}} = \frac{n^2F^2A\sqrt{\omega}\sqrt{D}C\Delta E}{4RT}$$

where ΔE is the amplitude. The peak width of $90.4/n$ mV (at 25 °C) is independent of the AC frequency, but this is only valid under conditions when the applied potential amplitude is less than $10/n$ mV, where n is the number of electrons associated with the process, in order to ensure system linearity. These parameters are convenient for checking the reversibility of the system.

The detection limit for analytical studies is about 10^{-7} M for reversible reactions. Substantial loss in sensitivity occurs for compounds with slow electron transfer kinetics. This may be advantageous for measurements of species with fast electron transfer kinetics in the presence of another that is irreversible, such as dissolved oxygen.

There is an intimate link between AC voltammetry and electrochemical impedance. In the latter, a fixed dc potential is applied and the frequency of the AC perturbation is scanned, whereas in AC voltammetry the frequency is fixed and the potential is scanned. Electrochemical impedance spectroscopy (EIS) is used mainly for characterisation, but there are some analytical applications, mainly in immunosensors.

AC voltammetry is little used at present owing mainly to the fact that control instruments are digitally based and it is thus simpler, from an instrumental point of view, to implement pulse techniques. Applications of AC voltammetry to pharmaceutically active compounds in the literature mostly use mercury working electrodes, with some examples at solid electrodes.

3.5 Stripping Techniques

The detection limits of DC voltammetry, CV and LSV are only about 10^{-5} M due to the contributions of charging and other background currents. They can be extended down to 10^{-7}–10^{-8} M using DPV and SWV, as described above.

Pre-concentration of an analyte on the electrode surface enables the detection limits to be lowered further down to 10^{-10}–10^{-12} M by using pulse voltammetric methods in the determination step, following the pre-concentration. Such methods are normally known as stripping voltammetry (SV). SV is composed of different related techniques, namely anodic, cathodic, adsorptive voltammetric and potentiometric stripping.

Stripping voltammetry has the following advantages:

1. Extremely low detection and determination limits (about 10^{-10}–10^{-12} M)
2. Multielement and speciation capabilities
3. Wide spectrum of drug compounds' electroanalysis
4. High precision and accuracy
5. Suitability for on-line and in-situ measurements
6. Relative simplicity and rapidity
7. Insignificant effect of the matrix from excipients (in drug dosage forms) or endogenous substances (in biological media)
8. Low equipment cost

Growing concerns about trace amounts of pharmaceutically active compounds and heavy-metal combinations have led to an increase in the monitoring of low amounts of compounds or metals in a variety of matrices such as dosage forms or biological samples. Traditionally, such measurements have been carried out in a centralised laboratory, which implies time-consuming sample preparation, transportation and storage steps.

Stripping analysis has a special place in trace drug analysis. The electroactive compounds are accumulated from the solution phase onto a solid electrode or a liquid mercury electrode before analysis, which increases sensitivity owing to the timescale of accumulation relative to the timescale of the determination step [1–20, 78–84]. The pre-concentration step before analysis is done either by forming an amalgam or complex with the particular analyte and the electrode material (mercury electrodes) or by adsorbing the substrate on the electrode surface [1–7, 11–18, 77–82]. The portable instrumentation and low power demands of stripping analysis satisfy many of the requirements for on-site and in-situ measurements of trace amounts of compounds.

SV consists of the following steps:

1. Accumulation or deposition of the compound on the electrode surface at open circuit or under controlled applied potential.
2. Equilibration. This period may be up to 30 s, but in many cases can be completely suppressed.
3. Determination, by oxidation or reduction (usually dissolution or stripping) of the compound from the electrode surface using a linear scan, pulse (DPV, SWV) technique or chemical reaction.

During the accumulation step, the solution is usually stirred or a rotating electrode is used, to increase solution flow and thus deposition rate. This step occurs during a preset time at a chosen applied potential. After equilibration, if needed, and after stopping stirring, the measurement (stripping) step is carried out [1–20, 78–84]. It is important that there is a linear relationship between the amount of accumulated substance and concentration, which can reach saturation at higher concentrations or for long deposition times. The length of the pre-concentration time determines the analyte concentration, which in turn determines the detection limit of the technique for the analyte. Thus, SV is used mainly for trace amounts of compounds.

For analytical purposes, the influence of deposition time and its optimisation should be investigated. The optimum deposition time is usually selected as a compromise between stripping peak height and the requirements of analysis time. At these low levels, the purity of the water and of the reagents used in the preparation of the supporting electrolyte is extremely important.

Mercury has often been used as an electrode material because of its chemical inertness in most aqueous solutions and because hydrogen is evolved only at very negative potentials, thus providing access to a wider range of negative potentials than other electrode materials. Thus, reductions of many compounds can be studied. Mercury drops can be renewed between experiments providing a new, reproducible surface for each determination. Also, mercury provides the opportunity for the element in question to be absorbed within the mercury, depending on the target compound, forming an amalgam, rather just adsorbing onto the surface [1–7, 77–82, 85–88]. Mercury electrodes may be configured as either a HMDE/SMDE or thin film. The thin film mercury electrode may yield a lower detection limit whilst the HMDE is simpler and less prone to contamination.

There are some advantages and disadvantages for each of these electrodes which are:

1. Reproducibility: The surface of the mercury electrode must remain constant from one experiment to another. This is relatively easy to achieve with a SMDE/

HMDE. A mercury thin film electrode on a carbon surface is, in fact, made up of discrete mercury droplets, and it is not always easy to obtain a reproducible surface.
2. Analyte accumulated: This depends on the surface area-to-volume ratio and the rate of stirring, during the pre-concentration step. The thin film mercury electrode has a larger surface area-to-volume ratio than the SMDE/HMDE, which is more advantageous for a given deposition time.
3. Stability: The SMDE/HMDE is more physically stable than the mercury thin film electrode. However, the combination of faster stirring and the higher surface area-to-volume ratio results in greater efficiency in the deposition step for the mercury thin film electrode, so that the deposition time is generally shorter.

In recent years, there has been growing concern about the use of pure mercury as an electrode material in stripping techniques, owing to its toxicity. Intensive research efforts have been devoted to the development of alternative electrode materials, with performance at negative potentials approaching that of mercury. If highly negative potentials are not needed, then carbon and metallic (gold, platinum, silver, bismuth, etc.) electrodes either of conventional size or as micro-electrodes can be used for the detection of trace levels of heavy metals, pharmaceuticals and other substances.

Bismuth film electrodes have been developed since the year 2000 as an alternative to the mercury thin film electrode and have a large negative potential range. The toxicity of bismuth is negligible, making these electrodes very attractive for routine applications in environmental, trace metal and pharmaceutical analysis, as reviewed in [126]. More recently, antimony film electrodes have been successfully developed with the same aim [127]. A variety of new and newly introduced materials such as dental amalgam, boron-doped diamond, carbon composite, carbon film and screen-printed carbon electrodes have also been used for stripping voltammetry.

In pharmaceutical analysis, SV techniques are widely used and very popular because of the low LOD and LOQ values, their accuracy and precision, as well as the low cost of equipment compared to other analytical methods. Different types of SV, classified primarily by the nature and scan direction of the pre-concentration and the stripping steps, respectively, are described in more detail in the following sections.

3.5.1 Anodic Stripping Voltammetry

Anodic stripping voltammetry (ASV) is the most widely used type of stripping analysis, especially for determination of metal ions such as lead, cadmium, copper and zinc. It involves the reduction of a chemically labile metal ion as the pre-concentration step to the metal on the electrode surface or within a mercury electrode as an amalgam, at a controlled, sufficiently negative potential. In the

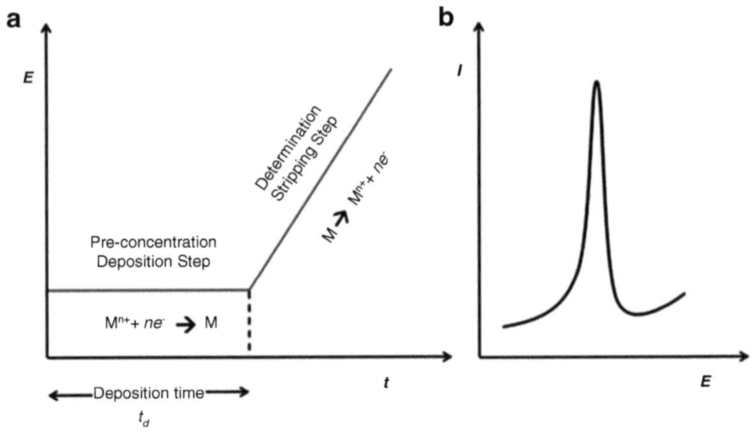

Fig. 3.10 Anodic stripping voltammetry (ASV): (**a**) potential–time profile and (**b**) anodic stripping voltammogram

determination step, the metal is reoxidised, i.e. "stripped", from the electrode surface by means of a positive potential scan giving rise to an anodic dissolution peak current (Fig. 3.10).

The process can be represented as:

$$\begin{array}{ll} \text{Pre-concentration}: & M^{n+} + ne^- \rightarrow M \\ \text{Determination}: & M \rightarrow M^{n+} + ne^- \end{array}$$

where, in the case of mercury, M is absorbed within the mercury as amalgam and in other solid electrode materials forms a solid aggregate on the electrode surface. After measurements, the electrode is cleaned of any remaining metal deposits by holding the potential at a sufficiently positive value during a period of time that is determined by subsequent running of a blank experiment under the same experimental conditions (and when no stripping peak should appear).

The very low LOD values of ASV, coupled with its accuracy, multiple simultaneous metal analysis and speciation capabilities, low cost and suitability for on-site and on-line measurements, make it one of the most powerful techniques for trace metal analysis. The peak potentials depend on the metal oxidation potential, and so mixtures of metal ions can be simultaneously identified and quantitatively determined, since the peak current is proportional to analyte concentration.

Generally, the deposition potential is about 0.3–0.4 V more negative than the reoxidation potential of the least easily reduced metal ion to be determined. Solution stirring or electrode rotation is often done to increase transport, the duration of this deposition step being selected according to the sample concentration and other requirements. Usually, between 1 and 10 min periods are sufficient for measurements in the range of 10^{-7}–10^{-9} M.

After an equilibration rest period, if needed, the potential is scanned positively, by linearly sweep, by DPV or by SWV [1–11, 85–88]. Using pulse waveforms, oxygen interferences can be reduced as well as giving higher sensitivity, particularly with SWV.

At mercury electrodes, the amalgam-forming pre-concentration step is

$$M^{n+} + ne^- + Hg \leftrightarrows M(Hg)$$

The concentration of the metal in the amalgam (C_{Hg}) is described by Faraday's law after a given preconcentration period:

$$C_{Hg} = \frac{I_L t_d}{nFV_{Hg}}$$

where t_d is the deposition time, I_L is the limiting current for the deposition of the metal and V_{Hg} is the volume of the mercury electrode [1–11, 85–88]. After the pre-concentration period, any forced convection is stopped. For reversible stripping reactions, the applied potential controls the concentration at the mercury–solution interface. In LSV of the reversible surface reaction at thin mercury film electrodes, the stripping peak current, I_p, is:

$$I_p = \frac{n^2F^2v^{1/2}AlC_{Hg}}{2.7RT}$$

where A is the electrode surface area and l the thickness, v is the potential scan rate and C_{Hg} is the concentration of the absorbed compound in the mercury film at the end of the accumulation period.

Microelectrode arrays of carbon, iridium, platinum and gold, fabricated by lithographic processes, offer a further advantage over individual macro-electrodes, when ASV assays of very low concentration levels of metal ions have to be performed. Developments in the technology of screen-printed electrodes (SPEs) have also impacted on ASV measurements. Mercury-coated SPEs have been developed for use as low-cost disposable metal-sensing devices.

The importance of ASV has increased awareness of the chronic toxic effects of metals and metal ions. Cadmium, lead and mercury are toxic even in trace amounts. However, zinc and copper are essential for living organisms in trace amounts, but high concentrations are toxic. ASV can detect and selectively determine a variety of these metals in environmental and physiological matrices.

One of the major advantages of ASV over other techniques used for trace analysis of metal ions is its multielement detection capability, so long as the stripping peaks do not overlap. Nevertheless, there may be formation of intermetallic compounds such as zinc–copper or surface blockage by surfactants or organic compounds. In this context, addition of a complexing agent that has different affinities for different metal ions may be beneficial.

ASV has been widely used for the determination of trace metal or organic compounds without interference from the matrix. A particular advantage of ASV, over the other stripping methods, is that measurement of the chemically labile fraction of the metal in solution is possible, whereas other methods such as atomic absorption spectroscopy usually only yield the total metal concentration. It can be used for the assay of trace metal ions in biological media such as blood, serum, plasma, urine and tissues. The absence of interference from alkali metals makes this method extremely powerful for the analysis of biological, pharmaceutical and environmental samples.

3.5.2 Cathodic Stripping Voltammetry

Cathodic stripping voltammetry (CSV) differs from ASV in that accumulation is performed by oxidation to a compound of the metal in question and stripping by reduction, i.e. CSV is the mirror image of ASV. This means that Hg electrodes can normally only be employed if the pre-concentration potential is negative, except for some anions (see below) so that solid electrode materials which have a positive potential range are often used. CSV can be employed to measure some anions as well as cations [1–11, 89–91]. However, solid electrodes can sometimes lead to poor reproducibility and low sensitivity, and irreversible reduction of hydroxides as a result of irregularities on the electrode surface may occur.

An example of a cation is Mn^{2+}, which is difficult to determine by ASV since accumulation potentials of the order of −1.8 V vs. SCE are necessary. Oxidation of Mn^{2+} on a carbon or platinum electrode leads to formation of a deposit of stoichiometric MnO_2 in the pre-concentration step:

$$Mn^{2+} + 2H_2O \rightarrow MnO_2 + 4H^+ + 2e^-$$

which is subsequently reduced to Mn^{2+} in the determination step.

For anions, exemplifying at mercury, the potential is held at an oxidising potential for Hg in order to form a sparingly soluble mercury compound on the electrode surface.

$$X^{n-} + Hg \overset{Deposition}{\rightarrow} HgX + ne^-$$

These species are stripped from the electrode surface by a negative-going potential scan that reduces the mercury salt to metallic Hg:

$$HgX + ne^- \overset{Stripping}{\rightarrow} Hg + X^{n-}$$

Pre-concentration occurs only on the electrode surface of the SMDE/HMDE or MFE. Hence, selectivity, precision and accuracy can be improved by isolating the analyte from the sample matrix such as removing inactive ingredients from pharmaceutical dosage forms and endogenous substances from biological samples.

CSV has been used for the detection and determination of inorganic ions such as halide ions (iodide, chloride and bromide), selenide and sulphide ions and a number of organic pharmaceutical compounds, nucleic acid bases and pesticides, etc. [91–95].

In general, interfacial complications, such as surface saturation, interferences between working analytes and other experimental difficulties have limited the use of CSV for routine analysis. Also, uncertainties exist with respect to the selectivity of CSV for the analysis of organic drug-active compounds.

3.5.3 Adsorptive Stripping Voltammetry

Adsorptive stripping voltammetry (AdSV) involves three main steps, which are similar to those employed in ASV and CSV. The difference is that the pre-concentration step is by adsorption accumulation either at open circuit or assisted by an applied potential (Fig. 3.11).

More specifically:

Step 1: Interfacial accumulation:

$$M^{n+} + nL^{-1} \rightarrow ML_n \quad \text{(aqueous)} \tag{3.1}$$

$$ML_n(\text{aqueous}) \rightarrow ML_n \quad \text{(adsorption)} \tag{3.2}$$

Addition of a complexing or chelating agent to enable the formation of an adsorbable metal complex or chelate in solution, usually at an optimised pH, is followed by pre-concentration of the metal complex or chelate by adsorption (3.2).

Step 2: Equilibration (rest) period: up to 30 s.

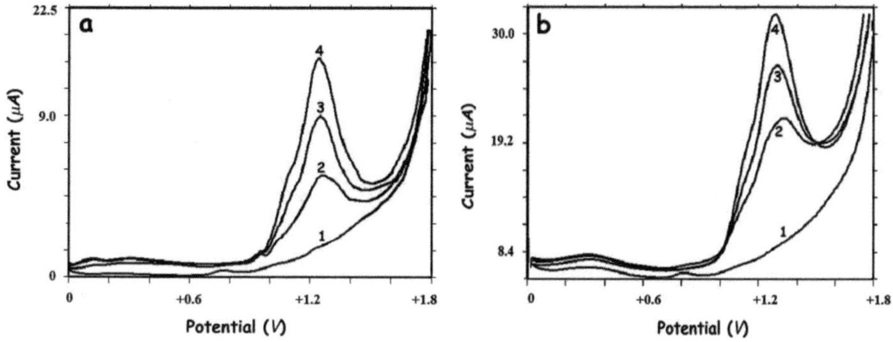

Fig. 3.11 The determination of carvedilol using (**a**) DPAdSV and (**b**) SWAdSV. (1) Supporting electrolyte [0.2 M H_2SO_4 with a constant amount of methanol (20 %)]; (2) 1×10^{-6} M; (3) 4×10^{-6} M; (4) 6×10^{-6} M. Ag/AgCl was used as reference electrode

Step 3: Stripping (or analysing) step:

$$ML_n(\text{adsorption}) + ne^- \rightarrow M^{n+} + nL^{-1} \quad (3.3)$$

This step involves an application of a negative or positive potential sweep to cause reduction or oxidation of the adsorbed molecule, complex or chelate. The resulting stripping peak currents are directly proportional to analyte concentration (Fig. 3.11).

The sensitivity of AdSV methods depends on the following electrochemical and chemical factors:

- Selected accumulation or adsorption potential
- Selection of the complexing or chelating agents and solution composition
- Reduction or oxidation potential of the adsorbed compound or complex

Thus, detailed and carefully controlled potential optimisation studies and of added ligand (in the case of complex formation) should be carried out. The analysis step can be done by linear scan or pulse techniques as with ASV or CSV.

Besides trace metals, AdSV techniques have been shown to be highly selective, sensitive and suitable for analysing organic compounds that exhibit surface-active properties. The most used working electrodes are mercury based such as HMDE, SMDE and MFE for measuring reducible species for the analysis of drugs. These electrodes offer the advantages of not needing surface cleaning, reproducible surface area and electrochemical response and can be used in a wide negative potential range [70–84, 115–123].

AdSV can be carried out at solid electrodes such as GC, BDD, CPE, wax-impregnated graphite, Pt, Au, Ru or at some modified electrodes. Microelectrodes and ultra-microelectrodes have been successfully used, especially for some metal assays, mostly after plating with mercury. Bi film electrodes have been recently used for AdSV methods for the determination of some metals. At chemically modified electrodes (CME), not only purely adsorptive accumulation, but also chemisorption can take place under controlled conditions. Metallic and carbon based solid electrodes are suitable for studying adsorbable substances that can be oxidised at the electrode [115–128].

A necessary requirement is that the reducible or oxidisable component must be close to the electrode surface and, as the adsorbed complex is usually formed as a single molecular layer on the electrode, that nearly all adsorbed compounds or complex are reduced during the potential scan. AdSV techniques can work over a wide concentration range between 10^{-11} and 10^{-3} M with approximately 10^{-12} M detection limit.

Lower detection limits can be accomplished by coupling the very efficient adsorptive accumulation of electroactive species or inorganic complexes on the electrode surface with different processes such as catalytic reactions. This is called catalytic adsorptive stripping voltammetry (CAdSV). Many metal complexes can be adsorbed onto electrode surfaces by catalytic reactions with electro-inactive oxidising agents or induce catalytic hydrogen evolution. In other cases, the addition of an oxidising agent can increase the sensitivity of AdSV significantly by enabling chemical oxidation of reduction products generated from the electrochemical reduction step. The resulting electrochemical responses are due to both adsorption and the catalytic reaction. Examples of CAdSV are for single or two-electron reduction reactions of inclusion complexes of metals or ions (e.g. Co, Cr, Fe, Pt, Ti, nitrite, nitrate, chloride, chlorate, bromated, peroxide and hydrogen ions) [70–84, 115–123].

The main disadvantage of AdSV is interference from other substances in the sample solution. Such compounds are usually organic but can also be inorganic such as halides; competitive complexation of other metals by the selected complexing agent can also occur. Interference effects can often be minimised by applying the right accumulation potential, shorter accumulation time and appropriate electrolyte parameters (e.g. pH, buffer, ionic strength).

AdSV techniques are well established and represent a large area with many possible applications in the analysis of pharmaceutical active compounds in their dosage forms or in biological media. AdSV with pulse techniques is highly selective, sensitive and precise for detecting ultra-trace amounts of pharmaceutically active compounds [102–111], important in research into drug design, bioavailability and safety.

3.5.4 Abrasive Stripping Voltammetry

Abrasive stripping voltammetry (AbSV) enables analysis of electroactive solid substances and was introduced by Scholz et al [112–119]. In this technique, a trace amount of a solid sample is mechanically transferred onto the electrode surface by abrasion, usually with paraffin-impregnated graphite electrode used for this technique. However, chemical vapour deposition diamond electrodes have also been used in this technique for analysing some metals [120, 121]. Subsequently, a suitable voltammetric measurement technique is used for the electrochemical stripping step in which the abraded material is electrochemically stripped off and measured by conventional electroanalytical methods such as DPV, SWV and CV. The transfer of solid compounds is only mechanical, by rubbing [112–120]. The tested materials are independent of the fraction properties and hardness of the electrode. Generally, soft alloys, minerals, pesticides and inorganic powders can be analysed [120]. After each measurement, the solid electrode surface is cleaned by rubbing the electrode onto a smooth filter paper [11, 112–120]. The AbSV technique allows easy and fast analysis, including fingerprint identification of inorganic and organic solid substances, and avoids the dissolution of the sample. However, the applications of AbSV in drug analysis are very limited [122–124].

3.5.5 Potentiometric Stripping Analysis

Potentiometric stripping analysis (PSA) was suggested in 1976 by Jagner and Graneli [96] as a novel electroanalytical technique for the determination of trace metal ions and organic compounds. It is a variant of the previously used "Chronopotentiometric Stripping Analysis (CPSA)" [97]. The CPSA method has several different versions. In the galvanostatic version of CPSA, after the pre-concentration of analyte at the electrode, a fixed current is applied to oxidise (or reduce) the accumulated species, and the *E*–*t* curve is recorded. During the stripping process, the potential remains close to a constant value, during a time known as the transition time, τ, which is directly proportional to the concentration of analyte in solution. CPSA can also be used without application of a current and takes place because of a redox reaction between the concentrated analyte and an oxidant in solution such as oxygen or permanganate [96, 98]. Both of these can be used in PSA.

PSA has been recently successfully applied in routine analysis due to the tremendous progresses in computer control allowing rapid control and data recording [26, 99–105]. After pre-concentration at a controlled fixed potential, determination can be done by an external current or by chemical oxidation, e.g. with oxygen or mercury ions present in solution:

$$\mathrm{M(Hg)} + \mathrm{Oxidant} \rightarrow \mathrm{M}^{n+} + \mathrm{Hg} + ne^-$$

Oxidation reactions follow the sequence corresponding to the equilibrium potentials of the accumulated species, making it possible to detect and determine them individually in the same experiment, Fig. 3.12a. Whilst each species is being oxidised the potential hardly changes, and when it has all been removed the potential changes rapidly to the potential corresponding to oxidation of the next species.

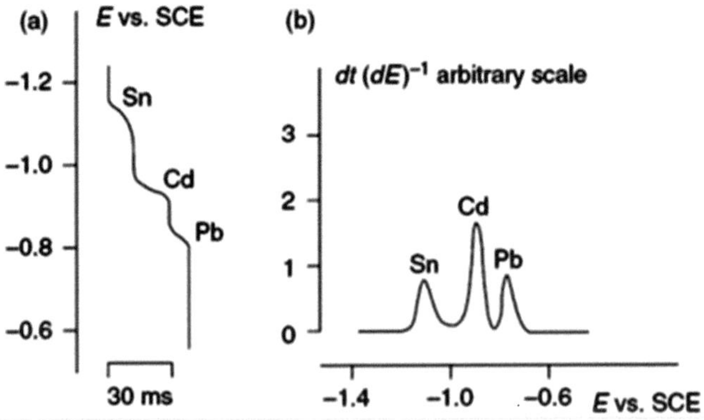

Fig. 3.12 PSA method: (**a**) Typical stripping potential–time steps and (**b**) d*t*/d*E* versus *E* curve. [Reprinted with permission from [102]]

The main innovation of PSA is to calculate the derivative dt/dE of the E–t curve and plot this vs. potential, Fig. 3.12b. A series of peaks is obtained, the position of which identifies the element, and the peak current is proportional to concentration. In this way, the information is displayed in a similar way as to other stripping techniques.

Adsorptive accumulation can also be used in PSA. This type of pre-concentration step allows determination of a large number of organic compounds such as pharmaceutically active compounds. As with AdSV, conditions are chosen so that the pre-concentration adsorption is a mass transport controlled process, with stirring or forced convection to reduce pre-concentration time.

Advantages of PSA are:

1. In PSA by chemical oxidation, there is no current during the stripping step, so there is no charging current effect on the measurements.
2. PSA has been proved to be feasible in samples with ionic strengths down to 10^{-4} M, as well as in polar organic solvents such as acetic acid, ethanol, methanol and propanol and in the presence of organic electroactive compounds, provided that they are not deposited on the electrode or are themselves oxidised.
3. Oxygen can be used as the oxidising agent; hence, deoxygenation is not necessary which may be required for ASV without pulse stripping.
4. The rate of transport of oxidants is affected similarly to that of adsorbable interferents, so that reproducibility is not affected.
5. The potential is controlled by an oxidation process, the "scan rate" is self-optimised and so signal discrimination in dt/dE vs. E plots is better, with less peak overlap than in ASV.

Disadvantages of PSA are:

1. The decrease in oxidant concentration during pre-electrolysis.
2. Direct PSA with adequate sensitivity is only feasible for a small number of analytes, especially when dissolved oxygen is used as the oxidant.

The applications of PSA and CPSA are identical to those of ASV for elements such as Bi, Cd, Cu, Ga, In, Pb, Mn, Tl, Zn, Co, Ni, Fe and Se at MFEs. Other elements such as Hg, Ag, As, Sb and Sn can be determined at gold electrodes. PSA is a non-destructive method which permits the simultaneous determination of trace levels of metals (ppb) in biological samples.

3.6 Conclusions

Modern electroanalytical techniques, especially voltammetric methods, provide reliable, sensitive, selective and reproducible data for the quantification of electroactive analytes such as drug-active compounds. Advances in electrochemical instrumentation, particularly involving pulse techniques, have allowed increased precision, accuracy, sensitivity and automation, which have found important applications in the electroanalysis of drugs. The improvement of the quality of life has stimulated considerable research in drug design, bioavailability and safety. Hence, for realising these targets highly sensitive, selective, rapid, accurate, precise and simple electroanalytical methods can be easily developed and applied for routine use.

The electroanalytical methods discussed in this chapter and more detailed on applications in Chapter 7 and 9 are effective for detecting and determining a number of drug-active compounds and metals in their dosage forms in a variety of matrices down to extremely low concentration levels, particularly by stripping techniques. A very broad field of applications of voltammetry is opening for the analysis of organic and inorganic drug-active compounds. There are thus many reasons to believe that electroanalytical methods for pharmaceutical assays should continue to develop considerably in the future. Electroanalytical methods represent an excellent tool for comparison purposes and for accuracy control in quality assurance systems and in all steps of validation studies.

References

1. Bockris J, Khan SUM (1993) Surface electrochemistry: a molecular level approach. Plenum, New York and London
2. Gileadi E, Kirowa-Eisner E, Penciner J (1975) Interfacial chemistry: an experimental approach. Addison-Wesley, Reading, MA
3. Bard AJ, Faulkner LR (2001) Electrochemical methods, fundamentals and applications, 2nd edn. Wiley, New York
4. Brett CMA, Oliveira-Brett AM (1993) Electrochemistry: principles methods and applications. Oxford University Press, Oxford
5. Frumkin A, Petrii OA, Damaskin BB (1980) In: Bockris JO'M, Conway BE, Yeager E (eds) Comprehensive treatise in electrochemistry, vol 1. Plenum, London, New York
6. Smyth MR, Vas JG (1992) Analytical voltammetry, comprehensive analytical chemistry. Elsevier, Amsterdam
7. Kissinger PT, Heineman WR (eds) (1996) Laboratory techniques in electroanalytical chemistry, 2nd edn. Marcell Dekker, New York
8. Brett CMA, Oliveira-Brett AMCF (1998) Electroanalysis. Oxford University Press, Oxford, UK
9. Bagotsky VS (ed) (2006) Fundamentals of electrochemistry, 2nd edn. Wiley Interscience, John Wiley & Sons Pub, New Jersey
10. Trasatti S (1977) In: Gerischer H, Tobias CW (eds) Advances in electrochemistry and electrochemical engineering, vol 10. Wiley-Interscience, New York, London
11. Wang J (2006) Analytical electrochemistry, 3rd edn. Wiley, New Jersey
12. Koryta J, Dvorak J, Kavan L (eds) (1993) Principles of electrochemistry, 2nd edn. Wiley, New York
13. Zoski CG (ed) (2007) Handbook of electrochemistry, 1st edn. Elsevier, Amsterdam
14. Greef R, Peat R, Peter LM, Pletcher D, Robinson J (1990) Instrumental methods in electrochemistry. Ellis Harvood Limites, New York
15. Brett CMA, Oliveira-Brett AM (2003) Step and pulse techniques. In: Unwin P (ed) Encyclopaedia of electrochemistry. Wiley, Weinheim
16. Kellner R, Mermet JM, Otto M, Valcercel M, Widmer HM (eds) (2004) Analytical chemistry: a modern approach to analytical science, 2nd edn. Wiley, Weinheim

17. Harvey D (2000) Modern analytical chemistry. Mc Graw Hill, Boston
18. Hart JP (1990) Electroanalysis of biologically important compounds. Ellis Harwood, London
19. Branina KH, Neyman E (1993) Electroanalytical stripping methods. In: Winefordner JD (ed) Chemical analysis: a series of monographs an analytical chemistry and its applications, vol 126. Wiley, New York
20. Ozkan SA, Uslu B, Aboul-Enein HY (2003) Crit Rev Anal Chem 33:155–181
21. Uslu B, Ozkan SA (2007) Anal Lett 40:817–853
22. Vire JC, Kauffmann JM (1994) Curr Top Electrochem 3:493–498
23. Barek J, Mejstrik V, Muck A, Zima J (2000) Crit Rev Anal Chem 30:37–57
24. Uslu B, Ozkan SA (2007) Comb Chem High Throughput Screen 10:495–513
25. Kauffmann JM, Vire JC (1993) Anal Chim Acta 273:329–334
26. Kauffmann JM, Guiberteau A (1994) Quim Anal 13:169–175
27. Ozkan SA, Uslu B, Senturk Z (2004) Electroanalysis 16:231–237
28. Uslu B, Ozkan SA (2004) Electrochim Acta 49:4321–4329
29. Suryanarayanan V, Zhang Y, Yoshihara S, Shirakashi T (2005) Electroanalysis 17:925–932
30. Dogan B, Uslu B, Suzen S, Ozkan SA (2005) Electroanalysis 17:1886–1894
31. Jain R, Jadon N, Radhapyari K (2006) Talanta 70:383–386
32. Demircan S, Kir S, Ozkan SA (2007) Anal Lett 40:1177–1195
33. Shahrokhian S, Karimi M, Khajehsharifi H (2005) Sens Actuators B 109:278–284
34. Ghoneim EM, El-Attar MA, Hamam E, Khashaba PY (2007) J Pharm Biomed Anal 43: 1465–1469
35. Yang G, Wang C, Zhang R, Wang C, Qu Q, Hu X (2008) Bioelectrochem 73:37–42
36. Raoof JB, Ojani R, Majidian M, Chekin F (2008) J Appl Electrochem 39:799–810
37. Wang F, Wu Y, Liu J, Ye B (2009) Electrochim Acta 54:1408–1413
38. Hammam E (2004) J Pharm Biomed Anal 34:1109–1116
39. Gao W, Song J, Wu N (2005) J Electroanal Chem 576:1–7
40. Shahrokhian S, Amiri M (2007) Microchim Acta 157:149–158
41. Raoof JB, Ojani R, Beitollahi H (2007) Electroanalysis 19:1822–1830
42. Nigovic B (2006) Anal Bioanal Chem 384:431–437
43. Ganjali MR, Norouzi P, Ghorbani M, Sepehri A (2005) Talanta 66:1225–1233
44. Muralidharan B, Gopu G, Vedhi C, Manisankar P (2009) J Appl Electrochem 39:1177–1184
45. Uslu B, Topal BD, Ozkan SA (2005) Anal Chim Acta 537:307–313
46. Torriero AAJ, Luco JM, Sereno L, Raba J (2004) Talanta 62:247–254
47. Ghoneim MM, Baumann W, Hammam E, Tawfik A (2004) Talanta 64:857–864
48. Zayed SIM, Issa YM (2009) Bioelectrochem 75:9–12
49. Semaan FS, Pinto EM, Cavalheiro ETG, Brett CMA (2008) Electroanalysis 20:2287–2293
50. Ribeiro FWP, Cardoso AS, Portela RR, Lima JES, Machado SAS, Lima-Neto P, Souza D, Correia AN (2008) Electroanalysis 20:2031–2039
51. Barker GC, Gardner AW (1958) Phil Trans R Lond A 5:302–315
52. Scholz F (ed) (2010) Electroanalytical methods, guide to experiments and applications, 2nd revised and extended ed. Springer, Berlin
53. Barker GC, Jenkins IL (1952) Analyst 77:685–696
54. Barker GC, Gardner AW (1960) Z Anal Chem 173:79–83
55. Oldham KB, Perry EP (1966) Anal Chem 38:867–872
56. Osteryoung J, Kirowa-Eisner E (1980) Anal Chem 52:62–66
57. Camacho L, Ruiz JJ, Serna C, Molina A, Ortiz Can FM (1994) J Chem 72:2369–2381
58. Molina A, Martinez-Ortiz F, Laborda E, Compton RG (2010) J Electroanal Chem 648:67–77
59. Laborda E, Rogers EI, Martinez-Ortiz F, Limon-Petersen JG, Rees NV, Molina A, Compton RG (2009) J Electroanal Chem 634:1–10
60. Bond AM (1980) Modern polarographic methods in analytical chemistry. Marcel Dekker, New York
61. Lovric M (2002) Square-wave voltammetry. In: Scholz F (ed) Electroanalytical methods. Springer, Berlin

62. Mirceski V, Komorsky-Lovric S, Lovric M (2007) In: Scholz F (ed) Square wave voltammetry theory and application. Berlin, Springer
63. O'Dea JJ, Osteryoung J, Osteryoung RA (1981) Anal Chem 53:695–701
64. Kalousek M (1948) Collect Czech Chem Commun 13:105–115
65. Osteryoung JG, Osteryoung RA (1985) Anal Chem 57:101A–110A
66. Souza D, Machado SAS, Avaca LA (2003) Quim Nova 26:81–89
67. Souza D, Codognoto L, Malagutti AR, Toledo RA, Pedrosa VA, Oliveira RTS, Mazo LH, Avaca LA, Machado SAS (2004) Quim Nova 27:790–797
68. Osteryoung J, O'Dea JJ (1986) Square wave voltammetry. In: Bard AJ (ed) Electroanalytical chemistry, vol 14. Marcel Dekker, New York
69. Eccles GN (1991) Crit Rev Anal Chem 22:345–380
70. Ozkan SA (ed) (2012) Electroanalytical methods in pharmaceutical analysis and their validation, 1st edn. New York, HNB
71. Ishibashi M, Fujinaga T (1952) Bull Chem Soc Jpn 25:68–72
72. Kinard WF, Philp RH, Propst RC (1967) Anal Chem 39:1557–1564
73. Radej J, Ruzic I, Konrad D, Branica M (1973) J Electroanal Chem 46:261–264
74. Barker GC (1966) Adv Polarogr 1:144–157
75. Anderson JE, Bond AM (1982) Anal Chem 54:1575–1586
76. Smith DE (1971) Crit Rev Anal Chem 2:247–343
77. Wang J (1985) Stripping analysis. Principles, instrumentation and applications. VCH, Weinheim, New York
78. Wang J, Tian B, Wang J, Lu J, Olsen C, Yarnitzky C, Olsen K, Hammerstorm D, Bennett W (1999) Anal Chim Acta 385:429–435
79. Abu Zuhri AZ, Voelter W (1998) Fresenius J Anal Chem 360:1–9
80. Brainina KZ, Malakhova NA, Stojka NY (2000) Fresenius J Anal Chem 368:307–325
81. Herzog G, Arrigan DWM (2005) TrAC Trends Anal Chem 24:208–217
82. Brown RJC, Milton MJT (2005) TrAC Trends Anal Chem 24:266–274
83. Ozkan SA (2009) Curr Pharm Anal 5:127–143
84. Radi AE (2006) Curr Pharm Anal 2:1–8
85. Florence TM (1970) J Electroanal Chem 27:273–281
86. Stulikova M (1973) J Electroanal Chem 48:33–45
87. Batley GE, Florence TM (1974) J Electroanal Chem 55:23–43
88. Kounaves SP (1992) Anal Chem 64:2998–3003
89. Brainina KZ (1971) Talanta 18:513–539
90. Dennis BL, Mayers JL, Wilson GS (1976) Anal Chem 48:1611–1616
91. Florence TM (1979) J Electroanal Chem 97:219–236
92. Moore WM, Gaylor VF (1977) Anal Chem 49:1386–1390
93. Palecek E (1980) Anal Lett 13:331–371
94. Palecek E, Osteryoung J, Osteryoung RA (1982) Anal Chem 54:1389–1394
95. Palecek E (1980) Anal Biochem 108:129–138
96. Jagner D, Graneli A (1976) Anal Chim Acta 83:19–26
97. Doronin AN, Kabanova OL (1965) Zh Anal Khim 20:1321–1326
98. Jagner D, Aren K (1978) Anal Chim Acta 100:375–388
99. Jagner D (1982) Analyst 107:593–599
100. Pinilla Gil E, Ostapczuk P (1994) Anal Chim Acta 293:55–65
101. Kauffmann JM, Pekli-Novak M, Nagy A (1996) Acta Pharm Hung 66:57–64
102. Jagner D (1983) TrAC Trends Anal Chem 2:53–58
103. Buffle J, Tercier-Weaber ML (2005) TrAC Trends Anal Chem 24:172–191
104. Hussam A, Coetzee JF (1985) Anal Chem 57:581–585
105. Estela JM, Tomas C, Cladera A, Cerda V (1995) Crit Rev Anal Chem 25:91–141
106. Barek J, Peckova K, Vyskocil V (2008) Curr Anal Chem 4:242–249
107. Vire JC, Kauffmann JM, Patriarche GJ (1989) J Pharm Biomed Anal 12:1323–1335
108. Alghamdi AH (2010) Arab J Chem 3:1–7

109. Economou A, Fielden PR (1997) TrAC Trends Anal Chem 16:286–292
110. Kaldova R, Kopanica M (1989) Pure Appl Chem 61:97–112
111. Gupta VK, Jain R, Radhapyari K, Jadon N, Agarwal S (2011) Anal Biochem 408:179–196
112. Scholz F, Lange B (1992) TrAC Trends Anal Chem 11:359–367
113. Scholz F, Lange B, Jaworski A, Pelzer J (1991) Fresenius J Anal Chem 340:140–144
114. Scholz F, Lange B (1990) Fresenius J Anal Chem 338:293–294
115. Scholz F, Müller WD, Nitschke L, Rabi F, Livanova L, Fleischfresser C, Thierfelder C (1990) Fresenius J Anal Chem 338:37–40
116. Scholz F, Nitschke L, Henrion G, Damas-Chun F (1989) Fresenius J Anal Chem 335: 189–194
117. Scholz F, Nitschke L, Henrion G (1989) Fresenius J Anal Chem 334:56–58
118. Scholz F, Nitschke L, Henrion G (1989) Naturwiss 76:71–72
119. Scholz F, Schroder U, Gulaboski R (eds) (2005) Electrochemistry of immobilized particles and droplets. Springer, Berlin
120. Scholz F, Meyer B (1998) Electroanal Chem 20:1–86
121. Chatterjee A, Wiltshire R, Holt KB, Compton RG, Foord JS, Marken F (2002) Diamond Relat Mater 11:646–655
122. Komorsky-Lovric S, Nigovic B (2004) J Pharm Biomed Anal 36:81–89
123. Komorsky-Lovric S, Nigovic B (2006) J Electroanal Chem 593:125–130
124. Nigovic B, Komorsky-Lovric S, Devcic D (2008) Croat Chem Acta 81:453–459
125. Osteryoung JG, Schreiner MM (1988) Crit Rev Anal Chem 19:S1–S27
126. Svancara I, Prior C, Hocevar SB, Wang J (2010) Electroanalysis 22:1405–1420
127. Slavec M, Hocevar SB, Baldrianova L, Tesarova E, Svancara I, Ogorevc B, Vytras K (2010) Electroanalysis 22:1617–1622
128. Dogan B, Ozkan SA (2005) Electroanalysis 17:2074–2083

Chapter 4
Solid Electrodes in Drug Analysis

Contents

4.1 Introduction

New platforms for electroanalytical research, and applications of electroanalysis involving electrochemical signal transduction at solid electrodes for detection and sensing of a great variety of chemically, biologically and environmentally important analytes, have been developed and employed for more than five decades [1].

Although the first electroanalytical applications used the dropping mercury electrode, experiments using solid electrodes are much easier to perform, either in stationary or in flow through systems.

As electroanalytical methods continue to develop, the range of solid working electrode materials is also expanding, and the analyst must become more aware of the materials available and their electrochemical properties. A great challenge for high-quality electroanalytical determinations is reproducibility, a low background current, an easily renewable electrode surface, and a fast electron transfer rate for the target analyte.

S.A. Ozkan et al., *Electroanalysis in Biomedical and Pharmaceutical Sciences*, Monographs in Electrochemistry, DOI 10.1007/978-3-662-47138-8_4

Applications of solid electrodes to the electroanalysis of pharmaceuticals in their dosage forms and biological samples are very wide. The electroanalytical determination of many drugs will be discussed in subsequent chapters.

4.2 Why Solid Electrode Materials?

Electrochemistry is concerned with phenomena associated with charge separation at an electrode surface. The choice of a solid electrode material depends to a great extent on the electrode's useful potential range, in a particular solvent, Table 4.1, and should always be chosen from the best quality and purity material. The potential range within which the electrode can be used is called the working potential window of the electrode, because only this potential range can be used for analytical measurements. The limits are usually determined by solvent oxidation and reduction, e.g. in water solvent to oxygen and hydrogen, respectively.

The increasing applications of solid electrodes are due to the fact that they have good, or at least reasonable, negative and positive potential ranges compared with the limited, almost non-existent, mercury electrode positive potential range [1–19]. Solid electrode surface reproducibility is more difficult to control than the renewable mercury electrode surface, but this is compensated by the advantage of the solid electrode being more mechanically stable as well as their larger positive potential range than mercury-based electrodes.

Table 4.1 Potential range of Hg, Pt and carbon electrodes

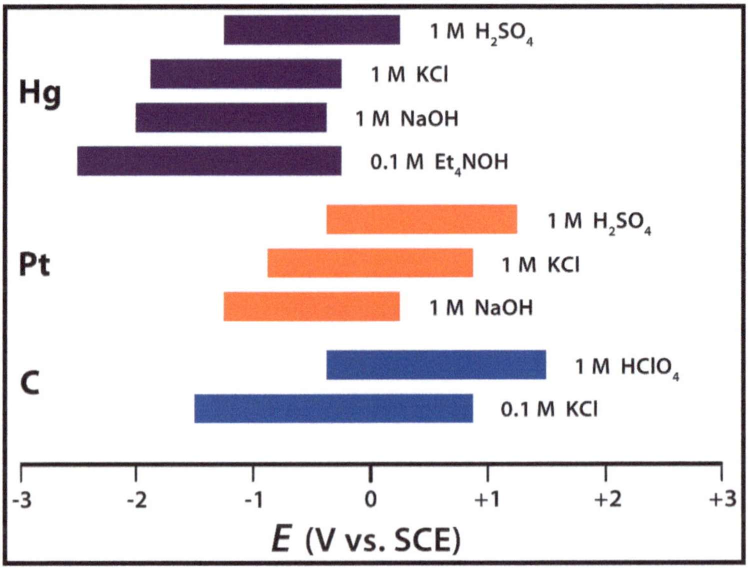

Hydrogen ion reduction on mercury electrodes is very slow and occurs at significantly more negative potentials than required by thermodynamics. This phenomenon is called hydrogen overpotential and makes mercury a very useful cathode. By contrast, mercury is easily oxidised at positive potentials.

Hydrogen ion reduction on Pt, Au, carbon, Bi and other solid electrode materials occur at much less negative potentials but are much better in the positive potential range (Table 4.1). In particular, working potential range of carbon electrodes is very wide because of their high hydrogen overpotential and good stability under anodic polarisation compared with most other solid electrode materials.

Research into the electron transfer oxidation reactions of many organic molecules can only be undertaken using solid electrodes with an appropriate positive potential range. In the selection of the solid electrode material, two factors have to be taken into consideration: the redox behaviour of the investigated analyte and the background current of the solid electrode over the potential region required for the measurements.

At the surface of the solid working electrode, dissolved electroactive analytes exchange one or more electrons. The solid electrode should have a reproducible response and a high signal-to-noise ratio. The electroanalytical method will measure the current, which is proportional to the analyte concentration, on applying a potential difference to the working electrode with respect to the reference electrode, the latter having a constant potential.

The composition and the geometry of the solid electrode must be considered since they will influence the performance and development of the electroanalytical method.

4.3 Properties of Solid Electrode Materials

In the selection of the solid electrode material, the redox behaviour of the analyte, the nature of the supporting electrolyte, the pH and the potential region required for the measurement of the redox processes of electroactive drugs have to be considered.

Solid electrodes can be made of different conducting materials and have different geometries, sizes, hydrodynamic conditions under which they can operate and active surface chemical modifications. Solid electrode surfaces are always much more rough than a liquid mercury surface, but this also means that solid electrodes can have a bigger electroactive surface area than the geometric area.

The performance of an electroanalytical method depends on the geometry and material composition of the solid electrode. Thus, chemical inertness and electrochemical stability over a wide range of conditions, electrical conductivity, fast electron transfer for a wide variety of redox systems, reproducible electrical and chemical properties, must be taken into account when it is being developed. The material should be easy to obtain and manufacture, non-toxic and inexpensive.

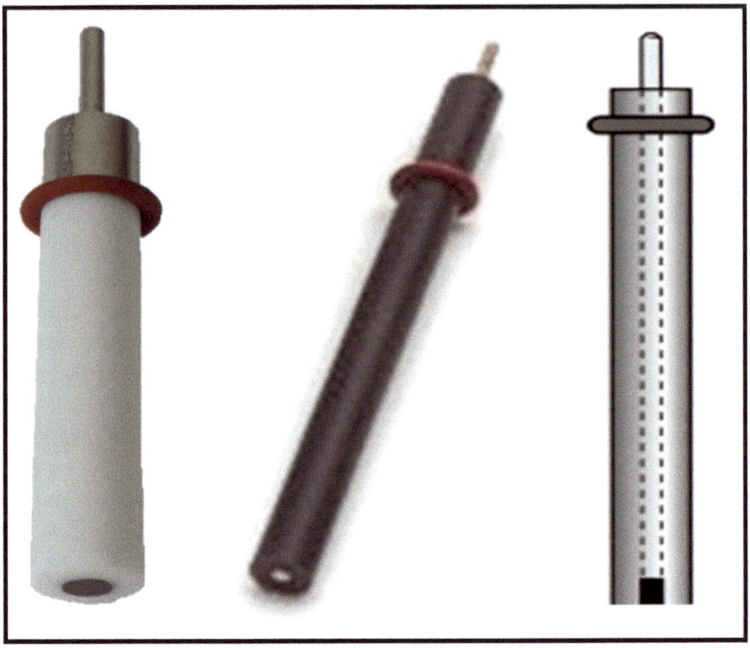

Fig. 4.1 Disc solid electrodes [Reprinted with permission from BAS Inc.]

Solid electrodes can be used in stationary, rotating and flow systems and have different configurations such as planar, rod or tube.

The most common solid electrode geometry consists of a disc of the solid conducting electrode material, attached by conducting glue to a short cylindrical conducting metal (such as brass) rod. This is inserted within a tightly fitting tube of insulating material that is chemically inert in aqueous and non-aqueous electrolytes, such as Teflon, PEEK, glass or epoxy resin, so that only the disc surface contacts with the solution, the external electrical contact being made at the back, Fig. 4.1. Such an inlaid disc is the most convenient form for cleaning and polishing. However, precautions should be taken to avoid contamination of the solid surface (e.g. with epoxy resin if this is used) during the preparation steps.

Diffusion of analyte to the solid electrode surface is influenced by the electrode size (macro- or micro-) and geometry. Disc electrodes with diameters greater than 100 μm are called macroelectrodes, and the measured currents are in the μA to mA range. Double working solid ring–disc electrodes have been used in rotating systems, Fig. 4.2, and in flow systems such as wall-jet or thin-layer electrochemical detectors.

Other solid electrode configurations include ultramicroelectrodes and micro-fabricated screen-printed electrodes or silicon-based thin-film electrodes. Micro-electrodes with dimensions less than 100 μm have currents in the pA to nA range. Although the currents at microelectrodes are small and may require specialised

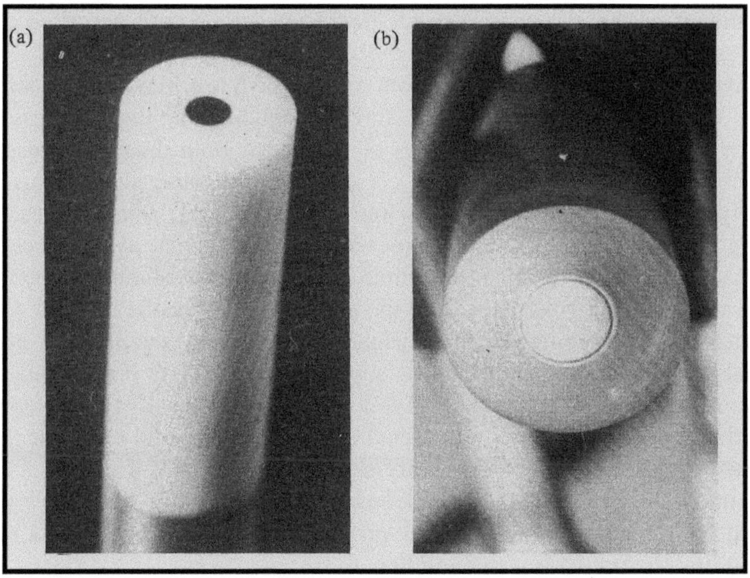

Fig. 4.2 Photographs: (**a**) rotating disc electrode (RDE), (**b**) rotating ring–disc electrode (RDE). (With permission from reference [20])

electrochemical equipment, they have a greater signal-to-background ratio owing to enhanced diffusion and thence higher current density and, being small, find use in applications where the small size is required.

The solid electrode surface can also be chemically modified in many ways enabling a large number of applications in industry, quality control of drugs and food, determination in pharmaceutical dosage forms, environmental monitoring, etc.

4.4 Surface Cleaning Pre-treatments

The application of solid electrodes to drug analysis needs to take into consideration the organic compound character of most medicinal drugs. Their redox processes are usually accompanied by strong adsorption of the drug, its redox products or both, in which case the solid electrode surface can be poisoned, leading to difficulties with reproducibility. For this reason, and to ensure good reproducibility, surface cleaning pre-treatment is usually required.

Electrode surface cleaning pre-treatments can be electrochemical and/or mechanical that will always to have to be adjusted to the electroanalytical procedure and drug analyte, and chemical pre-treatments with solid smooth electrode surfaces can also be employed. Solid metal electrodes are usually polished mechanically and are sometimes chemically etched with nitric acid or aqua regia. Suitable pre-treatment of the electrode surface should always precede any drug analysis using a solid electrode.

Mechanical pre-treatment to a smooth mirror finish is achieved using abrasives, such as diamond spray or alumina slurry, depending on the hardness of the electrode material, on polishing tables or on cloths. Both diamond and alumina abrasives are available in various particle sizes, and polishing should start using a relatively large particle size and following with successively smaller sizes, usually finishing with 1 micron particles. In some extreme cases, it may also be necessary to use ultrasound after mechanical polishing with diamond or alumina to completely remove all the abrasive particles from the pores of the electrode surface.

Electrochemical pre-treatment is usually achieved by cycling the electrode potential between chosen limits in the supporting electrolyte until a steady state baseline is obtained. The objective of this procedure is to activate and oxidise or reduce the impurities or undesirable functionalities on the solid electrode surface.

In the case of metal electrodes, care should be taken that the electrochemical pre-treatment only uses a cyclic voltammetry potential range between a positive potential limit that is before the formation of the metal oxide layer and a negative potential limit before hydrogen evolution.

A solid electrode surface conditioning cleaning procedure comprising mechanical followed by electrochemical pre-treatment ensures very reproducible experimental results.

4.5 Solid Electrode Materials

Many different solid materials are used for constructing working electrodes, such as metals, different forms of carbon and surface-modified electrodes, and less commonly metal oxides and conducting polymers. The physical properties of the solid electrode materials used in electroanalysis will now be reviewed.

4.5.1 Metal Electrodes

The most used solid metal electrode materials are Pt and Au. The first solid electrode used in voltammetry was Ag [1–4, 21]. However, this electrode has an extremely limited potential window and, therefore, restricted electroanalytical applications. Other solid metal electrodes such as Bi, Pd, Rh, Cu, Ru, Ni, Cd, Sn and In have also been used for specific electroanalytical applications. As an example, Cu, Ni and Ag can be used for some specific applications on polar aliphatic compounds such as carbohydrates and amino acid in different buffered media.

Metals are very important solid electrode materials, with the advantage of high conductivity and low background current. They are most often used for studying electron transfer kinetics, mechanism determination, thermodynamic parameters and seldom in electroanalytical applications, although forced convection easily increases their sensitivity and reproducibility. Metal electrodes usually have the form of disc, ring or short wire electrodes.

In general, solid metal electrodes have low background currents, surfaces suitable for various sensing and detection applications and provide reproducible results. Nevertheless, it should be kept in mind that their inertness is relative: at certain applied potentials, oxides are formed by reaction between the electrooxidised metal and oxygen, or hydrogen can adsorb on the metal surface. This can occur in aqueous and in some non-aqueous solutions. An important factor in using solid electrodes is the dependence of the response on the surface state of the electrode. Fundamental studies of the electrolyte solution/metal electrode interface can only be made by using well-defined single crystal electrodes.

Noble metals are commonly considered to be inert, but under certain electrochemical conditions electrochemical dissolution of the metal can occur; the metal electrodes' highly active surface can be covered by a surface oxide film over a broad range of positive potentials. This will influence the electrode oxidation processes since the electrochemical charge transfer reactions will occur at the surface oxide rather than at the metal surface. Their limited use in electroanalytical methods is also because of their rather low hydrogen overvoltage [1–4, 8–12]. Using non-aqueous media, these difficulties and problems are less severe, and noble metal electrodes are often an ideal choice.

For trace metal analysis, as an alternative to mercury film electrodes, Bi film working electrodes are successfully used in stripping voltammetric methods and they have been found to be more attractive and a good solid material to substitute mercury film electrodes [1–5, 10–19]. Bi is an environmentally friendly element and has low toxicity. Bi-modified electrodes are usually prepared in situ on a carbon disc substrate, the most common being glassy carbon; they have a high hydrogen overpotential, lead to wide linear dynamic ranges, low limits of detection and excellent resolution of neighbouring peaks.

4.5.1.1 Platinum Electrodes

Of the noble metals, platinum is the most widely used electrode material. The main advantage and widespread electrochemical use of Pt electrodes arises from its relatively high chemical inertness, electrocatalytic properties, clear separation of the potential regions for hydrogen and oxygen adsorption, excellent corrosion resistance and ease of fabrication in many forms. For quantitative electroanalytical studies, Pt electrodes of various geometries, Fig. 4.3, are mainly small discs, short straight or helical wires and nets; disposable screen-printed Pt electrodes are also available and will be described.

The biggest disadvantage in the use of Pt electrodes at positive potentials higher than +0.8 V vs. SCE is formation of a Pt oxide layer, Fig. 4.4. The Pt oxide film can strongly alter and slow down the kinetics of the electrode reaction or react with the

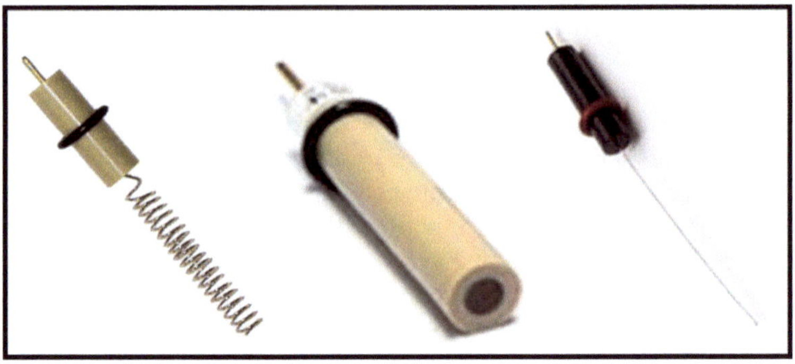

Fig. 4.3 Pt electrode different geometries [Reprinted with permission from BAS Inc.]

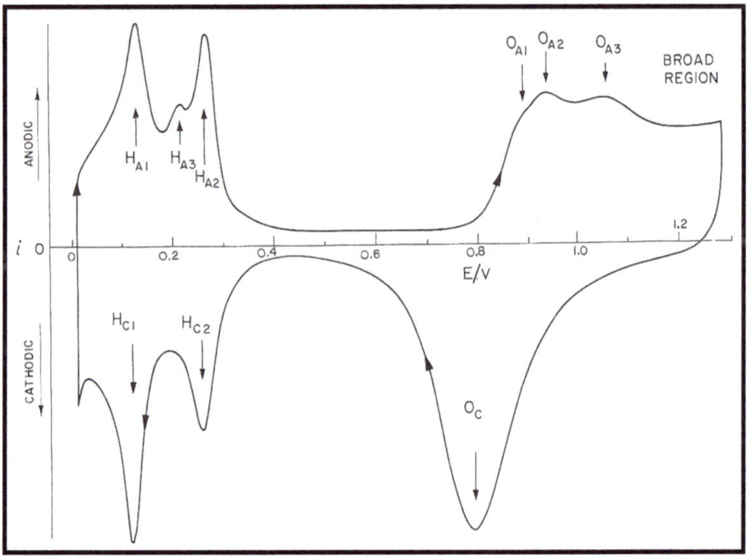

Fig. 4.4 Current–potential curve for platinum surface oxide formation and reduction in 0.5 M H_2SO_4 (Reprinted with permission from [22])

analyte giving rise to unexpected behaviour, leading to irreproducible data [1–4, 12, 19]. Care must be taken to only use Pt at less positive potentials in order to keep the electrode surface free from Pt oxide, Fig. 4.4 [22].

Another disadvantage of the use of Pt electrodes, besides its high cost, is that the presence of even small amounts of water or acid in the electrolyte leads to the reduction of hydrogen ion so that hydrogen evolution occurs at less negative potentials, and at these potentials the Pt electrode is partially covered by a layer of adsorbed hydrogen atoms, Fig. 4.4 [22].

The cyclic voltammogram for Pt in acid solution, Fig. 4.4, clearly shows the anodic peaks corresponding to Pt oxide formation and the very broad cathodic peak corresponding to subsequent Pt oxide reduction. At more negative potentials, cathodic peaks occur corresponding to the reduction of the adsorbed hydrogen on Pt followed on the reverse scan by the anodic peaks of adsorbed hydrogen on Pt.

Various pre-treatments can be used to clean the Pt electrode surface. Application of very negative potentials should be avoided because, when scanning in the anodic direction, anodic dissolution of adsorbed hydrogen can lead to incorrect results being obtained.

4.5.1.2 Gold Electrodes

The second most widely used solid metal electrode is gold, usually less expensive than platinum, but it is not as electrochemically inert. Similar to Pt, the Au electrode in aqueous electrolyte at potentials higher than +0.8 V is covered by an Au oxide film [1–4, 12, 19], making its use in the positive potential range reduced, Fig. 4.5.

Cyclic voltammograms of a monocrystalline Au electrode in 10 mM NaF, Fig. 4.5, clearly show anodic peaks corresponding to Au oxide formation and a very broad cathodic peak corresponding to subsequent Au oxide reduction [23].

Gold electrodes are much more useful than Pt electrodes in the negative potential region for reduction studies, since hydrogen is not adsorbed to an appreciable extent. After hydrogen evolution, Au electrodes show practically none of the hydrogen dissolution current associated with Pt electrodes but, on the other hand,

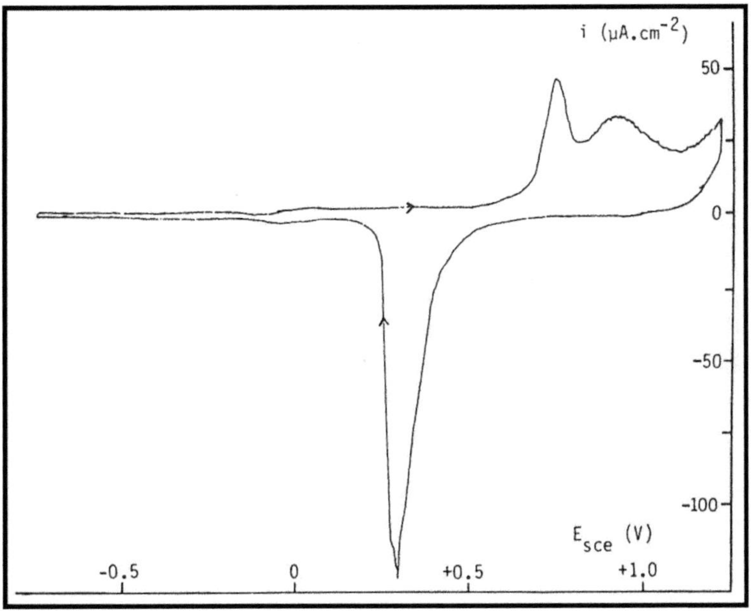

Fig. 4.5 Cyclic voltammogram of a Au (210) electrode in 10 mM NaF, sweep rate 20 mV/s (Reprinted with permission from [23])

the electrocatalytic activity for most charge transfer reactions is considerably lower than at Pt electrodes.

In chloride-containing media, such as hydrochloric acid, Au electrodes should not be used for anodic reaction studies due to the formation of a complex with chloride, limiting the useful potential limit to ~ +0.6 V.

The other drawback of Au electrodes is adsorption blocking of the surface by sulphur-containing species such as thiols, mercaptans and other inorganic sulphur species such as hydrogen sulphide. The Au–S bond is very strong, but such adsorption can be used to good effect to form self-assembled organosulphur monolayers. Au electrodes are also widely used as substrates for stripping analysis of trace metals.

Similarly to Pt electrodes for quantitative electroanalytical studies, Au electrodes of various geometries are used, mainly small discs or short wires. Various cleaning pre-treatments can be used to clean the Au electrode surface depending on the application [1–4, 12–19].

4.5.2 Carbon Electrodes

Carbon, in different forms, is the most widely used solid electrode material. All carbon allotropes have been used, but the way the building blocks are put together is different and the properties of the carbon allotropes themselves can vary widely. The electronic properties, surface microstructure, surface chemistry, microstructure and electronic properties are different, and this can influence redox system electrode reaction kinetics and mechanisms.

The carbon allotropes most used are reticulated vitreous carbon (RVC), glassy carbon (GC), graphite, highly oriented pyrolytic graphite (HOPG), carbon films, carbon paste (CP), wax impregnated graphite (WIP), pencil lead graphite, boron-doped diamond (BDD), carbon felt and cloth, carbon fibres, screen-printed carbon (SPC), fullerenes, carbon nanotubes (CNT), graphene, Kel-graph, whiskers, titanium carbide, etc. [1–4, 24–27]. All of these can be used for the electrochemical analysis of organic electroactive drugs by reduction or oxidation, even at high positive potentials by oxidation, and they are generally inexpensive.

Carbon fibres are the highest strength fibres, graphite one of the best lubricants, diamond the strongest crystal and hardest material and glassy carbon is macroscopically non-crystalline but one of the best helium gas barriers, besides new forms of carbon such as fullerene molecules and hexagonal polytypes of diamond.

Electrochemical reactions on carbon-based electrodes are normally slower than at noble metal electrodes, the electron transfer kinetics being dependent on structure and surface preparation, and some background current, but they have a wider potential range than metal electrodes in the negative and particularly so in the positive direction, generally showing chemical inertness, and are of low cost.

The carbon atom has six electrons of which four, in its valence shell, fill the sp, sp^2 and sp^3 hybrid orbitals which are responsible for the bonding structures in diamond, vitreous carbon, graphite, fullerene, pyrolytic carbon, nanotubes, etc. This explains the ability of carbon to bond with itself and with other atoms in very many combinations of chain and ring forms.

The study of the structure–function relationships at carbon-based electrodes has to consider the surface chemistry, especially, the variety of surface oxygen functional groups, particularly on sp^2-bonded carbon materials, such as carboxy, carbonyl and hydroxyl, that occur on the oxidised carbon surface; the structure of surface oxides on carbon is very complex. These influence reactivity, and thence also chemical derivatisations that are possible on, for example, carbon paste, glassy carbon, graphite and diamond surfaces. The high degree of delocalisation of π electrons, together with weak van der Waals forces, provide good electrical conductivity and influence the electrical double layer, molecular adsorption and electrode polarity.

The graphitic sp^2 atomic structure is present in different carbon materials such as pyrolytic graphite, carbon nanotubes, graphene, vitreous carbon and carbon fibres. Diamond, fullerene and allotropic lonsdaleite are sp^3 atomic structure forms. Within all used carbon electrode materials, graphite is by far the softest and diamond is one of the hardest known materials [1–4, 15, 24–39].

The high surface activity of carbon materials explains their susceptibility to poisoning by adsorption of organic compounds and especially by electroactive pharmaceutical compounds.

Many pre-treatment cleaning procedures have been developed, depending on the type of carbon, both renewing the surface and activating carbon-based electrodes, and play a role in increasing the π electron transfer rates. Carbon electrode surfaces do not interact with the accumulated or adsorbed compounds, a feature which rules out the appearance of a systematic error caused by such interactions, but the stability, surface area and reproducibility parameters in the analytical performance largely depend on the surface after cleaning.

The successful application of carbon electrodes in electroanalytical studies of electroactive pharmaceutical compounds is due to the high chemical and electrochemical stability of carbon materials, relatively high hydrogen and oxygen overpotential in different electrolytes, a broad working potential range, simplicity of mechanical renewal of the electrode surface and availability of different carbon materials.

4.5.2.1 Graphite Electrodes

Graphite is one of the carbon allotropes and is produced with different sizes and percentages of crystalline graphite, and consequently different conductivity and stability. Graphite is a thermodynamically stable form of elemental carbon; it is one of the softest minerals known to man and is a good conductor of electricity, Fig. 4.6 [1–4, 8–10, 15–19]. The electronically conductive carbons are derived from the hexagonal crystalline modification of graphite. Graphite materials, such as

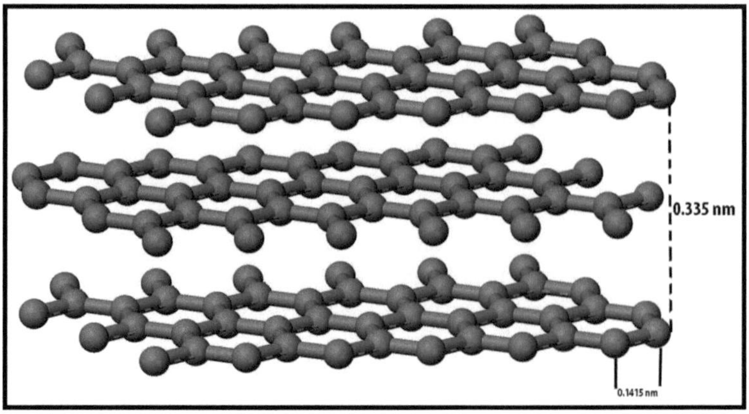

Fig. 4.6 Structure of graphite

pyrolytic graphite, carbon fibre, vitreous carbon, carbon black and others are aggregates of graphite crystallites, referred to as polycrystalline graphite.

A single sheet of graphite is called graphene, consisting of carbon atoms packed in a hexagonal lattice. Every carbon atom is bonded to three adjacent carbon atoms that lie at the vertices of equilateral triangles by four valence electrons available to participate in the formation of chemical bonds. Three of these electrons are used in forming strong covalent bonds with the adjacent carbon atoms packed in a hexagonal lattice in the sheet. This covalent bond has a short length and high strength due to electrons shared between atoms. The fourth electron is free to wander over the surface of the sheet making graphene an electrical conductor.

In graphite, the individual sheets of carbon atoms exist in parallel stacked layers, in a layered structure, connected by van der Waals forces to form crystallites, Fig. 4.6. The conductivity parallel to the carbon sheets is high and perpendicular to them is low. Each sheet of carbon atoms is translated (offset) by one-half of a unit such that alternate sheets are in the same position. The spacing between the layer planes is relatively large (0.335 nm) or more than twice the spacing between atoms within the basal plane (sheet) of 0.142 nm and approximately twice the van der Waals radius of carbon. Because these forces are weak, the sheets can easily slide past each other and graphite, unlike diamond, is therefore extremely soft, whereas the graphene sheet has strong covalent bonding between the carbon atoms.

Graphite is remarkable for the large variety of materials that can be produced from its basic form such as extremely strong fibres, easily sheared lubricants, gas-tight barriers and gas adsorbers, and the sliding sheets give graphite its softness for writing and its lubricating properties [1–4, 8–10, 15–19].

The choice of exposed basal or edge plane alters the electrochemical response owing to the different structure of the exposed surface, see Fig. 4.6. The graphite edge orientation is more reactive than is the graphite basal plane towards electron transfer and adsorption. Different edge-to-basal plane ratios of carbon-based materials display different observed electron transfer kinetics for the investigation of a redox active compound.

Large single crystals of graphite are rare; therefore, graphite is employed in polycrystalline form (solid blocks, paste electrodes, suspensions, etc.). Nevertheless, some carbons prepared by chemical vapour deposition (pyrolytic carbons) match the properties of a graphite single crystal [1–4, 8–10, 15–19].

Pores arising in graphite are sometimes impregnated with ceresin or paraffin under vacuum in order to impede the entry of solution into the graphite electrode lattice. When the in-plane dimension of the graphene sheet is small (i.e. high fraction of crystallite edges) and the spacing between them is large, this carbon is designated as amorphous (e.g. powders and glassy carbon). These materials are isotropic, as they have the same conducting properties in all directions.

The definition of the difference between "graphite" and "carbon" is somewhat subjective. It is normally considered that carbon has a non-ordered atomic structure, i.e. "amorphous", although most carbons have small crystallites of graphitic structure, and are isotropic, since they have the same conducting properties in all directions.

Spectroscopic graphite with a low level of metallic impurities is used in graphite rod electrodes for electroanalytical applications. Graphite composite electrodes are prepared from graphite powder mixed with a suitable binder and then bonded, either physically or chemically, to form a conductive solid composite.

Pyrolytic graphite is very pure and exhibits anisotropic properties, depending on the crystal orientation. The supported layer of pyrolytic graphite is actually polycrystalline, but the individual crystallites show a higher degree of preferred orientation with carbon hexagons parallel to the surface of the substrate. Such graphite can be either natural or artificial/synthetic.

The artificial type is called highly ordered pyrolytic graphite (HOPG), Fig. 4.7a, and is prepared by exposing pyrolytic graphite to high pressure and high temperature. It is a relatively new form of highly pure graphitic carbon and is an excellent atomically flat substrate with a renewable and smooth surface, especially for use in atomic force microscopy (AFM), Fig. 4.8.

An atomic resolution AFM image of HOPG shows a close packed array, with each atom surrounded by six nearest neighbours, Fig. 4.7b. The distance between any two of these atoms is 0.246 nm. This structure shows directly the hexagonal rings of the structure of graphite. In fact, HOPG has a polycrystalline structure, the size of which varies, the maximum being 10 mm for the highest quality. A freshly cleaved surface consists of atomic steps, 0.2−0.3 nm, and other bigger steps of several or dozens of atomic layers.

HOPG is completely non-polar. It is prepared from ordinary pyrolytic graphite by pressure annealing in a hot press at about 3,500 °C, and the smooth shiny basal surface is essentially free from surface oxygen functionalities and not very reactive. The basal faces, parallel to the plane of the carbon hexagons, show chemical and electrochemical properties significantly different from the perpendicular edge faces. The basal face is almost flat, with featureless voltammetric curves and relatively low charging currents in aqueous supporting electrolytes. Unlike Pt, the

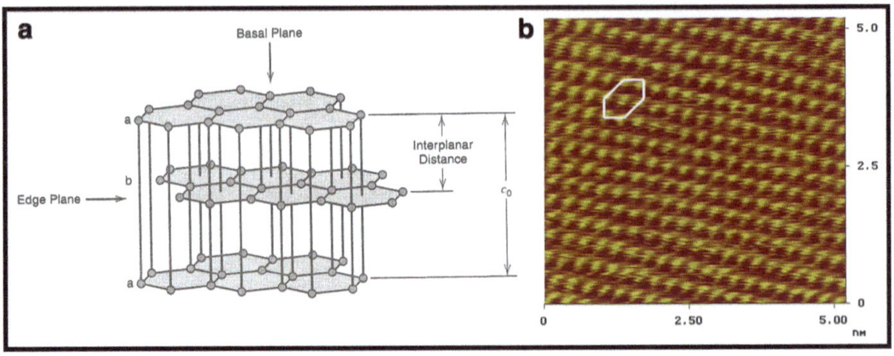

Fig. 4.7 Structure of highly oriented pyrolytic graphite (**a**) [Reprinted with permission from [40], (**b**) AFM (Reprinted with permission from [41])

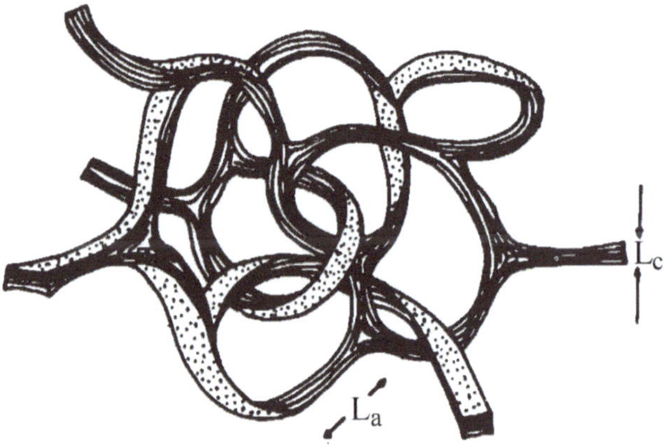

Fig. 4.8 Structure of glassy carbon. (Reprinted with permission from [46])

HOPG basal plane does not adsorb anions and most organic substances can be cleaved with adhesive tape, the new exposed clean surface remaining stable in air for many hours.

Due to their anisotropic structure, pyrolytic graphite and HOPG electrodes have differences in their preparation procedures. Both pyrolytic graphite and highly ordered pyrolytic graphite have weak interplanar bonds, and may be manually cleaved to expose basal surfaces. Teflon or epoxy resin are usually used to make the electrode holder material, a cylindrical recess drilled into a Teflon rod, and an electrical contact placed in the back of the recess, as for metal electrodes.

Some conventional cleaning pre-treatment procedures, such as mechanical polishing, cannot be employed because of the softness of HOPG and pyrolytic graphite. So, chemical cleaning in a convenient solvent, laser activation, heat treatment and electrochemical pre-treatment procedures have to be applied. For solvent cleaning, solvents such as acetonitrile, toluene and dichloromethane can be used. This pre-treatment works by dissolving materials adsorbed on the surface [3, 4, 10, 15, 39, 42–44]. Laser activation pre-treatment can be used to activate HOPG towards electron transfer. This pre-treatment cleans existing active sites and can create new ones, depending on the experimental conditions. The heat treatment procedure activates HOPG by desorbing contaminants or removing chemisorbed oxygen from the exposed edge plane sites, depending on the temperature. Electrochemical pre-treatment is the most used cleaning procedure.

Although, graphite-based electrodes are not as widely used as glassy carbon, boron-doped diamond, carbon paste or carbon nanotube electrodes [15–19, 39, 42–44]; they have also been used for the electroanalytical determination of drugs in their dosage forms and in biological samples.

4.5.2.2 Glassy Carbon Electrodes

Glassy carbon (GC), also referred to as vitreous carbon, a material which started being manufactured in the early 1960s, is a conductive form of carbon usually made by pyrolyzing precursors and is the most commonly used carbon electrode in electroanalysis since the 1980s [2–4, 15, 18, 19].

The structure of GC consists of graphitic planes randomly organised as tetrahedral domains in a complex topology linked by short oxygen-containing bridges, closely related to that of a vitreous material, with high lustre and glass-like fracture characteristics, hence the designations of glassy carbon, vitreous carbon or pyrolytic carbon. Unlike HOPG and other graphites, GC is microstructurally isotropic, the same electrical properties in all directions, and very hard, like diamond, thus making it moderately expensive compared with Pt and Au electrodes. It is difficult to machine in different electrode geometries, the most available being rods, discs and plates, and with a wide potential range, due to the high overpotentials for hydrogen and oxygen evolution.

The GC material is prepared from a pre-modelled polymeric (e.g. polyfurfurylalcohol or phenolic) resin body in an inert atmosphere by carefully applying a controlled heating programme (at temperatures between 1,000 and 3,000 °C) under pressure, and an increased electrical conductivity is obtained on increasing the preparation temperature [2–4, 15, 18, 19, 45]. The heat treatment is often applied slowly over days and causes release of atoms other than carbon, producing a conjugated graphitic microstructure. The original polymer backbone stays largely intact, preventing the formation of extended graphitic domains. The GC microstructure, Fig. 4.9, is a material of pure carbon combining glass-like high mechanical hardness characteristics due to the interwoven sp^2 carbon ribbons, (characterised by randomly oriented strips (lamellae) of pseudographitic layers of carbon hexagons) with the physical properties of graphite. The accepted model for the GC microstructure, Fig. 4.9, is that proposed by Jenkins and Kawamura [2–4, 15, 18, 19].

The random structure of GC, Fig. 4.9 [46], bears some resemblance to that of a polymer, in which the fibrils are very narrow curved and twisted ribbons of

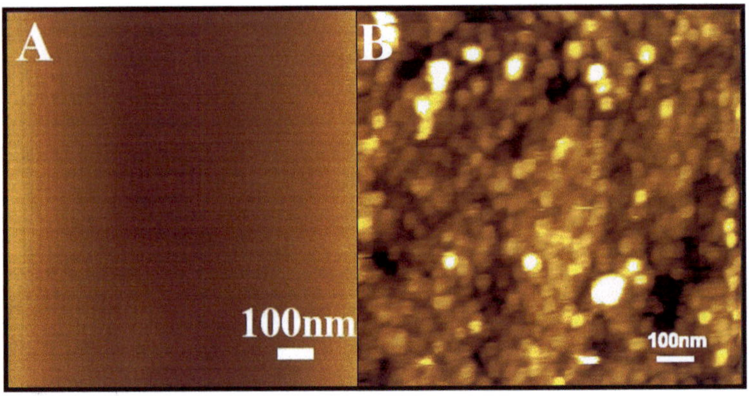

Fig. 4.9 AFM topographical images in air: (**A**) HOPG and (**B**) GC electrodes. (Reprinted with permission from [41])

graphitic carbon [2–4, 31, 35, 38, 47, 48]. These ribbon-like tangled aromatic molecules are cross-linked by carbon–carbon covalent bonds with varying bond energies. Some of these bonds may be highly strained. This extremely high density and small fine-pore size structure gives GC characteristics that enable the adsorption of some small molecules such as water, methanol and isopropanol.

In contrast to graphite, GC has isotropic electrical conductivity and has remarkable physico-chemical properties, such as high strength and high resistance to chemical attack. It is mechanically polishable and is extremely impermeable to gases and liquids.

GC can be produced in three basic types which have essentially the same microstructure, but different macrostructures: solid monolithic vitreous, foam reticulated, spheres or particles, the most common being vitreous carbon, Fig. 4.9, and reticulated carbon. The density of GC is less than HOPG, indicating that it is a nanoporous material and contains some void spaces, Fig. 4.8.

GC is extremely resistant to many chemicals especially when prepared at very high temperatures. Only strong oxidising agents like oxygen at elevated temperature, or hot melts, or strong oxidising acids can attack glassy carbon. But even under these circumstances, it is probably the most inert carbon-based material [2–4, 15, 31, 47, 48]. According to the general specifications, GC is similar to polycrystalline graphite in composition, bonding and resistance, but differs greatly in porosity, low density, high hardness, high strength, good electrical conductivity and mechanical properties, all due to the difference in structure.

Electrochemical experiments normally give rise to similar redox behaviour at GC and HOPG electrodes. However, the HOPG surface is used as a substrate in AFM studies, because it is atomically flat with less than 0.06 nm of root-mean-square (r.m.s.) roughness for a $1{,}000 \times 1{,}000\ \text{nm}^2$ surface area, Fig. 4.9A, whereas the GC electrode is only used for voltammetric characterisation as it is much rougher, with 2.10 nm r.m.s. roughness for the same surface area, Fig. 4.9B, therefore unsuitable for AFM surface characterisation.

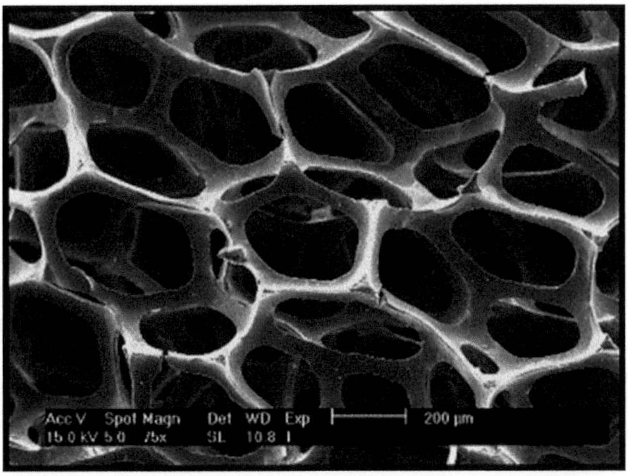

Fig. 4.10 Reticulated vitreous carbon structure. (Reprinted with permission from [49])

Reticulated vitreous carbon (RVC), also called polymeric carbon since it derives mostly from the carbonisation of polymeric precursors, is an open-pore foam material of honeycomb structure composed solely of vitreous carbon, one of the most chemically inert forms of carbon known which is resistant to oxidation, Fig. 4.10.

RVC has an exceptionally high void volume (90−97 %), high porosity, high surface area, rigid structure, low density and low resistance to fluid flow, very good mechanical properties, high corrosion resistance and lower thermal conductivity than other forms of carbon but has a very high resistance to temperature change in non-oxidising structures [1–4, 15, 31, 50].

However, the skeletal structure of RVC is fragile and brittle and needs physical support, and the low volumetric carbon content per area means that care has to be taken to ensure a uniform potential and current distribution through the material. The RVC electrode surface can easily become poisoned when used for organic drug oxidation studies, and it is difficult to avoid the drugs and oxidation products being accumulated in the pores, even using chemical solvents. This limits the use of RVC for the reproducible electroanalytical determination and screening of pharmaceutical dosage forms and biological fluids.

4.5.2.3 Carbon Paste and Carbon Composite Electrodes

Carbon paste electrodes (CPE) were introduced in 1958 by R.N. Adams for the electroanalytical determinations of drugs [34]. They are prepared using a dispersion of carbon particles mixed in a pasting liquid, with good positive and negative potential range, and have since attained very broad applications in electroanalysis [18, 19, 30–38]. The electroactive surface area is very similar to that of a single-phase electrode of the same size. The reason is that the surface is composed of tiny micron-sized carbon particles, which acts as a microelectrode assembly with overlapping diffusion fields.

CPEs are a type of carbon composite electrodes, which all use electrically conductive carbon particle powder mixed with an insulating binder. In the case of CPEs, examples of pasting liquids are water-immiscible non-conducting organic binders such as Nujol, uvasol, paraffin oil, silicon oil, bromonaphthalene and bromoform. CPEs offer an easily renewable and modified surface, are of low cost and normally have a very low background current over the entire positive and negative potential range [18, 19, 30–38]. Classical graphite particles are now often replaced by particles of other forms of carbon such as fullerenes, carbon nanotubes, graphene and carbon fibres. Pasting liquids are being replaced by binders which become solid such as polyurethane, Kel-F and Kelgraph, in this case forming a conductive solid composite electrode, the surface of which can be polished in the same way as other solid electrodes.

The basic requirements for an organic binder are its insolubility in the solution under measurement, a low vapour pressure to ensure both mechanical stability and long lifetime and inertness in the working potential range in electroanalytical studies [18, 19, 51–55]. It is also important to use an organic binder that does not cause the electron transfer rates to decrease on increasing the amount of pasting liquid.

Graphite can also be impregnated with resins, in order to produce tight materials of high chemical stability. This type of polycrystalline graphite electrode is called wax-impregnated graphite electrode (WIGE). The modifier is either dissolved in pasting liquid or physically mixed with the paste when preparing a modified carbon paste electrode. The modifier itself could be electroactive or may be a complexing agent which can extract an electroactive analyte into the surface layers of the paste electrode. After evaporating the solvent, the impregnated carbon powder is ready to use [2–4, 18, 19, 55–62].

The nature and behaviour of common carbon pastes can be described by means of physico-chemical properties such as composite character, hydrophobicity, conductivity, disintegration and lifetime. Some surface treatments can also improve polarisation characteristics of carbon pastes. The potential range of commonly used carbon pastes also depends on the pH of the working electrolyte solutions in a way that can differ from bulk carbon to some extent, but this is usually minor.

Fullerenes are a new class of carbon-only molecules, discovered in 1985 by Kroto [63], the most well-known being composed of 60 carbon atoms (C_{60}) arranged in a soccer ball structure, Fig. 4.11. They are carbon clusters, whose surface is formed by 12 pentagons and any number of hexagons. As can be deduced from Fig. 4.11, fullerene is based on sp^2 hybridised carbon like glassy carbon and graphite, but differs in its topology. Despite the presence of fused hexagons of sp^2 carbons, fullerenes do not create planar sheets as graphite does. Their large surface area, stability and wide positive potential range have prompted their use in CPEs after mixture with appropriate pastes.

Carbon paste is typically ready within a few minutes and can also be stored in a suitable container for later use or packed immediately into the electrode body, but long-time storage is not recommended because of their limited lifetime. It is recommended to prepare only the necessary amount of paste and use it immediately.

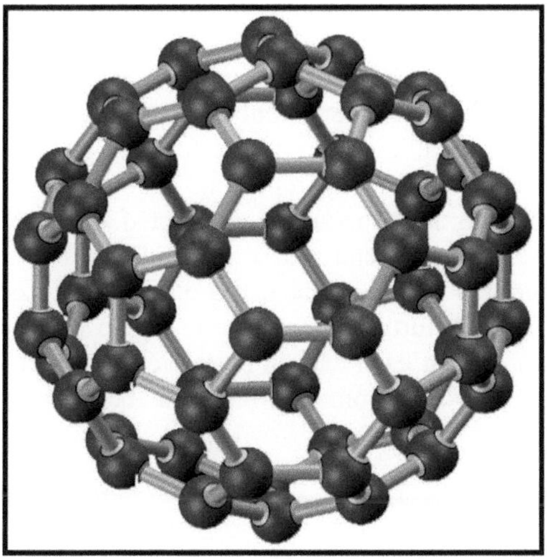

Fig. 4.11 Buckminsterfullerene, C_{60} [63] (Reprinted with permission from [64])

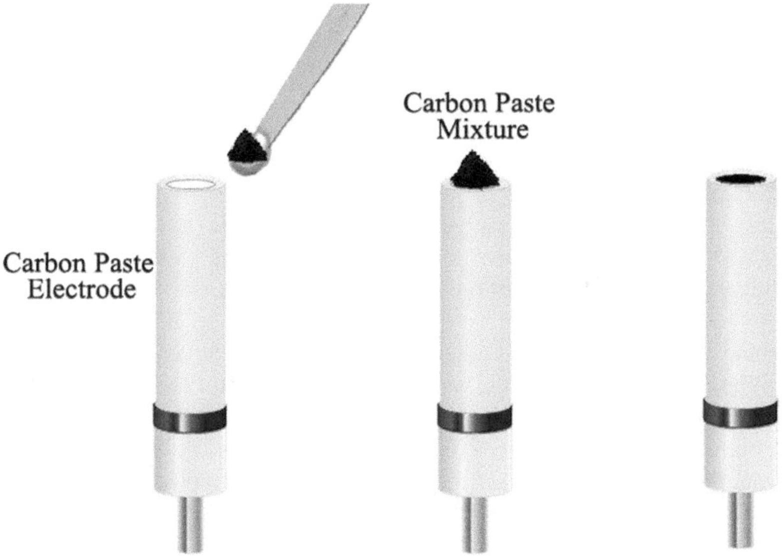

Fig. 4.12 Preparation of carbon paste disc electrodes [Reprinted with permission from BAS Inc.]

Disc CPEs are prepared by inserting the carbon paste in a shallow hole drilled into a short Teflon rod, polyethylene syringe or glass tube [1, 55], which is in electrical contact with a wire at the back, Fig. 4.12. The surface is best smoothed by rubbing the electrode several times and slowly polishing on a mildly abrasive surface.

CPEs are one of the most simple and easy to prepare carbon-based electrodes, with precision in current measurements, only surpassed by noble metal electrodes, and with a large surface area, important to use for drug electroanalysis.

Surface renewal can be done by replacing an outer layer of the paste and re-smoothing it, providing a fresh surface unaffected by the electrode history. In some assemblies, the internal metal rod is attached to a screw thread so that the paste can be extruded and then smoothed. If a plastic sheath is used then renewal can be done by cutting. A further alternative is to use it in disposable electrode form. This demonstrates that each carbon paste electrode surface is an individual surface, and there may be some differences between surfaces, affecting reproducibility.

4.5.2.4 Carbon Fibre Microelectrodes

Carbon fibres are a type of high strength material, often stronger than steel but much lighter, with very wide applications and have been available for more than 40 years. Like all forms of conductive carbon are produced by high-temperature pyrolysis, at several hundred or thousand degrees Celsius, of a carbon-based precursor, such as a polymer textile, or via catalytic chemical vapour deposition [1–4, 13–19, 64–68]. The heat treatment process is similar to that for the preparation of glassy carbon. Typical dimensions of the carbon fibre tip are in the range of 5–20 μm diameter and 5–15 mm length.

Highly oriented polyacrylonitrile (PAN)-based fibres have an “onion” type arrangement of the “concentric” or “random” graphite layers, whereas pitch-based fibres have a cross section of “radial” or ”PanAm” graphitic lamellae, Fig. 4.13 [69].

Carbon fibre electrodes have at least one dimension that is micron size, the disc end, so they are microelectrodes, with a radial diffusion of the solution analyte to the carbon fibre microelectrode surface.

They are very strong in the axis direction, have small ohmic drops and low background current [67, 68], due to their well-ordered graphite-like structure and low porosity. Carbon fibres are normally cut to a length of ~0.5 mm prior to use and typically mounted at the tip of a pulled glass capillary with epoxy adhesive, Fig. 4.14 [70].

Microelectrodes offer some advantages over conventional solid and micro-electrodes in electrochemical studies in terms of a faster response time; a lower background current and a smaller size that enables measurements to small amount of samples; etc.

A major advantage of the carbon fibre microelectrodes is the possibility of using high scan rates. Because of their small electrode area, reduced capacitance and fast time constants can be obtained and thence discrimination against charging currents

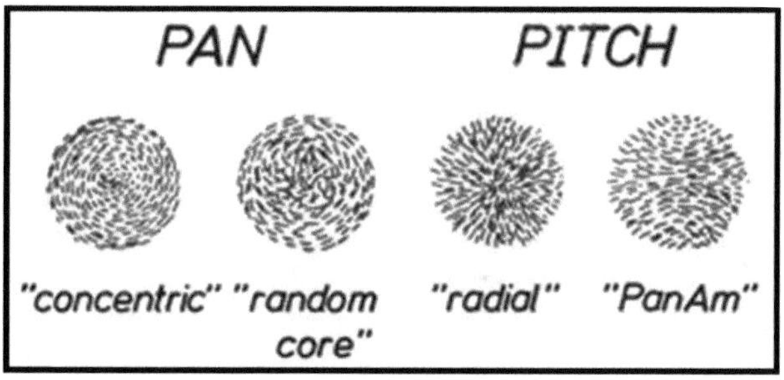

Fig. 4.13 Typical schematic cross sections of highly oriented carbon fibres. (Reprinted with permission from [69])

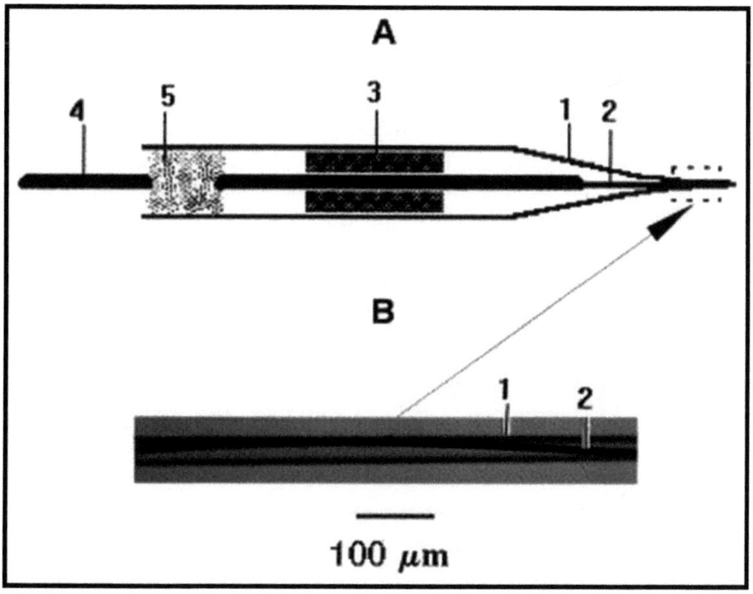

Fig. 4.14 (**a**) A scheme of a fine carbon-fibre microelectrode. (1) Micropipette; (2) carbon-fibre; (3) thick-walled capillary; (4) copper leading-out wire; (5) Wood's alloy. (**b**) Microphotograph of a part of the carbon-fibre situated within the micropipette's tip. In process of introducing the fibre into the micropipette, this part of the fibre bends slightly under the applied pressure. (1) Glass; (2) carbon fibre. (Reprinted with permission from [70])

[66–68], and scan rates exceeding 1 V/s can be used. This makes them ideal for use in fast-scan cyclic voltammetry.

Carbon fibres have been used in anodic measurements in various micro-environments, such as neuroelectrochemistry research in brain slices or in vivo in rat brains, in single cells or in vesicular volumes. They have the ability to monitor in a subsecond time frame and record in real time and, additionally, other devices cannot be used because they are too big.

4.5.2.5 Carbon Nanotube Electrodes

Nanotubules of graphite were observed in 1991 by Iijima deposited on the cathode during the direct current arcing of graphite for the preparation of fullerenes [71], and since then different methods were developed for carbon nanotube (CNT) preparation, such as chemical deposition from the gas phase, arc discharge between graphite electrodes or laser evaporation, etc. [72–77].

CNTs are the only form of carbon with extended bonding and yet with no dangling bonds and represent an increasingly important group of nanomaterials with unique geometric, mechanical, electronic and chemical properties. They are one of the most important classes of "new" carbon materials consisting of hollow cylinders made of carbon.

A CNT can be viewed as a graphene sheet which is rolled and the edges joined to make a tube. This can be done in three ways, to give armchair, zig-zag or chiral nanotubes, Fig. 4.15 a. In the zig-zag type, the graphene layer is rolled up in a way to make the ideal ends of an open tube give a zig-zagged edge. In comparison to the zig–zag tubes, the graphene sheet is turned by 30° before rolling up to give armchair type nanotubes. If the angle of turning the graphene layer before rolling up is between 0° and 30°, chiral nanotubes are obtained. Only armchair nanotubes are electrical conductors; other nanotubes are semiconducting. These structures have been established from symmetry considerations and from determination of the band structure of CNTs produced in different ways [71–79]. The CNTs are usually closed at each end due to the presence of five-membered rings.

CNTs are entirely composed of sp^2 bonds, similar to those of graphite, providing the molecules with their unique strength, and can be single-walled carbon nanotubes (SWCNT), with one graphene cylinder, Fig. 4.15a, or multiwalled carbon nanotubes (MWCNT), with many nested cylinders having an interlayer

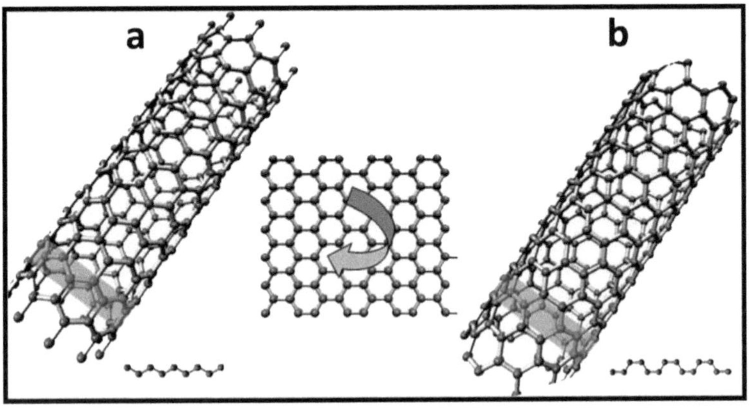

Fig. 4.15 (**a**) A single-walled carbon nanotube can be pictured as a rolled graphene sheet. Nanotubes are classified according to the Rolling direction as zigzag, armchair and chiral. (**b**) Multiwalled nanotubes can be pictured as a set of concentric single-walled nanotubes differing from them in their properties. (Reprinted with permission from [64])

spacing approximately that of graphite, Fig. 4.15b. This bonding structure, which is stronger than the sp^3 bonds found in diamond, provides CNT with their unique strength but using hexagons alone, carbon cannot yield closed three-dimensional structures.

Thus, SWCNT have diameters in the range of 2–3 nm whereas MWCNTs are made of concentric tubes, with external diameters that can vary up to 100 nm [76–79]. Their unique structure results in unique macroscopic properties, including high tensile strength, high electrical conductivity, high resistance to heat and lack of chemical reactivity in many circumstances. The inner layer tubes of the MWCNT have π-orbitals in their structure similar to graphite, and the interactions between neighbouring tubes within the MWCNT lead to additional stabilisation compared to SWCNT.

CNTs with different diameters can be prepared by various methods, including electrochemical synthesis and pyrolysis of precursor organic molecules, and some of these tubes may fit one into another to make a MWCNT. Electron micrographs show that the space between individual tubes is usually ~0.34 nm, Fig. 4.15.

In order to make CNT-modified electrodes, it is often necessary to functionalise them first, in order to introduce surface functional groups and increase the hydrophilic character. The functionalised CNTs are deposited onto a conducting surface like glassy carbon or graphite, where they will be immobilised in both oriented and non-oriented positions, increasing the electroactive surface area of the substrate electrode. Sometimes Nafion or a conducting polymer is added to improve the physical robustness of the modifier layer.

Covalent functionalisation of the CNTs structure is achieved after purification of the nanotubes obtained from different production methods. Reaction with hot, concentrated oxidising mineral acids like nitric or sulphuric acids introduces carboxyl groups at the ends of the tubes and defects on the side walls. For MWCNT, the reaction has to be carefully controlled and the conditions adjusted to determine whether or not the outer walls of the tubes are oxidatively removed, allowing a variation not only of the length, but even of external tube diameter. Carboxyl derivatives obtained from the oxidative opening of nanotubes can be further modified by classical organic chemistry methods. The attachment products from long alkyl chains exhibit a markedly increased solubility in organic solvents. The debundling of SWCNT can also be promoted by the opening and functionalising the end of tubes as the intertubular van der Waals exchange decreases due to the modification.

CNTs are one of the most commonly used building blocks in nanotechnology. Their attractiveness as electrochemical nanoprobes is due to their small size, with larger surface area, high sensitivity, fast and reproducible responses, good conductivity and high chemical stability. The signal is increased compared with other traditional carbon-based electrodes, such as carbon paste or glassy carbon, due to the CNT high surface area [75–90]. Additionally, there may be electrocatalytic effects compared to other forms of carbon, owing to active sites associated with CNT defects. Additional defects can be introduced by nitrogen doping during CNT synthesis, nitrogen atoms substituting carbon atoms, which can lead to enhanced electrocatalytic effects [81]. The chemical and electrochemical properties of carbon nanotubes can be particularly considered as suitable for the design of a variety of sensors, and CNTs have been used to build novel electrodes with interest for electroanalytical applications.

4.5.2.6 Boron-Doped Diamond Electrodes

Diamond and graphite are two of the most interesting minerals and are both carbon allotropes. Diamonds are transparent and brilliant, whereas graphite is opaque and metallic. Another important physical difference is their hardness. Diamond is the hardest known substance in nature, is used to cut glass and in industrial drill bits, is widely sought after because of its rarity and unique crystalline structure and is used in jewellery. Graphite is very soft. The reason for their extreme differences in hardness and other physical properties is explained by their molecular crystal structure.

Diamond is typically crystallised in the "cubic" system, consisting of sp^3 hybridised tetrahedrally bonded carbon atoms, Fig. 4.16. Micro-structurally, the atoms arrange themselves in stacked, six-membered rings in a puckered rather than a planar conformation. Diamond has an extraordinary chemical stability, is insoluble in water and organic solvents, has a very high melting point (about 4,000 °C) and is therefore a very poor electrical conductor with no surface redox processes as that occur for other carbon electrodes.

Boron-doped diamond (BDD), Fig. 4.17, has been developed as a new conductive material in which an industrial diamond film is doped with boron, becoming electrically conducting. BDD electrodes have found many applications as electrode material in electroanalysis.

BDD thin films can be produced synthetically by one of several established deposition protocols, the most popular being hot filament and microwave-assisted chemical vapour deposition, using boron-containing compounds such as trimethylboron, trimethyl borate or B_2H_6. [92–96]. Boron is easily incorporated in the diamond film, substituting carbon atoms in the lattice. Heavily doped films are produced in this way making synthesised diamond semi-conducting, but also affecting factors such as growth rate, morphology and crystallinity, as well as conductivity of the diamond particles, Fig. 4.17e [97].

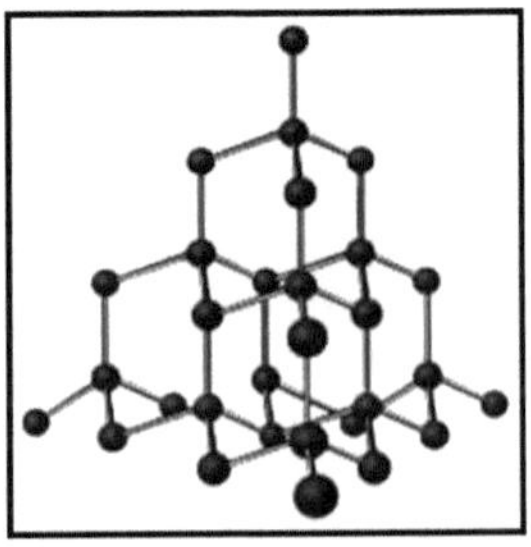

Fig. 4.16 Diamond structure

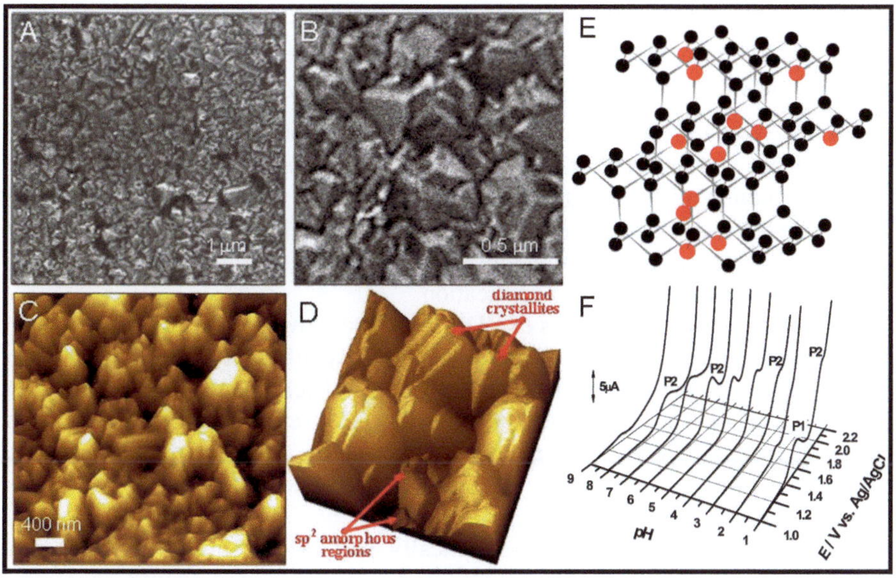

Fig. 4.17 Morphological characterisation of BDD electrode surface: (**a**, **b**) SEM images, (**c**, **d**) AFM images, (**E**) boron atoms randomly distributed in a diamond lattice and (**F**) Differential pulse for a wide pH range. (Adapted with permission from [91])

BDD electrodes are thus semiconductor electrodes with microcrystalline structure and relatively rough surfaces on the micrometric scale, with exceptional chemical inertness and mechanical strength, negligible adsorption of organic compounds, extreme hardness and applications in aggressive media such as strong acids. BDD is an excellent electrode material with a very wide potential range in aqueous solution and low background current, due to the absence of carbon–oxygen functionalities and low sensitivity to dissolved oxygen, [92–104].

Among the electrochemical transducers, the BDD electrode presents unique properties. The BDD electrode has an inert character, weak adsorption properties, with an excellent extended positive potential window. The BDD electrode behaviour is strongly related to the controlled in situ electrochemical generation (by water oxidation) of hydroxyl radicals (OH) and their subsequent reactions. However, the BDD surface is not completely inert, and the OH electrochemically generated on a BDD electrode has been investigated in different electrolyte media, Fig. 4.17f [91].

The effect of anodic and cathodic pre-treatments on the electrochemical response of a BDD electrode and the influence of the pre-treatment in different supporting electrolytes on the final surface termination were investigated [105]. The BDD anodic and cathodic pre-treatments consist of:

1. Anodic pre-treatment—a positive potential $E_{ap} = +3.0$ V is applied for 30 min, and the BDD surface is oxidised. The anodic pre-treatment causes the BDD hydrogen terminations to be converted to oxygen terminations.

2. Cathodic pre-treatment—a negative potential $E_{ap} = -3.0$ V applied for 30 min, and the BDD surface is reduced. The cathodic pre-treatment causes an increase in the thickness of the adsorption layer formed by hydrogen termination.

The electrochemical response of the BDD surface varies as a function of surface pre-treatment and also with the electrolyte used in the pre-treatment, significant oxygen evolution always occurring at +1.4−1.6 V.

Despite the many advantages of BDD electrodes, their high cost and difficulties in finding an appropriate substrate on which to deposit the thin diamond layer are major drawbacks to widespread application.

Nevertheless, the applications of BDD electrodes have demonstrated their superior electrochemical properties that are significantly different from those of other carbon electrodes, e.g. GC and HOPG. The very low background current is less than that at metal and GC electrodes, making the BDD electrode superior to other electrode materials, enhancing the sensitivity for the detection of a number of pharmaceutically and biologically important compounds that exist at nanomolar or picomolar concentrations.

The BDD electrode surface is mechanically inert against polishing or cavitation during exposure to ultrasound radiation; it is chemically inert against aggressive chemicals and under extreme polarisation.

4.5.3 Screen-Printed Electrodes

Screen-printing technology is widely used for the mass production of disposable electrochemical sensors, Fig. 4.18. Since the 1990s, screen-printing technology, adapted from the microelectronics industry, has offered high-volume production of extremely low cost and yet highly reproducible and reliable single-use disposable SPE sensors; a technique which holds great promise for on-site monitoring.

The complete electrochemical electrode set configuration, i.e. working, reference and auxiliary electrode, is contained in a disposable screen-printed electrode (SPE) that is highly suitable for working with micro-volumes and decentralised assays or to develop specific sensors by modifying their surface with various materials, usually in a strip form, Fig. 4.18. SPEs have been used successfully in the development of sensors and biosensors in various configurations for a wide range of analytical applications [107–111].

SPEs can be made of carbon, Au, Pt, Bi, AGr or CNT inks. The support is coated on defined areas with a conducting ink that contains the electrode active material. Part is then covered by an insulating film to define an external electrical contact at one extremity and an electrode surface at the other, Fig. 4.18. Typical dimensions of the SPE strip are H33 × L10 × W0.5 mm. The shape and the surface of SPEs are not nearly as smooth as more traditional electrodes such as metal, pyrolytic graphite or GC electrodes [107–119].

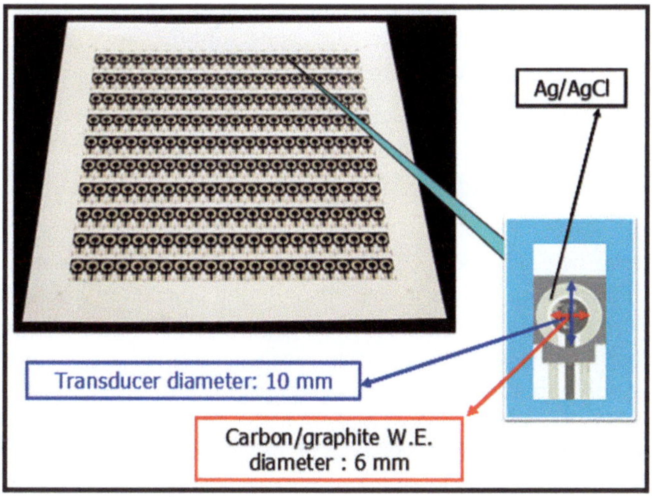

Fig. 4.18 Design of screen-printed electrochemical transducers. (Reprinted with permission from the authors [106])

Stable reference electrodes are printed from Ag/AgCl inks, Fig. 4.18, but it is important to allow the electrode to equilibrate with the test solution for at least 60 s before performing an experiment.

Both reference electrode and electrical contacts are often made of silver, chloride ion being added to the test solution if necessary to form the Ag/AgCl reference electrodes. Disposable SPEs are designed to work with microlitre-size samples, ideal for quality control or research purposes. A wide variety of SPEs are commercially available [107–119].

The great versatility of SPEs is due to the wide range of ways in which a large number of SPEs can be printed on different types of substrates, such as plastic or ceramic and to the use of possible modifiers mixed with the ink. The ink composition used for printing the SPEs can be modified by the addition of very different substances, such as metal microparticles, enzymes, polymers, complexing agents and which may enhance the selectivity and sensitivity of a particular analysis.

SPEs are designated according to the application and working electrode, such as unmodified SPE, film coated SP electrodes, metal-based SPEs, chemically modified SP carbon electrodes, ion selective SP electrodes, SP metal film electrodes, enzyme-modified SPE and antigen/antibody-modified SPE.

The exact determination of carbon SPE area is difficult, since the special binder compounds in the carbon ink may introduce unexpected features in the voltammograms, and the heterogeneous kinetics at the electrode surface can be sluggish. However, the main advantage of SPEs is being disposable and, as they are used only once and then discarded, surface fouling is not a problem.

SPEs' small portable instrumentation can be easily used outside the laboratory, especially for environmental measurements. The analyte sample can be placed on the exposed electrodes in the form of small drop in situ. In the laboratory, they can be used in a larger volume of analyte-containing solution, appropriate for analysis of pharmaceutical or biological samples.

4.5.4 Chemically Modified Electrodes

Chemically modified electrodes (CMEs) consist of conducting or semiconducting materials modified using different reagents with the objective of obtaining specific surface properties for electroanalytical applications. Such manipulation of the molecular composition of the electrode surface aims at improving sensitivity, selectivity and/or stability, tailoring the response in order to meet new analytical needs [120–132].

The enhanced selectivity of CMEs can be obtained by either an analyte-specific pre-concentration effect or by selecting a layer with properties that catalyse a specific reaction [115, 121, 133–139]. Whilst conventional voltammetric and amperometric electrodes serve mainly for carrying the electrical current, powerful sensing devices can be designed relying on specific modification of their surfaces. The unmodified electrodes are called "bare", "native" or "virgin" electrodes [33, 59].

CMEs are made by physisorption/chemisorption, covalent bonding, sol–gel encapsulation, coating the surface by a polymer, by polynuclear or mixed films or by mixing the modifier with the electrode matrix material, i.e. composite modification. Good control of electrode characteristics and reactivity is achieved by the surface modification [1–4, 115, 121–143].

The main objectives of chemical modification are:

- Acceleration of electron transfer reactions
- Enabling chemical reactions at the electrode surface such as pre-concentration reactions by attaching ligands
- Changing transport properties to the electrode surface
- Excluding interferents
- Creating selective membrane permeability

Chemical modifiers are permselective membranes, organic ligands, microparticles, organometallic or inorganic catalysts, ion exchangers, biological materials, zeolites, clays and silica-based materials, and most polymers can be applied to the electrode surface by a combination of electrostatic interaction and low solubility in the electrolyte solution [1–4, 115, 121, 133–138]. Immobilisation of a chemical modifier on the electrode surface should maintain its electrochemical and physical properties.

The simplest modification procedure is physical or chemical adsorption, with the advantage that it does not require any special reagents. In an ideal case, a monolayer coverage electroactive mediator, such as an aromatic organic compound, is attached by adsorption onto a flat surface. In this case, the adsorptive strength increases with the increase in the number of aromatic rings. The adsorption is due to the strong interaction between the electrode surface and the extended π-electron systems of these redox molecules. The immobilised redox couple shows a significantly different electrochemical behaviour from that arising due to diffusion of a species from bulk electrolyte solution to the electrode surface. However, its lifetime on the surface can be relatively short as it may gradually desorb and escape into solution.

Covalently bonded-modified electrodes are usually much more stable than electrodes modified by physi/chemisorption. However, their preparation can be much more difficult and the reaction conditions must be rigorously controlled. This procedure has been particularly useful for attaching enzymes directly to electrode surfaces, hence they have been designated enzyme-modified electrodes or simply electrochemical biosensors.

Covalent bonding requires a chemical reaction with the substrate electrode, so the electrode surface must be pre-treated. The surfaces of carbonaceous materials contain, after special treatment, quinonoid, carboxy and phenolic groups, which are capable of forming covalent bonds. Carbon-based electrodes offer more functionality not only for groups like hydroxyl, carboxyl or carbonyl but also for quinone and lactone groups. The surface of metal-based electrodes such as Pt, Pb and Ni is often covered with a hydroxide layer or an oxide layer.

Chemisorption and low solubility in the contacting solution or physical anchoring can held surface organic, organometallic or inorganic polymer films on porous electrodes which can form electronically conductive and non-conductive polymer film coatings.

Polymer film coating modification consists in spreading and evaporating a polymer solution on the electrode surface or carrying out electropolymerisation on the electrode substrate surface. The former can be obtained using different procedures such as dip coating, solvent evaporation or spin coating. The latter can be carried out by electrochemical polymerisation/deposition, radio frequency polymerisation and cross-linking. The thickness of the polymer layer, type of attachment and stability are important parameters of the modified electrode. Mono or multiple layers can be obtained in the same way until the desired thickness is obtained.

The most uniform films are obtained by spin coating, which means evaporating on a rotating electrode. This gives precise control of film thickness and is particularly attractive for miniaturised sensor surfaces.

Dip coating consists of immersing the electrode in a solution of the polymer for a period of time sufficient for spontaneous film formation by adsorption to occur. Multiple film layers can be obtained by withdrawing the electrode from the solution, allowing the film of polymer solution to dry and repeat the procedure.

Solvent evaporation consists in applying a droplet of a polymer solution on an upward-facing electrode surface and allowing the solvent to evaporate. A major advantage is that the polymer coverage can be immediately calculated from the original polymer solution concentration and droplet volume.

Electrochemical polymerisation of appropriate monomers can be used to form polymer films that are electronically conductive (conducting polymers or redox polymers). Normally, this is done by potential cycling in which the potential is first cycled to a sufficiently positive potential to form monomer cation radicals at the electrode surface, usually around +1.0 V vs SCE, which initiates polymer formation. The potential sweep is then inverted, usually to a slightly negative potential, and the cycles are then repeated.

The rate of polymer formation, film thickness and morphology depends on the potential cycling conditions, number of cycles and solution composition. Examples of electronically conducting polymers prepared in this way are polyaniline (PANI), polypyrrole (PPy) and poly(3,4-ethylenedioxythiophene) (PEDOT) and of redox polymers, polyphenazines [144]. Sometimes metal nanoparticles are incorporated in the polymer film to give electrocatalytic effects.

Radio frequency polymerisation is another procedure for obtaining polymer CMEs. Vapours of the monomer are exposed to radiofrequency plasma discharge to induce polymer film formation on the electrode substrate. The disadvantage may be that the high energy of the RF discharge may result in chemical damage, thereby producing unknown functionalities and structural modifications to the polymer.

Cross-linked films are often formed by copolymerisation of bifunctional and polyfunctional monomers. Cross-linking may be activated chemically, electrochemically, photolytically, radiolytically or thermally. A chemical step is designed to couple chemical components of a film on an electrode to impart a desired property such as increased stability, decreased permeability or altered electron transport characteristics.

Sol–gel polymerisation is the formation of a polymer network from suitable precursors, such as oxysilanes, by condensation reactions, with elimination of water or alcohols [145]. Molecules such as enzymes can be encapsulated in these networks, the size of the nanocages within the network being tuned by choice of the oxysilane precursor. These precursors can also contain groups such as amines so that electrostatic or even cross-linking reactions with encapsulated molecules can take place.

CMEs can contain multiple chemical modifiers that can be dissolved in a binder or admixed mechanically into the matrix during homogenisation. The composite modification procedure is based on mixing a slightly soluble modifier with a conductive matrix.

Considerable progress has been made in the development of "tailor-made" electrode surfaces by chemical modification of electrode surfaces with electroactive polymer films with new electrocatalytic properties. Electrochemical polymerisation has proven most useful for the preparation of CMEs. Electrochemical deposition using potentiostatic or galvanostatic methods relies on the dependence of polymer solubility on ionic state, so that film formation will occur, often irreversibly, when a polymer is oxidised or reduced to its less soluble state, and conducting and non-conducting polymers can be prepared.

The chemical modification process has to improve the analytical performance either by increasing their sensitivity or protecting the surface from undesired reactions.

Successful electroanalytical applications of CMEs depend on their ability to detect the identity of analyte easily, free from interferences, with stability and reproducibility, for a wide concentration range, using easy preparation techniques and be comparable or superior to a non-modified electrode.

CPEs as well as CNTs, Fig. 4.15, undoubtedly represent one of the most convenient possibilities for preparation of CMEs [115, 121, 133–141]. The coupling of electrochemical biosensors with nanoscale pastes is a tool for improved sensitivity, longer stability of the bioelement in biosensors and new detection possibilities, the main problem being to obtain a long term stability paste.

Enzyme CMEs detect the products of reaction between an immobilised enzyme layer and a reaction substrate in amperometric and potentiometric measurements, Fig. 4.19 [32, 51–55]. Thus, the searches for CMEs lead to rapidly responding reagent-less biosensors [122–135]. Conductive carbon pastes and inks have been employed for many years in a variety of biosensor applications, most notably in disposable blood glucose detectors [107–113]. A typical strategy involves placing an enzyme on the working electrode, either by coating the electrode or pre-mixing the ink with the enzyme, in order to enhance the CME electrochemical response for a given analyte.

CMEs are integrated chemical systems with applications in chemical sensors, electrocatalysis, molecular electronics, electrosynthesis, energy conversion and as molecular wires in amperometric enzyme electrodes.

Electroanalytical applications of CMEs can involve the advantages of increased sensitivity, electrocatalysis (acceleration of electron transfer rate), preferential accumulation of the analyte on the surface or selective membrane permeation.

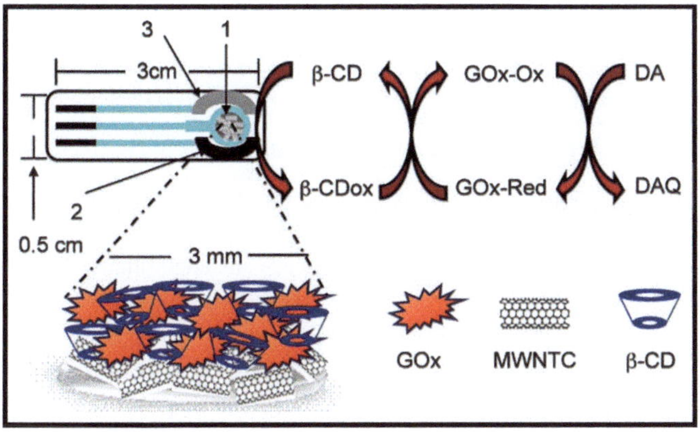

Fig. 4.19 Design of carbon nanotube-modified screen-printed electrode. (Reprinted with permission from [146])

4.6 Conclusion

Solid electrodes for electrochemistry and electroanalysis have been reviewed. The versatility and low cost of solid electrode technology are responsible for their continuous development and wide applications in research and in electroanalysis. Solid electrode materials, by themselves or after modification as modified electrodes, can be used in many configurations. New modifiers and mediators can enhance the reproducibility and sensitivity of the solid electrode surface and are leading to new and wider electroanalytical applications for the determination of pharmaceuticals in their dosage forms and in biological samples.

References

1. Adams RN (ed) (1969) Electrochemistry at solid electrodes. Marcel Dekker, New York
2. Wang J (ed) (2006) Analytical electrochemistry, 3rd edn. Wiley-VCH, New York
3. Kissinger PT, Heineman WR (eds) (1996) Laboratory techniques in electroanalytical Chemistry, 2nd edn. Marcel Dekker, New York
4. Smyth MR, Vos JG (eds) (1992) Analytical voltammetry, vol XXVII. Elsevier, Amsterdam
5. Bond AM (1980) Modern polarographic methods in analytical chemistry. Marcel Dekker, New York
6. Kellner R, Mermet JM, Otto M, Valcarcel M, Widmer HM (eds) (2004) Analytical chemistry, a modern approach to analytical science, 2nd edn. Wiley-VCH, New York
7. Christian GD (ed) (2004) Analytical chemistry, 6th edn. Wiley, New Jersey
8. Bard AJ, Faulkner LR (2001) Electrochemical methods, fundamentals and applications, 2nd edn. Wiley, New York
9. Brett CMA, Oliveira-Brett AM (eds) (1993) Electrochemistry, principles, methods and applications. Oxford University Press, New York
10. Branina KH, Neyman E (1993) Electroanalytical stripping methods. In: Winefordner JD (ed) Chemical analysis: a series of monographs an analytical chemistry and its applications, vol 126. Wiley, New York
11. Hart JP (ed) (1990) Electroanalysis of biologically important compounds. Ellis Harwood, New York
12. Trasatti S (ed) (1980) Electrodes of conductive metallic oxides, Part A. Elsevier, Amsterdam
13. Wang J (ed) (1988) Electroanalytical techniques in clinical chemistry and laboratory medicine. VCH, New York
14. Koryta J, Dvorak J, Kavan L (eds) (1993) Principles of electrochemistry, 2nd edn. Wiley, New York
15. Pierson HO (ed) (1993) Handbook of carbon, graphite, diamond and fullerenes Properties, processing and applications. Noyes, New Jersey
16. Meyyapan M (ed) (2005) Carbon nanotubes, science and applications. CRC, London
17. Bagotsky VS (ed) (2006) Fundamentals of electrochemistry, 2nd edn. Wiley-Interscience, New Jersey
18. Uslu B, Ozkan SA (2007) Anal Lett 40:817–853
19. Uslu B, Ozkan SA (2007) Comb Chem High Throughput Screen 10:495–513
20. Greef R, Peat R, Peter LM, Pletcher D, Robinson J (1985) Instrumental methods in electrochemistry. Ellis Horwood Ltd, New York
21. Solomen E (1987) Z Phys Chem 24:55–62
22. Angerstein-Kozlowska H, Conway BE, Sharp WVA (1973) J Electroanal Chem 43:9–36

23. Hamelin A (1985) Double layer properties at sp and sd metal single–crystal electrodes. In: Conway BE, White RE, Bockris JO (eds) Modern aspects of electrochemistry. Plenum, New York
24. Lipson H, Stokes AR (1942) Math Phys Sci 181:101–105
25. Bas B, Piech R, Niewiara E, Ziemnicka M, Stobierski L, Kubiak WW (2008) Electroanalysis 20:1655–1660
26. Bauman F, Shain I (1957) Anal Chem 29:303–306
27. Yang ZP, Enquist I, Kauffmann JM, Liedberg B (1996) Langmuir 12:1704–1707
28. He P, Ye J, Fang Y, Anzai J, Osa T (1997) Talanta 44:885–890
29. Wang J, Lu J, Hocevar S, Farias P, Ogorevc B (2000) Anal Chem 72:3218–3222
30. Wang J (2005) Electroanalysis 17:1341–1346
31. Zoski CG (ed) (2007) Handbook of electrochemistry, 1st edn. Elsevier, Amsterdam
32. Kalcher K, Svancara I, Metelka R, Vytras K, Walcarius A (2008) Heterogeneous carbon electrochemical sensors. In: Grimes CA, Dickey EC, Pishko MV (eds) Encyclopedia of sensors. American Scientific Publishers, New York
33. Kalcher K (1990) Electroanalysis 2:419–433
34. Adams RN (1958) Anal Chem 30:1576–1576
35. Olsen C, Adams RN (1960) Anal Chim Acta 22:528–536
36. Urbaniczky C, Lundstorm K (1984) J Electroanal Chem 176:169–182
37. Rice M, Galus Z, Adams RN (1983) J Electroanal Chem 143:89–102
38. Radi AE, Abd-Elghany N, Wahdan T (2007) Chem Pharm Bull 55:1379–1382
39. Kuwana T, French WG (1964) Anal Chem 36:241–242
40. Bard AJ, Faulkner LR (2001) Electrochemical Methods 13:560
41. Chiorcea AM, Oliveira-Brett AM (2004) Bioelectrochem 63:229–232
42. Schultz FA, Kuwana T (1965) J Electroanal Chem 10:95–103
43. Bauer D, Gaillochet MP (1974) Electrochim Acta 19:597–606
44. Lamache M, Bauer D (1979) Anal Chem 51:1320–1322
45. Maleki H, Cojocaru CD, Brett CMA, Jenkins GM, Selman JR (1998) J Electrochem Soc 145: 721–730
46. Jenkins GM, Kawamura K (1971) Nature 231:175–176
47. Harris PJF (ed) (1999) Carbon nanotubes and related structures, New materials for the twenty first century. Cambridge University Press, United Kingdom
48. Harris PJF (2005) Crit Rev Solid State Mater Sci 30:235–253
49. Trinidad P, Ponce de León C, Walsh FC (2006) Electrochim Acta 52:604–613
50. Friedrich JM, Ponce-de-Leon C, Reade GW, Walsh FC (2004) J Electroanal Chem 561: 203–217
51. Svancara I, Vytras K, Kalcher K, Walcarius A, Wang J (2009) Electroanalysis 21:7–28
52. Zima J, Svancara I, Barek J, Vytras K (2009) Crit Rev Anal Chem 39:204–227
53. Svancara I, Walcarius A, Kalcher K, Vytras K (2009) Cent Eur J Chem 7:598–656
54. Svancara I, Vytras K, Barek J, Zima J (2001) Crit Rev Anal Chem 31:311–345
55. Vytras K, Svancara I, Metelka R (2009) J Serb Chem Soc 74:1021–1033
56. Jarbawi TB, Heineman WR (1986) Anal Chim Acta 186:11–19
57. Brunetti B, Ugo P, Moretto LM, Martin CR (2000) J Electroanal Chem 491:166–167
58. Kalcher K, Kauffmann JM, Wang J, Svancara I, Vytras K, Yang Z (1995) Electroanalysis 7:5–22
59. Wilson GS (2002) Bioelectrochemistry. In: Bard AJ, Stratmann M (eds) Encyclopedia of electrochemistry, vol 9. Wiley-VCH, Weinheim
60. Sebkova S, Navratil T, Kopanica M (2005) Anal Lett 38:1747–1758
61. Gaylor VF, Conrad AL, Landerl JH (1957) Anal Chem 29:224–228
62. Elving PJ, Smith DL (1960) Anal Chem 32:1849–1854
63. Kroto HW, Heath JR, O'Brien SC, Curl RF, Smalley RE (1985) Nature 318:162–163
64. Guldi DM, Martun N (2010) Carbon nanotubes and related structures: synthesis, characterization, functionalization, and applications. Wiley-VCH, GmbH & Co. KGaA, Weinheim

65. Edmonds T (1985) Anal Chim Acta 175:1–22
66. Huffman ML, Venton BJ (2008) Electroanalysis 22:2422–2428
67. Howell JO, Wightman RM (1984) Anal Chem 56:524–529
68. Engstrom RC, Wightman RM, Kristensen EW (1988) Anal Chem 60:652–656
69. Besenhard JO, Schultze A, Schur K, Jannakoudakis PD (1991) Preparation of voltammetric and potentiometric carbon fibre microelectrodes. In: Montenegro MI, Queirós MA, Daschbach JL (eds) Microelectrodes: theory and applications, NATO ASI Series E. Springer, Netherlands
70. Kuras A, Gutmaniene N (2000) J Neurosci Meth 96:143–146
71. Iijima S (1991) Nature 354:56–58
72. Sherigara BS, Kutner W, D'Souza F (2003) Electroanalysis 15:753–772
73. Gogotsi Y (ed) (2006) Carbon nanomaterials. CRC Taylor Francis Group, New York
74. Dai H (2002) Acc Chem Res 35:1035–1044
75. Lawrence NS, Deo RD, Wang J (2005) Electroanalysis 17:65–72
76. Merkoci A (2006) Microchim Acta 152:157–174
77. Trojanowicz M (2006) TrAC Trends Anal Chem 25:480–489
78. Wang J (2005) Electroanalysis 17:7–14
79. Iijima S, Ichihashi T (1993) Nature 363:603–605
80. Dr. Robert R. Johnson http://astro.temple.edu/~rjohnson
81. Barsan MM, Carvalho RC, Zhong Y, Sun X, Brett CMA (2012) Electrochim Acta 85: 203–209
82. Li J, Cassell A, Delzeit L, Han J, Meyyappan M (2002) J Phys Chem B 106:9299–9305
83. Zhao Q, Gan Z, Zhuang Q (2002) Electroanalysis 14:1606–1607
84. Ge M, Sattler K (1994) Appl Phys Lett 65:2284–2286
85. Kwon YK, Lee YH, Kim SG, Jund P, Tomanek D, Smalley RE (1997) Phys Rev Lett 79: 2065–2068
86. Eftekhari A (ed) (2008) Nanostructured materials in electrochemistry. Wiley-VCH Verlag GmbH & Co KGaA, Weinheim
87. Loiseau A, Launois P, Petit P, Roche S, Salvetat JP (eds) (2006) Understanding carbon nanotubes. Springer, Berlin
88. Jorio A, Dresselhaus MS, Dresselhaus G (eds) (2008) Carbon nanotubes, advanced topics in the synthesis, structure, properties and applications. Springer, Berlin
89. Hodes G (ed) (2001) Electrochemistry of nanomaterials. Wiley-VCH Verlag GmbH, Weinheim
90. Balasubramanian K, Burghard M (eds) (2010) Carbon nanotubes, methods and protocols. Humana Press a part of Springer Science + Business Media, Totowa, NJ
91. Enache TA, Chiorcea-Paquim AM, Fatibello-Filho O, Oliveira-Brett AM (2009) Electrochem Commun 11:1342–1345
92. Yamanaka S, Watanabe H, Masai S, Takeuchi D, Okushi H, Kajimura K (1998) Jpn J Appl Phys 37:L1129
93. Sakharove AY, Peleskov YV, Quarto FD, Piazza S, Sunseri C, Teremetskaya IG, Varnin VP (1995) J Electrochem Soc 142:2704–2709
94. Lawrence NS, Pagels M, Meredith A, Jones TGJ, Hall CE, Pickles CSJ, Godfried HP, Banks CE, Compton RG, Jiang L (2006) Talanta 69:829–834
95. Ramamurti R, Becker M, Schuelke T, Grotjohn T, Reinhard D, Swain G, Asmussen J (2008) Diam Relat Mater 17:481–485
96. Lowe MA, Fischer AE, Swain GM (2006) J Electrochem Soc 153:B506–B511
97. Kapalka A, Fóti G, Comninellis C (2009) Electrochim Acta 54:2018–2023
98. Peleskov YV (2002) Russ J Electrochem 38:1275–1291
99. Compton RG, Foord JS, Marken F (2003) Electroanalysis 15:1349–1363
100. Fortin E, Chane-Tune J, Delabouglise D, Bouvier P, Livache T, Mailley P, Marcus B, Mermoux M, Petit JP, Szunerits S, Vieil E (2005) Electroanalysis 17:517–526
101. Codognoto L, Machado SAS, Avaca LA (2003) Diam Relat Mater 33:951–957
102 Rao TN, Fujishima A (2000) Diam Relat Mater 9:384–289
103. Hupert M, Muck A, Wang J, Stotter J, Cvackova Z, Haymond S, Show Y, Swain GM (2003) Diam Relat Mater 12:1940–1949

104. Xu J, Granger MC, Chen Q, Lister TE, Strojek JW, Swain GM (1997) Anal Chem 69: 591A–597A
105. Oliveira SCB, Oliveira-Brett AM (2010) J Electroanal Chem 648:60–66
106. Creanga C, Serban S, Pittson RW, El Murr N (2011) "No Calibration" type sensor in routine amperometric bio-sensing—an example of a disposable hydrogen peroxide biosensor, biomedical engineering. In: Serra PA (ed) Biosensors-emerging materials and applications. InTech, Croatia
107. Dominguez Renedo O, Alonso-Lomillo MA, Arcos Martinez MJ (2007) Talanta 73:202–219
108. Branina KZ, Kositzina AN, Ivanova A (2007) Electrochemical sensor analysis. In: Alegret S, Merkoci A, Barcelo D (eds) Comprehensive analytical chemistry, 1st edn. Elsevier, Amsterdam
109. Sljukic B, Malakhova NA, Branina KZ, Banks CE, Compton RG (2006) Electroanalysis 18: 928–930
110. Hart JP, Crew A, Crouch E, Honeychurch KC, Pemberton RM (2004) Anal Lett 37:789–830
111. Hart JP, Wring SA (1997) TrAC-Trends Anal Chem 16:89–103
112. Honeychurch KC, Hart JP (2003) TrAC Trends Anal Chem 22:456–469
113. Choi JY, Seo K, Cho SR, Oh JR, Kahng SH, Park J (2001) Anal Chim Acta 443:241–247
114. Radi A-E, Eissa A, Nassef HM (2014) J Electroanal Chem 717–718:24–28
115. Svancara I, Kotzian P, Bartos M, Vytras K (2005) Electrochem Commun 7:657–662
116. Materon EM, Wong A, Klein SI, Liu J, Sotomayor MDPT (2015) Electrochim Acta 158: 271–276
117. Hart JP, Wring SA (1994) Electroanalysis 6:617–624
118. Brett AM, Serrano SH, Gutz IG, La-Scalea MA (1997) Electroanalysis 9:110–114
119. Demetriades D, Economou A, Volugaropous A (2001) Anal Chim Acta 43:434–441
120. Murray RW (1981) Chemically modified electrodes. In: Bard AJ (ed) Electroanalytical chemistry. CRC, Boca Baton, FL
121. Eggins BR (ed) (2002) Chemical sensors and biosensors. Wiley-VCH, New York
122. Radi AE (2006) Curr Pharm Anal 2:1–8
123. Durst RA, Baumer AJ, Murray RW, Buck RP, Andrieux CP (1997) Pure Appl Chem 69: 1317–1323
124. Arrigan DW (1994) Analyst 119:1953–1966
125. Faulkner LR (1989) Electrochim Acta 34:1699–1706
126. Guadelupe AR, Abruna HD (1985) Anal Chem 57:142–149
127. Murray RW (1980) Acc Chem Res 13:135–141
128. Murray RW (1992) Molecular design of electrode surfaces, vol 22. Wiley and Sons, New York
129. Murray RW (1984) Chemically modified electrodes. In: Bard AJ (ed) Electroanalytical chemistry, A series of advances. Marcel Dekker, New York
130. Nikolelis DP, Krull UJ (1993) Electroanalysis 5:539–545
131. Odashima K, Sugawara M, Umezawa Y (1991) Trends Anal Chem 10:207–215
132. Radi A (2010) Comb Chem High Throughput Screen 13:728–752
133. Fujihira M, Rusling JF, Rubinstein I (2007) Modified electrodes. In: Bard AJ, Stratmann M (eds) Encyclopedia of electrochemistry, vol 10. Wiley-VCH, Weinheim
134. Stradiotto NR, Yamanaka H, Zanoni MVB (2003) J Braz Chem Soc 14:159–173
135. Zhang JD, Oyama M (2004) J Microchem 78:217–222
136. Zhao GH, Qi Y, Tian Y (2006) Electroanalysis 18:830–834
137. Wang J, Tian B, Rogers KR (1998) Anal Chem 70:1682–1685
138. Baldwin RP, Thomsen KN (1991) Talanta 38:645–652
139. Walcarius A (2001) Electroanalysis 13:701–718
140. Kauffmann JM, Guiberteau A (1994) Quim Anal 13:169–175
141. Wang J (1991) Electroanalysis 3:255–259
142. Budnikov GK, Labuda J (1992) Russ Chem Rev 61:1491–1514
143. Kutner W, Wang J, L'Her M, Buck RP (1998) Pure Appl Chem 70:1301–1318

144 Pauliukaite R, Ghica ME, Barsan MM, Brett CMA (2010) Anal Lett 43:1588–1608
145. Pierre AC (2004) Biocatal Biotransform 22:145–170
146. Alarcón G, Guix M, Silva WC, Ramírez AM, Palomar M, Merkoçi A (2010) Biosens Bioelectron 26:1768–1773

Chapter 5
Screen-Printed Electrodes (SPE) for Drug Compounds Determination

Contents

5.1 Introduction

The proper medication dosage in pharmaceutical formulations and in physiological fluids is of primary importance in health care. Qualitative and quantitative data are needed for pharmaceuticals in order to check for appropriate drug content and to detect any counterfeit or industrial production error. The determination of the active drug compound in plasma is of interest regarding its bioavailability and its therapeutic efficacy. The determination of drug residues and growth factors in animal products (milk, meat, honey...) and in surface water is also essential in terms of environmental and health concerns.

The analysis of a drug compound can be accurately realized by numerous more or less sophisticated methods employing separation steps such as in liquid chromatography [1, 2], and capillary electrophoresis [3, 4], by direct spectroscopic or electrochemical methods [5–7] and by classical volumetric (titrimetry) methods depending on the complexity of the studied sample. When dealing with drug determination in biological fluids besides the above cited instrumental methods, immunoassays may be advantageously applied at first glance for drug screening, e.g., by competitive optical or electrochemical enzyme immunoassays [8, 9], surface plasmon resonance-based inhibition immunoassays [10], chemiluminescence enzyme immunoassay [11], indirect competitive inhibition enzyme

S.A. Ozkan et al., *Electroanalysis in Biomedical and Pharmaceutical Sciences*, Monographs in Electrochemistry, DOI 10.1007/978-3-662-47138-8_5

immunoassays [12], or indirect competitive enzyme-linked immunosorbent assays (ELISA) [13].

Biosensors thanks to their hybrid structure, with a biological element retained in close proximity of a physical transducer, can advantageously be considered in drug assays since they may offer high selectivity and sensitivity due to signal amplification [14, 15]. Due to the inherent fragile nature of the biocomponent, however, biosensors have not yet reached the expected success in routine use for drug compound assays.

With the commercial launch, in the early nineties, of sensing tools with disposable tips based on screen-printed electrodes (SPE) for glucose assay in blood samples, a new era started in the field of (bio)sensor technology. At first focused on biosensors development, the trend logically oriented progressively to the application of SPEs as planar devices for single or multiple uses in electroanalytical applications. The attractiveness of an SPE is due to the fact that it comprises the (two) three electrodes system in a miniaturized planar configuration. Any electroanalytical waveforms can be applied with the working electrode material based on carbon or noble metals printed on a plastic, glass, or ceramic substrate [16]. The SPEs may be reused or may serve for a single use, and thus contamination and memory effects are avoided. They may be applied for the analysis of volumes as low as a few ten of microliters. SPEs are mass produced with a reproducibility of the working electrode area of less than 5 %. The working electrode surface of a SPE can be readily modified, which opens many possibilities for extended applications. These modifications include the immobilization of a biological component, enzyme or antibody, a thin polymer layer, an organometallic catalyst, nanometallic particles, carbon nanotubes, etc. Several recent review articles have illustrated the potential of application of SPEs [17–21].

5.2 SPE Properties

The screen printing technology exits since a long time ago; it is used in art and textile industry. It has been judiciously implemented to create electrodes. The SPE manufacturing uses several steps with screen printing of conductive inks onto a flat substrate material [22]. Two different types of substrates are commonly employed: ceramic and plastic-based materials [23]. The ink is applied with a floodbar through a patterned stencil or mask on this substrate. Then, the ink undergoes a thermal curing in order to solidify the ink. The curing temperature has an effect on the physical and electrochemical characteristics of the carbon material which is in the ink. At temperatures up to 80 °C graphite active edges are formed. At temperatures in excess of 120 °C, cracks began to appear in the carbon paste leading to a higher effective surface area and improved SPE performances [23]. Silver ink is commonly used as underlying conductive track. Finally, a protective isolating ink coating allows to define the SPE active area [18]. In case of ceramic substrate, a step of firing at high temperature is required in the production. A large range of inks

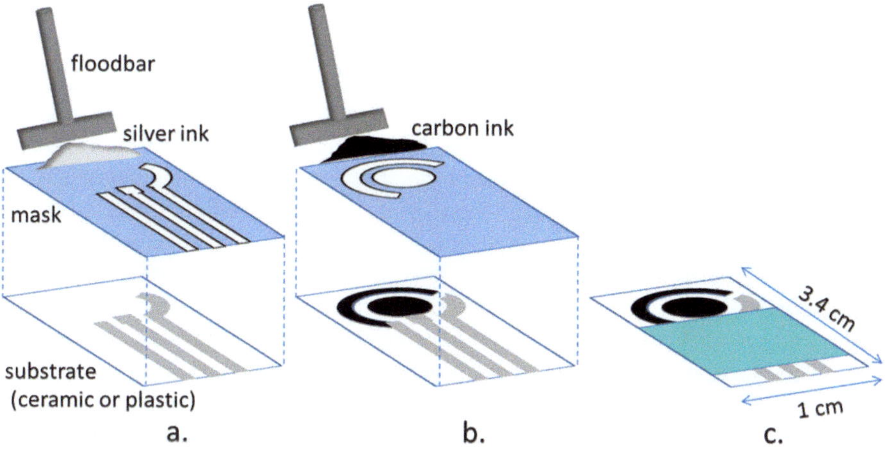

Fig. 5.1 Representation of SPE production. (**a**) Application of silver ink through the mask to print the conductive track and the reference electrode (*circular arc*) on the ceramic or plastic substrate. (**b**) Application of carbon ink to print counter (*circular arc*) and working electrode (disk). (**c**) SPE with the protective ink coating

can be used: carbon and metal (Pt or Au)-based inks are commonly used for printing the working electrode, whereas silver—(silver/silver chloride)-based inks are used for patterning the reference electrode [21] (Fig. 5.1).

Carbon-based inks are frequently used because carbon is relatively inexpensive, easy to modify, and chemically inert. Carbon inks contain a binder (polymer mix), an organic solvent, and graphite particles [24]. The graphite is used as conducting material. It is of interest to compare the characteristics of a SPE and the well-known carbon paste electrode (CPE). The carbon paste in a CPE is a mix of graphite and mineral oil (immiscible with water and non-conductive). The CPE was developed in the late fifties by R.N. Adams [25]. This electrode has been extensively investigated for its potential use in electroanalysis [26, 27]. It offers a low background current in a broad potential domain, the surface can be readily renewed, the CP can be modified by various (bio)reagents, the analyte may be preconcentrated onto the electrode, etc. CPEs are generally employed at the academic level; they have not attracted interest in the pharmaceutical industry and drug control laboratories likely due to the manual surface renewing and surface smoothing requirements which affect the reproducibility of the results. Alternatively, carbon composite electrodes (CCE) have been designed by mixing graphite with epoxy, Teflon, or another hard binding agent [28, 29]. The latter are more robust and suitable for hydroorganic solvents and more stable in hydrodynamic conditions comparing to the CPE [30]. Contrary to glassy carbon or boron doped diamond electrodes, the CCEs have, however, not yet reached commercial success. The working electrode of SPEs can be regarded as a miniaturized version of the CCE designed in a three-electrode configuration.

5.3 SPE Modifications

The surface of the working electrode (WE) of a SPE can be modified for selectivity and sensitivity improvements. Single- (SW) or multi-walled (MW) carbon nanotubes (CNT) and graphene, all of the three being carbon allotropes, contribute to enhance the conductivity of the electrode. The graphite can advantageously be replaced by carbon nanotubes or graphene in the ink composition. Gold, silver, or platinum nanoparticles can be mixed within or electrodeposited onto the WE to improve the electronic transfer properties. The nanoparticles confer the metal characteristics to the electrode and increase the active surface area. An oxidation of the carbon surface can also be realized to generate hydrophilic groups with electrocatalytical properties and for subsequent immobilization reactions (Fig. 5.2).

Bioactive components such as enzymes, antibodies, antigens, DNA, or complexing agent such as cyclodextrins can be retained physically or chemically onto the electrode surface (Fig. 5.3).

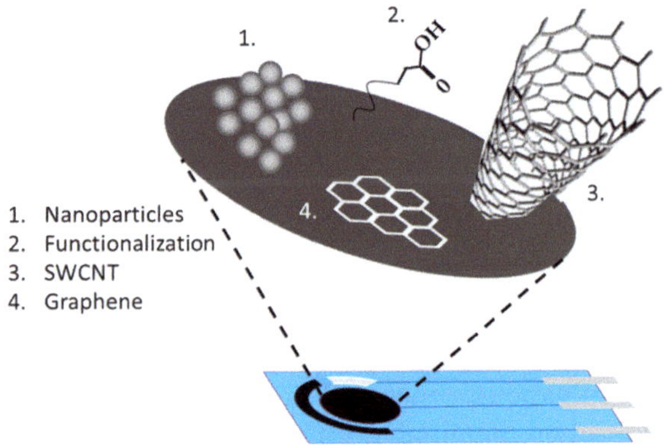

Fig. 5.2 Cartoon of SPE modifications to increase the reactivity of the electrode

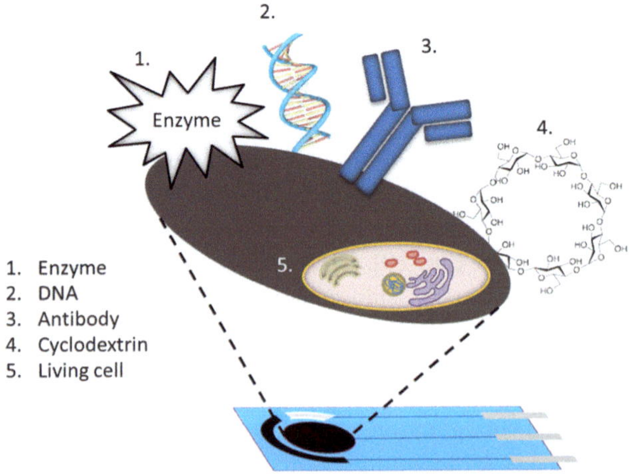

Fig. 5.3 Cartoon of SPE modifications to confer selectivity at the electrode

Adsorption is the simplest physical immobilization method. The SPE is placed in contact with an aqueous solution of the component for a fixed time period. Van der Waal's forces, electrostatic interactions (in case of charged species), and other weak bonds appear between the SPE surface and the component. The unadsorbed component is removed by gently washing the SPE with a buffer solution [31]. The component can also be physically entrapped within a thin polymer layer onto the SPE. The species can be trapped in a three-dimensional matrix such as a polymer or clay material [32]. Biocomponents that are thermoresistant can be mixed with the ink [33]. Such immobilization methods provide a surface with randomly oriented modifiers.

Covalent links such as peptide bonds can be realized between surface carboxylic groups and the amino groups of a biocomponent by activation of the carboxylic groups by carbodiimide. *N*-hydroxysuccinimide can be associated to carbodiimide in order to increase the reaction efficiency. This procedure can be used to immobilize a protein (enzyme, antibody...) onto a carbon SPE. Carboxylic groups can be generated at the carbon SPE surface by chemical or electrochemical oxidation [34]. Amino functional groups can also be linked onto the SPE. Glutaraldehyde is often used as bifunctional activating agent in order to link two amino groups [35]. Amino groups present at a SPE can eventually be transformed with nitrous acid into a diazonium salt and reacted with a protein to form a diazo bond [36]. Proteins can be chemically derivatized with aniline derivatives with subsequent diazotation and electro-addressing of the modified protein onto a carbon-based SPE [37]. Thiols (RSH) can be directly immobilized on gold SPE or carbon SPE modified by gold nanoparticles because the SH group self-assembled spontaneously onto gold with formation of a stable chemical link. Gold-based inks have the advantage that they can be used to immobilize DNA, enzyme, or other proteins by exploiting this S–Au relatively strong bond formation [18]. A mercaptoacid can be used to introduce carboxylic groups at the surface of a gold SPE [36]. Immobilization of proteins by cross-linking is also possible. A cross-linking agent is used generally in the presence of an inert protein such as bovine serum albumin to obtain a co-reticulation. The immobilization of a biocomponent can be achieved via avidin adsorbed onto the SPE and subsequent use of the strong link between (strept)avidin and biotin or via an antibody linked onto the SPE [38]. A recent review has described several immobilization strategies to develop enzymatic biosensors [39] (Fig. 5.4).

5.4 SPE Applications (Table 5.1)

Electroanalytical methods using SPEs for drug determination reported in the literature are mainly intended to check for the active ingredient content in pharmaceuticals or in plasma (serum). Some papers describe the development of methods to detect antibiotics in animal products because antimicrobial agents are administered to animals to promote growth and for therapeutic and prophylactic reasons.

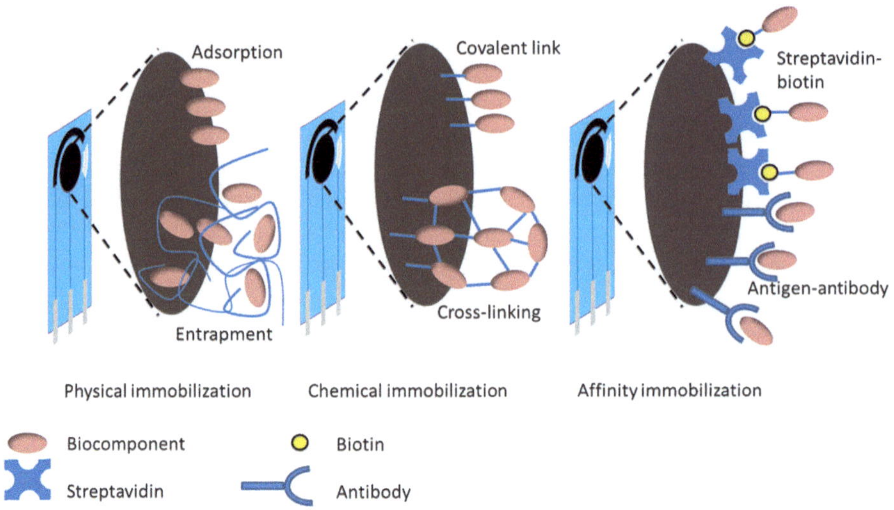

Fig. 5.4 Cartoon of several methods for a biocomponent immobilization onto a SPE

Many regulatory authorities have stipulated maximum residue levels for antibiotics in foods of animal origin. The antibiotic presence in milk or meat can be toxic, can cause allergic reactions in some hypersensitive to individuals, or can lead to problems regarding the spread of drug-resistant microorganisms [40–42]. Small-sized devices using the SPE technology can be regarded as of particular interest for rapid and accurate assays of such samples.

5.4.1 *Unmodified SPE*

The unmodified carbon- or metal-based SPEs can be dedicated to the determination of drug compounds in a simple matrix such as tablet or food after solid phase extraction. Direct determination with minimum sample treatment other than dilution with buffer and filtration (centrifugation) of the sample can be readily performed with a SPE provided that selectivity criteria are achieved. Usually differential pulse (DPV) or square wave (SWV) voltammetry or amperometry (under hydrodynamic conditions) can be implemented, and selectivity is based on the distinct redox potential of the analyte of interest.

A rapid and simple method for procaine determination was developed using flow injection analysis (FIA) with a carbon SPE serving as amperometric detector. Procaine is an injectable local anesthetic often used due to its quick action and lower toxicity than other anesthetics. The recorded signal corresponded to the irreversible oxidation of an amino group to hydroxylamine, in a process involving two protons and two electrons. The proposed FIA amperometric procedure was

Table 5.1 Examples of applications of SPEs for drug compounds determination

Analyte	Electrochemical technique	Working electrode	Surface modification	Analyzed sample	Method	LOD	References
Chloramphenicol succinate	Voltammetry		Molecularly imprinted film	Milk	Calibration curve	2 nM	[40]
Gentamicin	Amperometry (FIA)	Rhodinized carbon	Antibody against gentamicin (free and labeled by GOx)	Milk	Competitive immunoassay		[41]
Lincomycin	Solid phase extraction HPL coupled with a electrochemical sensor	Carbon	Preanodization	Feeds honey Milk Urine	Standard addition	0.08 μM	[42]
Procaine	Amperometry (FIA)	Carbon	–	Pharmaceutical formulation	Standard addition	6 μM	[43]
Levodopa	Amperometry (FIA)	Gold		Tablets Capsule	Standard addition	68 μM 0.99 μM	[44]
Tetracycline Chlortetracycline Oxytetracycline	Amperometry (FIA)	Gold	–	Capsule Chicken Shrimps (solid phase extraction for feeds)	Calibration curve (caps) Standard addition (feeds)	0.96 μM 0.58 μM 0.35 μM	[45]
Colchicine	Anodic DPV	Carbon		Tablets	Standard addition	0.103 μM	[46]
Kojic acid	SWV	Carbon	Preanodization	Cosmetic bleaching products	Standard addition	0.17 μM	[48]
Levetiracetam	Amperometry	Carbon	HRP immobilized in polypyrrole	Serum Tablets	Standard addition	9.81 μM	[49]
Levetiracetam	Amperometry	Carbon	HRP immobilized by a aryl diazonium salt	Tablets	Standard addition	17.5 μM	[50]
Phenobarbital	Chronoamperometry	Gold Carbon	CYP4502B4	Tablets	Standard addition	–	[36]
Phenobarbital	Chronoamperometry	Gold	CYP450 2B4	Tablets	Standard addition	0.289 μM	[51]
Paracetamol	Amperommetry	Carbon	HRP	Tablets	Standard addition	0.0621 μM	[53]
Ecstasy	Amperometry	Carbon	Antibody against free analyte and HRP Labeled species	Saliva Urine	Competitive immunoassay	2 nM 0.2 nM	[54]

(continued)

Table 5.1 (continued)

Analyte	Electrochemical technique	Working electrode	Surface modification	Analyzed sample	Method	LOD	References
Heptacaine Carbisocaine Pantacaine	DPV	Carbon	DNA	Serum	Standard addition	0.03 μM 0.03 μM 0.02 μM	[55]
1-methyl-2-piperidi-noethylester of 2-hexoxyphenyl-carbamic Acid	DPV	Carbon	DNA	Rabbit serum Human serum	Standard addition	0.06 μM 0.016 μM	[56]
Azepines Phenothiazine	DPV (preconcentration in the film)		Film of the polymers of β cyclodextrin (β-CDP)or carboxymethyl β cyclodextrin (β-CDPA)	Serum	Standard addition	0.04 to 0.5 μM (β-CDP) 0.01 to 0.11 μM (β-CDPA)	[57]
Serotonin	DPV	Colloidal gold	Nafion membrane	Brain Platelet-rich plasma	Standard addition	5 nM	[58]
Isoniazid	Linear sweep voltammetry DPV SQW	Carbon	Film of poly-L-histidine	Urine	Standard addition	0.5 μM 0.17 μM 0.25 μM	[59]
Paracetamol	Amperometry (FIA)	Carbon	Biocomposite film (MWCNT-PANI-polyFAD	Oral solution Effervescent tablets		–	[60]
Paracetamol Paracetamol Paracetamol	CV DPV Amperometry (FIA) Amperometry CV	Carbon Carbon	PEDOT film Cellulose acetate Polyurethane resin composite	Tablets Urine Pharmaceutical formulations	Standard addition Calibration curve	0.16 μM 13 μM 0.818 μM	[61–63]
Salbutamol Ractopamine	Conductimetry DPV	Carbon Silver Carbon	MIM MWCNT MIM	Pig urine Pig urine	Calibration curve Calibration curve	13.5 nM 6 nM,	[64, 65]

Hesperidin	CV AdS SWV	Carbon	MWCNT	Orange juice	Standard addition	0.61 μM 7 nM	[66]
Methimazole	Amperometry (FIA)	Carbon	MWCNT Tyrosinase immobilized onto a rotating disk	Tablet	Comparaison of the analyte-excipient response with pure ana-lyte response	0.056 μM	[67]
Paracetamol	Amperometry (FIA)	Carbon	MWCNT	Tablets	Calibration curve	0.1 μM	[68]
Silybin	AdS SWV	Carbon (single-sided heated)	MWCNT	Tablets	Standard addition	0.5 nM	[69]
Paracetamol Ascorbic acid	Amperometry	Carbon	SWCNT	Tablets	Calibration curve	0.2 nM 0.5 nM	[70]
Salbutamol	CV	Carbon	Inkjet-printed Graphene-PEDOT: PSS	Tablets Syrup	Calibration curve	1.25 μM	[72]
Arbutin	SWV	Carbon	Nontronite clay	Cosmetic bleaching products	Standard addition	0.18 μM	[73]
Codeine	SWV Amperometry (FIA)	Carbon	Nontronite clay	Urine Tablets	Standard addition	20 nM (SWV) 0.2 nM (FIA)	[74]
Lamotrigine	DPAdSV	Carbon	Silver nanoparticles	Capsules	Standard addition	0.372 μM	[75]
Isoniazid	Amperometry (FIA)	Carbon	Silver nanoparticles	Simulated Human urine	Calibration curve	2.6 μM	[76]
Dapsone	SWV	Carbon	Platinum nanoparticles	Tablets	STANDARD addition	0.76 μM	[77]

applied for the determination of procaine in the pharmaceutical formulation Timpanol®. The procaine content was determined by the standard addition method, and the results were compared with the official method proposed in the United States Pharmacopoeia: extraction-spectrophotometric method based on its maximum absorbance signal at 280 nm. The results were in agreement with the results obtained by the extraction-spectrophotometric method, as well as the label values described by the industrial laboratory [43].

Another example is the application of a gold-based SPE as an amperometric disposable sensor for levodopa (L-dopa) determination. L-dopa, the medication of choice for the treatment of Parkinson's disease, is principally metabolized by an enzymatic reaction (dopa-decarboxylase) to dopamine compensating for the deficiency of dopamine in the brain. In order to improve the sensitivity of the SPE, the determination of L-dopa was conducted using hydrodynamic conditions. Indeed, the obtained anodic peak in the flow cell was five times higher than when using the gold-based SPE under stationary condition. The proposed method was applied to the determination of L-dopa in two pharmaceutical formulations using the standard addition method without prior chemical/electrochemical treatment. The results and the statistical treatments of data confirmed that there were no significant differences between the proposed method and the United States Pharmacopoeia method [44].

The electrochemical analysis of tetracyclines was reported using a gold-based SPE by cyclic voltammetry and FIA with direct current (dc) amperometry. The proposed method was successfully applied to the determination of tetracycline, chlortetracycline, and oxytetracycline in pharmaceutical products and spiked foods. Tetracyclines are an important class of antibiotics with a broad spectrum. They are used in a variety of food-producing animals, so it was of interest to develop methods applicable to food matrices [45].

Bodoki et al. determined colchicine, using a graphite-based SPE. Colcichine is used as a specific anti-inflammatory agent in acute attacks of gout by inhibiting the migration of leucocytes to inflammatory areas. Both the dosage of the drug in tablets and the interaction of colchicine with physiological molecules were realized. For quantitative assays, an anodic response was obtained by DPV using the unmodified graphite-based SPE. Standard addition method was employed to determine the drug in tablets. The sensor was also used for the DNA-colchicine, β-tubulin-colchicine, and bovine serum albumin (BSA)-colchicine interaction studies. A double-stranded calf thymus DNA was immobilized onto the SPE surface. The drug interaction with DNA was monitored by the decrease of the guanine residue oxidation peak. No current decrease after the exposure to colchicine to the DNA biosensor was observed confirming a lack of DNA–colchicine interaction, a result also supported by the different antimitotic mechanisms of colchicine. The colchicine inhibits the microtubules polymerization by binding with high affinity to tubulin, a negatively charged protein formed by two globular polypeptidic subunits (α-tubulin and β-tubulin). The β-tubulin was immobilized onto the SPE surface.

The β-tubulin and colchicine interaction caused a shift of colchicine oxidation peak and a decrease of the anodic current. The interaction between BSA and colchicine was investigated by using a BSA-modified SPE. A potential shift and an anodic current decrease were observed due to the strong interaction of colchicine with the BSA resulting in a less favorable conformation of the colchicine molecule for the heterogeneous electron transfer [46].

5.4.2 Modified SPE Surface

5.4.2.1 Preanodization

The preanodization technique involves applying a positive voltage for a fixed period of time at the carbon electrode in the presence of a buffer. This produces high background currents (pseudocapacitive origin) and often permits to obtain a shift of the oxidation peak potential of the studied analyte to lower positive potentials. This is particularly interesting for compounds oxidized at potential close to the solvent evolution [47].

Lincomycin, an antibiotic drug used in human and veterinary medicine, can be determined in honey, milk, urine, or in tablets by FIA coupled with liquid chromatography (LC) and amperometric detection with a preanodized carbon-based SPE. The lincomycin oxidation peak potential shifted from 1.1 V to 0.9 V with the preanodized SPE versus the SPE. The current magnitude increased because during preanodization, the SPE became more porous with active edge plane sites and surface carbonyl functionalities. The obtained results for lincomycin determination in real samples confirmed the utility of using a preanodized SPCE coupled to LC [42].

Kojic acid, a tyrosinase inhibitor, was determined in cosmetic bleaching products thanks to a preanodized carbon-based SPE. The preanodization allowed the oxidation of the carbon-based SPE introducing the C=O group on the electrode surface resulting in kojic acid accumulation. This study demonstrated that the preanodized carbon-based SPE can be used to determine kojic acid in cosmetic products with an excellent sensitivity and selectivity by square-wave voltammetry (SWV) [48].

5.4.2.2 Biocomponent

5.4.2.2.1 Enzyme

The SPE surface modification by an enzyme combines the advantage of the SPE with the biocomponent selectivity and the enzymatic amplification of the signal. Horseradish peroxydase (HRP) can be advantageously used because of its commercial availability, stability, and high turnover [49, 50]. Another family of enzymes used for the SPE modification is the cytochrome P450 enzyme family

(CYP450) [36, 51]. CYP450 is particularly important in medical sciences since most CYP450 catalyze the in vivo oxidation of organic xenobiotics namely drug compounds into more hydrophilic species: $RH + O_2 + 2H^+ + 2e^- \rightarrow ROH + H_2O$.

Two carbon-based SPE-based biosensors for the determination of levetiracetam have been developed by Alonso-Lomillo et al. HRP was used as biological component. The HRP immobilization process was different for each biosensor. HRP immobilization was accomplished by pyrrole electropolymerization onto a carbon-based SPE [49] or onto a carbon-based SPE premodified with an aryl diazonium salt [50]. HRP biosensors are based on the enzymatic oxidation of the analyte in the presence of hydrogen peroxide with the signal corresponding to the reduction of the oxidized substrate at the electrode surface. In the presence of H_2O_2, the immobilized peroxidase (HRP Fe^{3+}) is oxidized to compound I (oxyferryl iron and a porphyrin cation radical). The latter is reduced first by a one electron transfer to an intermediate redox state (compound II) and subsequently by a second electron transfer to give the native resting state (Fe^{3+}) in the presence of the analyte, i.e., the electron donor [52].

Alonso-Lomillo et al. have developed a CYP450 2B4 biosensor for the determination of phenobarbital, a first generation anticonvulsant drug widely used to treat epilepsy. The enzyme was covalently immobilized with a diazonium salt (4-nitrobenzenediazonium tetrafluoroborate) or by the formation of a thiol monolayer onto the SPE. The carbon-based SPEs are functionalized by electrochemical grafting of a diazonium salt solution and subsequent generation of the amine form by electrochemical reduction of the nitro group by cyclic voltammetry. The gold-based SPEs and the carbon-based SPE modified by Au nanoparticles were functionalized by mercaptopropionic acid and thioctic acid through self-assembling formation of Au–S bonds. The carboxylic group was activated by *N*-(3-dimethylaminopropyl)-*N*'-ethylcarbodiimide (EDC) and *N*-hydroxysuccinimide (NHS) for immobilization. The three sensors used chronoamperometry to determine phenobarbital in tablets after dissolution in acetone–water and centrifugation [36]. Another CYP450 2B4 biosensor was developed for phenobarbital determination with another enzyme immobilization method. The CYP450 2B4 (CYP450) was immobilized into a polypyrrole matrix onto a gold-based SPE [51].

A biosensor based on the immobilization of HRP in a polyethyleneimine-zirconium alcoxide porous gel film onto a SPE for paracetamol was developed by V. Sima et al. The biosensor principle was based on monitoring the amperometric signal produced by electrochemical reduction of the enzymatically oxidized paracetamol. This biosensor enabled paracetamol determination in tablets [53].

5.4.2.2.2 Antibody

The SPE surface can readily be modified by an antibody (Ab) for highly selective assays. The antibody can be immobilized directly onto the SPE surface [54] or localized online in close proximity to the SPE as shown in a FIA setup [41]. A competitive immunoassay for amphetamine and analogue recreational drugs was developed for the drug determination in saliva and urine. The antibody was immobilized by passive adsorption onto the carbon-based SPE surface. Care was taken to cover the Ab-SPE with a blocking layer of casein for minimizing nonspecific adsorption. The SPE served as solid support for the immunoreaction between 3,4-methylenedioxyamphetamine (MDA) or 3,4-methylenedioxymethamphetamine (MDMA) linked to HRP and the analytes MDA and MDMA. The enzymatic substrate was tetramethylbenzidine (TMB)/ H_2O_2. The TMB oxidized by HRP was reduced at the SPE, and the resulting current was monitored. The correlation obtained with saliva sample was better than with urine sample because in saliva the drug is not metabolized whereas, in urine, a lot of metabolites are present which may be undetectable depending on the antibody specificity [54].

A competitive immunoassay coupled to a FIA and an electrochemical detection was developed by Van Es et al. for the quantification of gentamicin, an antibiotic, in milk. The antibody against gentamicin was immobilized onto a nitrocellulose strip. Free gentamycin and gentamicin covalently linked to glucose oxidase were in competition for the antibody adsorbed onto the strip. The analyte level was quantified according to the extent of glucose oxidase activity, measured electrochemically by means of hydrogen peroxide determination. The latter was amperometrically detected in a flow cell located right next to the site of the immunoreaction. Thanks to the use of the nitrocellulose strip to immobilize the antibody, the SPE was not modified so that the amperometric signal was not lowered due to problems of electrode accessibility [41].

5.4.2.2.3 DNA

A drug capable of accumulation within dsDNA without DNA damage and which is easily removed by dilution can be determined at a DNA-modified SPE. DNA-based biosensors have been developed to determine by DPV the local anesthesic drugs heptacaine, carbisocaine, and pentacaine [55]. The accumulation of the drug in the DNA layer can be increased at an applied potential of −0.20 V for 60s. This accumulation step was also implemented to determine another local anesthesic drug, namely the 1-methyl-2-piperidinoethylester of 2-hexoxyphenylcarbamic in spiked serum sample by DPV. Matrix interference was suppressed by protein precipitation with acetonitrile. The DNA biosensor permitted the investigation of the drug stability in spiked serum samples [56].

5.4.2.3 Membrane/Film

The SPE surface can be modified by a polymer film to improve the selectivity of the assay. The film can also be used for a pre-concentration step. β-cyclodextrin (β-CDP) or carboxymethylated β-cyclodextrin (β-CDPA) polymer films were used for determination of tricyclic antidepressive drugs such as azepines and phenothiazines [57]. The sensors were fabricated by casting a carbon-based SPE with thin films of β-CDP or β-CDPA. The electroanalytical two-step procedure involved the drug preconcentration at a selected constant potential followed by quantification by DPV.

A selective and sensitive electrochemical method based on a Nafion membrane-coated colloidal gold screen-printed electrode (Nafion/CGSPE) was described for the determination of serotonin (5-HT) in the presence of high concentrations of dopamine, ascorbic acid, and uric acid [58]. Serotonin is a monoamine neurotransmitter, and its level in the brain can reflect depression symptoms. The electrocatalytic response of serotonin on the Nafion/CGSPE was evaluated by cyclic voltammetry, and a linear calibration curve was obtained by DPV. Nafion is a good cation exchanger which adhered well onto the electrode surface. Serotonin can be adsorbed on Nafion-modified electrodes via ion-exchange interaction. The interferences of uric acid and ascorbic acid to 5-HT were successfully avoided because of their opposite electric charge in physiological conditions.

The electrode modification by a thin polymeric film can have as goal a change in oxidation or reduction potential of the analyte and the increase of peak current. L-histidine was electropolymerized into a film onto the surface of a carbon-based SPE for the determination of isoniazide (INZ), a tuberculostatic agent effective against mycobacterium strains. The electrode exhibited a catalytic effect with the electrochemical reduction of INZ occurring at less negative potentials and with an increased peak current compared to the non-modified SPE. The electrochemical behavior of the modified carbon-based SPE was investigated by cyclic, linear sweep (LSV), DPV, SWV, and electrochemical impedance spectroscopy (EIS) [59].

A conductive biocomposite film which contained multi-walled carbon nanotubes (MWCNTs) along with the incorporation of poly(aniline) and poly(flavinadenine dinucleotide) copolymer (PANIFAD) has been synthesized on a gold electrode and carbon-based SPE by potentiostatic methods and used for the determination of paracetamol in tablets. The use of MWNT enhanced the surface coverage concentration of the analyte in the film and increased the electron transfer rate constant with concomitant increase in sensitivity [60].

Another study showed that the paracetamol determination was possible by oxidation at a SPE modified with electrogenerated poly(3,4-ethylenedioxythiophene) (PEDOT) film. The PEDOT is a conducting polymer that is capable of catching anionic analytes based on electrostatic attractions. Moreover, the polymer hydrophobic nature facilitated accumulation of neutral, nonelectrolytic organic analytes. This work studied the electrochemical behavior of the non-charged paracetamol drug at SPE/PEDOT. The PEDOT film efficiently accelerated the paracetamol electron transfer rate, and the proposed method was applied to paracetamol determination in pharmaceutical tablet by FIA coupled to amperometric detection [61]. Paracetamol was also determined at carbon-based SPE with the working electrode covered by a cellulose acetate layer. The latter afforded an effective interferent screening for the amperometric assay of the drug in urine [62]. Another study used a carbon-based SPE blended with a polyurethane resin for the cyclic voltammetric assay of paracetamol in pharmaceuticals without interference from ascorbic acid [63].

Molecularly imprinted polymers (MIPs) are three-dimensional structures that can be formed in the presence of a template onto the SPE surface. After the polymerization step, the template is removed leaving open cavities for selective sensing. A sensor based on a MIP film for the detection of chloramphenicol succinate was described. The antimicrobial agent was successfully detected in milk samples [40]. Chai et al. have developed a conductimetric probe based on a molecularly imprinted membrane (MIM) for the detection of salbutamol. The sensor consisted of two parallel SPEs. One of the SPEs was coated with a MIM using salbutamol as the template, and the other was modified with a non-molecularly imprinted membrane. A change in the electrical conductivity in response to salbutamol was recorded and allowed the assay of salbutamol in pig urine. Salbutamol is a selective β 2-adrenoceptor agonist used as a bronchodilator in asthmatic patients, but it is used illegally as growth-promoting agents in animals [64]. The same group developed an amperometric sensor for the determination of ractopamine, a β-agonist used illegally to promote animal growth. The SPE was modified with MWCNTs, and a MIM was directly synthesized on these modified electrodes [65].

5.4.2.4 Nano-carbon Material

Carbon nanotubes (CNT) are allotropes of carbon with a cylindrical nanostructure. They have unique electrical properties and have a large specific surface. There are different types of CNT including single-walled (SWCNT) and multi-walled carbon nanotube (MWCNT). CNT-based SPEs have been developed to determine hesperidin, methimazol, and paracetamol [66–68].

Hesperidin is a flavonoid found in citrus fruit and used for the treatment of fragile capillaries. The sensor, developed to determine the hesperidin in orange juice, was based on MWCNT-SPEs using the technique of adsorptive stripping voltammetry (AdSV). The much larger electroactive surface area of these MWCNT-SPEs allowed for a greater degree of hesperidin adsorption. AdSV experiments using the MWCNT-SPEs were performed without accumulation in a droplet of solution containing hesperidin [66].

Methimazole is a drug used in the therapy of hyperthyroidism. Methimazole inhibits the addition of iodine by the thyroperoxidase, a necessary step in the synthesis of thyroid hormones. Tyrosinase, immobilized on a rotating disk, catalyzed the oxidation of catechols to o-benzoquinone, whose back electrochemical reduction was detected on a graphite-based SPE modified with MWCNT at

−150 mV. When methimazole was added to the solution, this thiol-containing compound participated in a Michael type reaction with o-benzoquinone to form the corresponding thioquinone derivative. This reaction suppressed the substrate recycling process between tyrosinase and the electrode, decreasing the peak current in direct relation to the increase of methimazole concentration. The sensor was used to determine methimazole in different pharmaceutical samples [67].

Fanjul-Bolado et al. have used a CNT-SPE for the study of the electrochemical oxidation mechanism of paracetamol. The paracetamol was oxidized to *N*-acetyl-*p*-quinoneimine. The study permitted the determination of paracetamol in effervescent tablets and oral solutions. A FIA system was used with the aim of automating the analysis. The amperometric detection combined with FIA was an attractive solution to determine the paracetamol at a high analysis throughput [68].

Silybin, the biologically active component of *silibum marinum*, a hepatoprotective plant, can be determined by a disposable MWCNT-modified single-sided heated carbon-based SPE (MWNT/ss-HSPCE). The method consisted of two steps (1) accumulation of silybin at the electrode surface. (2) SWV measurement. During the 5 min of accumulation, the electrode temperature was elevated at 50 °C improving the square wave voltammetry stripping peak current. The temperature difference between the heated electrode (50 °C) and the unheated solution (20 °C) induced a forced convection which was favorable to transport the electroactive substances from the solution to the electrode surface. The silybin content in pharmaceutical tablets can be determined under the optimum conditions by the standard addition method [69].

Paracetamol and ascorbic acid are frequently combined in pharmaceutical formulations. The oxidation potentials of ascorbic acid and paracetamol are rather close precluding the accurate paracetamol determination without separation or electrode modification (see above). Sarakbi et al. have developed a rapid and sensitive method for simultaneous quantification of these two components by LC using a monolithic column coupled to a SWCNT-based SPE housed in a wall-jet flow cell configuration. The linear calibration range for both compounds was comprised between 2.10^{-7}–1.10^{-4} M. Limit of detection (*LOD*) for APAP and AA was 2.10^{-7} and 5.10^{-7} M, respectively, and the limit of quantification (*LOQ*) was $5.10^{-7}$and 1.10^{-6} M [70].

Graphene is a carbon allotrope that has a two-dimensional honeycomb lattice structure. It is highly advantageous for electrochemistry due to its very large two-dimensional electrical conductivity, excellent electron transfer rate, and huge specific surface area [71]. Graphene exhibited a better conductivity, density of surface negative charge present, sensitivity, signal-to-noise ratio, and stability than SWCNT [51]. A inkjet-printed graphene-poly(3,4-ethylenedioxythiophene): poly (styrene-sulfonate) (GP-PEDOT:PSS) on a carbon-based SPE was developed for detection of salbutamol in syrup and tablets. An acceptably low matrice interference was noticed on the electrochemical oxidation of salbutamol at this GP-PEDOT:PSS-modified carbon SPE [72].

5.4.2.5 Clay

Clays are inorganic materials which can be deposited as a thin film onto an electrode surface. There are cationic clays that have negatively charged alumino silicate layers and anionic clays, with positively charged hydroxide layers. Electrochemical species can be accumulated in different places of the clay, yet only 10–30 % display a redox activity. The charge transport in the clay is due to their diffusion in a channel or/and to charge hopping. To minimize the intercalation of the electrochemical species and the loss of redox activity, a delamination process can be applied at the clay by using smaller size particles, pillaring agents, or surfactant agents. The clay matrix can be used to entrap a biomolecule or a redox mediator at the SPE surface. The target molecule can be also concentrated in the clay to enhance the sensitivity [32].

Arbutin, a hydroquinone glycoside, is the active component of the bearberry (*uvae ursi folium*). This plant is used as urinary antiseptic. Arbutine has a depigmenting effect by reducing the cellular activity of tyrosinase, an essential enzyme for melanin formation. Arbutin was determined in cosmetic bleaching products by SWV with the use of a preanodized nontronite clay-coated SPE. Nontronite clay (Fe_2^{3+} $(Si_{4-x}Al_xO_{10})(OH)_2$ $(Na^+_x$, $nH_2O)$ contains a high amount of trivalent iron and is electrochemically active. During the preanodization, hydrophilic functional groups were produced onto the SPE, and the iron in the clay was converted to a higher amount of oxidized iron. A substantial interaction between arbutin and oxidized iron can explain the further increase in peak current. The arbutin was determined in cosmetics by the method of standard addition [73].

Another nontronite clay-modified carbon-based SPE was developed for codeine determination. The drug was quantified in urine and commercial pharmaceuticals by amperometric FIA and square wave voltammetry [74].

5.4.2.6 Nanoparticles

With the aim that obtain a better limit of detection compared to carbon-based SPEs or mercury film carbon SPEs, Burgoa Calvo et al. developed a silver nanoparticles-modified carbon SPE for the determination of lamotrigine (LTG). LTG was determined in pharmaceutical formulations by differential pulse adsorptive stripping voltammetry (DPAdSV) using the modified SPE. LTG is an antiepileptic drug which inhibits the release of the excitatory amino acid neurotransmitters (glutamate and aspartate) which play an important role in epilepsy [75].

More recently, a FIA system was developed for INZ determination using a SPE modified with silver hexacyanoferrate nanoparticles (NPAg-HCF) as amperometric detector. This method permitted a rapid INZ determination in human urine spiked without significant influence of ascorbic acid, dopamine, glucose, and rifampicin (co-administrated with INZ in pharmaceutical formulations) [76].

Dodecanethiol-stabilized platinum nanoparticles were applied as modifier onto SPE for the determination of dapsone, a drug used for the treatment of various skin diseases such as leprosy among others. The sensor mechanism required first the platinum electro-oxidation and subsequently the dapsone eletrooxidation. This oxidation was assumed to be chemically catalyzed by the preformed Pt oxide nanoparticles [77].

5.5 Screen-Printed Electrode and Ascorbic Acid Determination

Ascorbic acid (AA), the vitamin C, is an essential nutriment because it is not synthesized by the humans. AA is an antioxidant and has a protective function against oxidative stress. Vitamin C deficiency causes scurvy, a common disease among the sailors when fruits and vegetables could not be stored for long periods. AA is naturally present in fresh vegetables and fruits, but also in dietary supplements and pharmaceutical formulations, alone or in association in multivitamin complex or with paracetamol.

Many sensors for AA determination by oxidation at low potential were developed. The low potential minimized the effect of most currently electrochemical interferents such as paracetamol, uric acid, and citric acid. Lower the oxidation potential of AA can be achieved by modification of the SPE surface. A study described a carbon-based SPE modified with a conductive polyaniline (PANI) layer [78]. Another team constructed carbon-based SPE impregnated with the electrocatalyst 2,6-dichlorophenolindophenol lanthanum salt ([DCPI]3La) [79]. The electrode surface must be modified by multilayer films containing MWCNTs and the redox polymer. The modification allowed a negatively shift of the AA oxidation potential and high electrocatalytic activity towards AA in the presence of uric acid and dopamine [80]. A SPE with a RuO_2 working electrode, an Ag/AgCl reference electrode, and a RuO_2 counter electrode permitted the determination of AA at low oxidation potential. The interference of uric acid and hydrogen peroxide was very low at the applied potential [81]. SPE using an ink containing graphene allowed the simultaneous determination of ascorbic acid, dopamine, and uric acid [82].

A FIA system with a heated carbon-based SPE was developed to improve the detection sensitivity without decomposing the substrates. When the working solution was preheated and then AA was added, the peak current increases considerably. Yet, if the AA solution was directly heated, AA was decomposed to an electro-inactive species. The SPE was heated at 46 °C with a light beam. The authors suggested that the increase in the interaction of AA with the SPE at the interface was the reason for the observed higher current at elevated temperatures [83] (Table 5.2).

Table 5.2 Examples of application of SPEs for ascorbic acid determination

Analyte	Electrochemical technique	Working electrode	Surface modification	Analyzed sample	Method	LOD	References
Ascorbic acid	Amperometry	Carbon	PANI film	Tablets Juice	Calibration curve	2.45 μM	[78]
	CV	Carbon	2,6-dichlorophenol Indophenol lanthanum salt ([DCPI]3La)	Standard solution		10 μM	[79]
	Amperometry	Carbon	MWCNT + redox Polymer	Tablets	Standard addition	0.3 μM	[80]
	Amperometry DPV	Ruthenium dioxide Graphene		Injection Human urine	Standard addition	0.95 μM	[81, 82]
	Amperometry	Carbon	Heated SPE	Commercial orange juice	Standard addition	28.7 nM	[83]

5.6 Conclusion

Several research groups have shown the possibility of applying screen-printed electrodes for drug compounds analysis in pharmaceutical forms and in biological matrices. Carbon-based SPEs are most popular but would need modification of the working electrode surface for selective and accurate determination. It is anticipated that robust modified SPEs applied with a proper validation procedure could gain success in many application areas where rapid drug control is required.

References

1. Fahmy OT, Korany MA, Maher HM (2004) J Pharm Biomed Anal 34:1099–1107
2. Rodila R, Kim GE, Fan L, Chang MS, Zhang J, Wu H, El-Shourbagy TA (2008) J Chromatogr B 872:128–132
3. Boone CM, Waterval JCM, Lingeman H, Ensing K, Underberg WJM (1999) J Pharm Biomed Anal 20:831–863
4. Morzunova TG (2006) Pharm Chem J 40:158–170
5. Patriarche JG, Zhang H (1990) Electroanalysis 2:573–579
6. Xu QX, Yuan AJ, Zhang R, Bian X, Chen D, Hu X (2009) Current Pharm Anal 5:144–155
7. Ozkan SA (ed) (2012) Electroanalytical methods in pharmaceutical analysis and their validation, 1st edn. HNB Pub, New York
8. Wang CJ, Tian Z, Byrnes K, Lin CC (1992) J Pharm Biomed Anal 10:473–479
9. Hu L, Zuo P, Ye BC (2010) Anal Biochem 405:89–95
10. Fitzpatrick B, O'Kennedy R (2004) J Immunol Methods 291:11–25
11. Kim JK, Adam A, Loo JCK, Ong H (1995) J Pharm Biomed Anal 13:885–891
12. Grothaus PG, Raybould TJG, Bignami GS, Lazo CB, Brynes JB (1993) J Immunol Methods 158:5–15
13. Svojanovsky SR, Egodage KL, Wu J, Slavik M, Wilson GS (1999) J Pharm Biomed Anal 20: 549–555
14. Yu D, Blankert B, Viré JC, Kauffmann JM (2005) Anal Lett 38:1687–1701
15. Kauffmann JM, El Kaoutit M, Parsajoo C (2012) Biosensors for drug testing and discovery. In: Higson S (ed) Biosensors for medical applications. Woodhead, Cambridge
16. Dropsens (2013) http://www.dropsens.com. Accessed 30 May 2013
17. Domínguez-Renedo O, Alonso-Lomillo MA, Ferreira-Gonçalves L, Arcos-Martínez MJ (2007) Talanta 73:202–219
18. Li M, Li YT, Li DW, Long YT (2012) Anal Chim Acta 13:31–44
19. Jubete E, Loaiza OA, Ochoteco E, Pomposo JA, Grande H, Rodriguez J (2009) J Sens 2009: 1–13
20. Honeychurch KC (2012) Insciences J 2:1–51
21. Avramescu A, Andreescu S, Noguer T, Bala C, Andreescu D, Marty J-L (2002) Anal Bioanal Chem 374:25–32
22. Wang J, Tian B, Nascimento VB, Angnes L (1998) Electrochim Acta 43:3459–3465
23. Grennan K, Killard AJ, Smyth MR (2001) Electroanalysis 13:745–50
24. Morrin A, Killard AJ, Smyth MR (2003) Anal Lett 36:2021–2039
25. Adams RN (1958) Anal Chem 30:1576–1576
26. Kalcher K, Kauffmann JM, Wang J, Švancara I, Vytřas K, Neuhold C, Yang Z (1995) Electroanalysis 7:5–22
27. Švancara I, Vytřas K, Kalcher K, Walcarius A, Wang J (2009) Electroanalysis 21:7–28

28. Ramirez-Garcia S, Alegret S, Céspedes F, Forster RJ (2002) Analyst 127:1512–1519
29. Corb I, Manea F, Radovan C, Pop A, Burtica G, Malchev P, Picken S, Schoonman J (2007) Sensors 7:2626–2635
30. Maleki N, Safavi A, Tajabadi F (2006) Anal Chem 78:3820–3826
31. Rico MA, Olivares-Marín M, Gil EP (2009) Talanta 80:631–635
32. Mousty C (2004) Appl Clay Sci 27:159–177
33. Ricci F, Amine A, Tuta CS, Ciucu AA, Lucarelli F, Palleschi G, Moscone D (2003) Anal Chim Acta 485:111–120
34. Susmel S, Guilbault GG, O'Sullivan CK (2003) Biosens Bioelectron 18:881–889
35. Lia YG, Zhoua YX, Jiangb ZH, Ma LR (1999) Anal Chim Acta 382:277–282
36. Alonso-Lomillo MA, Yardimci C, Domínguez-Renedo O, Arcos-Martínez MJ (2009) Anal Chim Acta 633:51–56
37. Corgier B, Marquette CA, Blum LJ (2005) J Am Chem Soc 127:18328–183332
38. Hernández-Santos D, Díaz-González M, González-García MB, Costa-García A (2004) Anal Chem 76:6887–6893
39. Sassolas A, Blum LJ, Leca-Bouvier BD (2012) Biotechnol Adv 30:489–511
40. Zhang NW, Ding MX, Liu GY, Song WW, Chai CY (2008) Chin J Anal Chem 36:1380–1384
41. Van Es RM, Setford SJ, Blankwater YJ, Meijer D (2001) Anal Chim Acta 429:37–47
42. Chiu MH, Yang HH, Liu CH, Zen JM (2009) J Chromatogr B 877:991–994
43. Bergamini MF, Santos AL, Stradiotto NR, Zanoni MVB (2007) J Pharm Biomed Anal 43: 315–319
44. Bergamini MF, Santos AL, Stradiotto NR, Zanoni MVB (2005) J Pharm Biomed Anal 39: 54–59
45. Masawat P, Slater JM (2007) Sens Actuators B 124:127–132
46. Bodoki E, Laschi S, Palchetti I, Săndulescu R, Mascini M (2008) Talanta 76:288–294
47. Gregg MR (1986) Chromatographia 21:705–707
48. Shih Y, Zen JM (1999) Electroanalysis 11:229–233
49. Alonso-Lomillo MA, Domínguez-Renedo O, Matos P, Arcos-Martínez MJ (2009) Bioelectrochemistry 74:306–309
50. Alonso-Lomillo MA, Domínguez-Renedo O, Hernández-Martín A, Arcos-Martínez MJ (2009) Anal Biochem 395:86–90
51. Alonso-Lomillo MA, Gonzalo-Ruiz J, Domínguez-Renedo O, Muñoz FJ, Arcos-Martínez MJ (2008) Biosens Bioelectron 23:1733–1737
52. Petit C, Murakami K, Erdem A, Kilinc E, Ortiz Borondo G, Liégeois JF, Kauffmann JM (1998) Electroanalysis 10:1241–1248
53. Sima V, Cristea C, Bodoki E, Duţu G, Săndulescu R (2010) Cent Eur J Chem 8:1034–1040
54. Butler D, Pravda M, Guilbault GG (2006) Anal Chim Acta 556:333–339
55. Vanícková M, Jakubková E, Bucková M, Čižmárik J, Labuda J (2004) Electroanalysis 16:328–330
56. Vanícková M, Lehotay J, Čižmárik J, Labuda J (2005) Bioelectrochemistry 66:125–127
57. Ferancová A, Korgová E, Buzinkaiová T, Kutner W, Štěpánek I, Labuda J (2001) Anal Chim Acta 447:47–54
58. Liu M, Xiang J, Zhou J, Ding H (2010) J Electroanal Chem 640:1–7
59. Bergamini MF, Santos DP, Zanoni MVB (2010) Bioelectrochemistry 77:133–138
60. Li Y, Umasankar Y, Chen SM (2009) Talanta 79:486–492
61. Su WY, Cheng SH (2010) Electroanalysis 22:707–714
62. Gilmartin MAT, Hart PJ (1994) Analyst 119:2431–2437
63. Saciloto TR, Cervini P, Gomes Cavalheiro ET (2013) Anal Lett 46:312–22
64. Chai C, Liu G, Li F, Liu X, Yao B, Wang L (2010) Anal Chim Acta 675:185–190
65. Zhang H, Liu G, Chai C (2012) Sens Actuators B 168:103–110
66. Sims MJ, Li Q, Kachoosangi RT, Wildgoose GG, Compton RG (2009) Electrochim Acta 54:5030–5034

67. Martinez NA, Messina GA, Bertolino FA, Salinas E, Raba J (2008) Sens Actuators B 133: 256–262
68. Fanjul-Bolado P, Lamas-Ardisana PJ, Hernández-Santos D, Costa-García A (2009) Anal Chim Acta 638:133–138
69. Wu SH, Nie FH, Chen QZ, Sun JJ (2011) Anal Chim Acta 687:43–49
70. Sarakbi A, Aydogmus Z, Sidali T, Gokce G, Kauffmann JM (2011) Electroanalysis 23:29–36
71. Alwarappan S, Erdem A, Liu C, Li CZ (2009) J Phys Chem C 113:8853–8857
72. Karuwan C, Sriprachuabwong C, Wisitsorrat A, Phokharatkul D, Sritongkham P, Tuantranont A (2012) Sens Actuators B 161:549–555
73. Shih Y, Zen JM (2000) Anal Chim Acta 412:63–68
74. Shih Y, Zen JM, Yang HH (2002) J Pharm Biomed Anal 29:827–833
75. Burgoa Calvo ME, Domínguez Renedo O, Arcos Martínez MJ (2007) Talanta 74:59–64
76. de Oliveira PR, Oliveira MM, Zarbin AJG, Marcolino-Junior LH, Bergamini MF (2012) Sens Actuators B 171–172:795–802
77. Caetano FR, Gevaerd A, Castro EG, Bergamini MF, Zarbin AJG, Marcolino-Junior LH (2012) Electrochim Acta 66:265–270
78. O'Connell PJ, Gormally C, Pravda M, Guilbault GG (2001) Anal Chim Acta 43:239–247
79. Florou AB, Prodromidis MI, Tzouwara-Karayanni SM, Karayannis MI (2000) Anal Chim Acta 423:107–114
80. Sha Y, Qian L, Ma Y, Bai H, Yang X (2006) Talanta 70:556–560
81. Wu J, Suls J, Sansen W (2000) Electrochem Commun 2:90–93
82. Ping J, Wu J, Wang Y, Ying Y (2012) Biosens Bioelectron 34:70–76
83. Ke JH, Tseng HJ, Hsu CT, Chen JC, Muthuraman G, Zen JM (2008) Sens Actuators B 130: 614–619

Chapter 6
Electrochemical Biosensors for Drug Analysis

Contents

List of Abbreviations

3-APTS	3-Aminopropyltriethoxysilane
AChE	Acetylcholinesterase
AISM	Ammonium ion selective membranes
ALP	Alkaline phosphatase
AME	Ammonium membrane electrode
AO	Ascorbate oxidase
AP	Acid phosphatase
APCPG	3-Aminopropyl-modified-controlled-pore glass
ASA	Anthracene sulfonic acid
ATC	Acetylthiocholine chloride
ATOE	Antimony-doped-tin oxide Electrode
AuE	Gold electrode
AuNPs	Gold nanoparticles
AuSPE	Screen printed gold-based electrode
BB	A four-channel thick-film transducer
BLM	Planar bilayer lipid membrane
BMIHFP	1-butyl-3-methylimidazolium hexafluorophosphate
BMIHFP	1-butyl-3-methylimidazolium hexafluorophosphate
BMITMS	1-butyl-3-methylimidazoliumbis (trifluoromethylsulfonyl) imide
BSA	Bovine serum albumin

S.A. Ozkan et al., *Electroanalysis in Biomedical and Pharmaceutical Sciences*,
Monographs in Electrochemistry, DOI 10.1007/978-3-662-47138-8_6

BuChE	Butyrylcholinesterase
CA	Cellulose acetate
CBS	Carbon-based strip
CFE	Carbon fiber electrode
CHIT	Chitosan
CL	Cardiolipin
CM	Cellophane membrane
CNT	Carbon nanotube
COD	Choline oxidase
COX	Cyclooxygenase
$\{Co(Sep)_3\}$	Cobalt sepulchrate
CPE	Carbon paste electrode
CPSA	Constant-current chronopotentiometric stripping analysis
CRE	Carbon rod electrode
CSPE	Screen printed carbon-based electrode
CTC	Cytochrome *C*
CTM	Cellulose triacetate membrane
CV	Cyclic voltammetry
CYP450	Cytochrome P450
D-AAOx	D-amino acid oxidase
DDAB	Didodecyldimethylammonium bromide
DL	Detection limit
DM	Dialysis membrane
DNA	Deoxyribonucleic acid
DOE	Dissolved oxygen electrode
DPV	Differential pulse voltammetric
dsDNA	Double strain DNA
EDA	Ethylenediamine
EDAC	Carbodiimide hydrochloride
EIS	Capacitive electrolyte–insulator–semiconductor
EnFET	Enzyme field-effect transistor
Fc	Ferrocene
FI	Flow injection system
FMCA	Ferrocenemonocarboxylic acid
GA	Glutaraldehyde
GCE	Glassy carbon electrode
GDH	Glucose dehydrogenase
GEE	Graphite/epoxy electrode
Gel-k-cM	Gel-like k-carrageenan membrane
GME	Graphite-methacrylate electrode
GOX	Glucose oxidase
GrE	Graphite electrode
GSD	α-glucosidase
HAS	Human serum albumin

HCF	Hexacyanoferrate
HRP	Horseradish peroxidase
ITO	Indium tin oxide
KCG	Kappa-Carrageenan gel
L-AAOx	L-amino acid oxidase
LAC	Laccase
LAPS	Light-addressable potentiometric sensors
LOX	Lipoxygenase
LOX	Lactate oxidase
L-T3	L-triiodothyronine
L-T4	L-thyroxine
LYO	Lysine oxidase
mAb	Monoclonal antibody
MAO	Monoamine oxidase
MB	Methylene blue
MDH	Morphine dehydrogenase
MMPs	Magnetized nanoporous silica-based microparticles
MPA	3-mercaptopropionic acid
MSP	Scaffold proteins
MUT	Mutarotase
MWCNT	Multiwall carbon nanotube
NADH	Nicotinamide adenine dinucleotide reduced form
NADPH	Nicotinamide adenine dinucleotide phosphate
NAu	Nanogold
NHS	*N*-hydroxysuccinimide
NM	Nylon membrane
NMP	Phenazinemethosulfate
NPt	Platinum nanoparticles
NYBM	Nylon Biodyne membrane
Os-polymer	Osmium-complex redox polymer
PAA	Polyacrylamide microparticles
PAAm	Polyacrylamide microgels
PANI	Polyaniline
pAPA	Poly(amphiphilic pyrrole ammonium)
PCS	Poly(carbamoyl) sulfonate (PCS) hydrogel
PDDA	Poly-(dimethyldiallylammonium chloride)
PEI	Polyethylenimine
PEVA	Poly(ethyleneco-vinyl acetate)
PGE	Pencil graphite electrode
PGE	Pyrolytic graphite electrode
pHE	pH electrodes
POX	Peroxidase
PPO	Polyphenol oxidase
PPY	Polypyrrole

PQQ	Pyrroloquinolinequinone
PtE	Pelatinelectrode
pt-PLL	Platinic chloride, poly-l-lysine
PtSPE	Screen printed platinum-based electrode
pTTBA	Poly-TTBA(5, 2′:5′, 2″-terthiophene-3′-carboxylic acid)
PVA-PE	Poly (vinylacetate)-polyethylene-copolymer
PVC	Polyvinyl chloride
PVF	Poly(vinylferrocenium)
Rh-GrE	Rhodium-graphite
RNA	Ribonucleic acid
SAH	Salicylate hydroxylase
SAM	Self-assembled monolayer
sCPE	Solid carbon paste electrode
SDS	Sodium dodecyl sulfate
SiF	Silk filament
SiSG	Silica sol gel
SOX	Sarcosine oxidase
SSAE	All-solid-state ammonium electrode
ssDNA	Single strain DNA
SWV	Square wave voltammetry
TCNQ	Tetracyanoquinodimethane
TCP	Toray carbon paper
TOX	Theophylline oxidase
TYR	Tyrosinase
XOX	Xanthine oxidase
ZrO	Zirconium alkoxide
ZrO–PEI	Zirconium alkoxide-polyethylenimine

6.1 Introduction

A vast number of instrumental techniques are currently available for the selective and sensitive determination of pharmacologically active compounds. Spectroscopic and electrochemical methods and hyphenated techniques such as liquid chromatography and capillary electrophoresis are nowadays routinely used at the academic level and in the pharmaceutical industry.

Biosensors occupy a specific niche in the arsenal of analytical methods for drug compounds analysis. They provide a different conceptual approach with a biocomponent intimately connected to a physical transducer sensitive to the physicochemical interaction between the biocomponent and the analyte of interest [1]. Chemical sensors used, e.g., for monitoring biological processes such as the in vivo pH or oxygen electrode are not considered as biosensors.

Biosensors are well suited for studying the interaction of an immobilized biological element (enzyme, antibody, oligonucleotide, DNA, cell population)

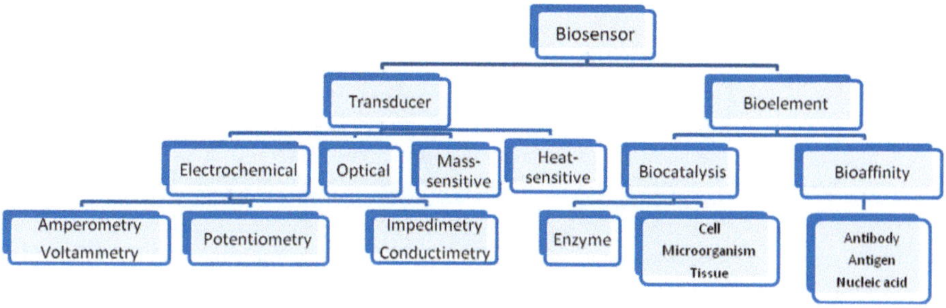

Fig. 6.1 Biosensors classification

with a given analyte (glucose, lactate, glutamate...) or a family of related substrates (phenols, catechols, phenothiazines...). Depending on the biological element and the nature of the interaction with the substrate, bio-affinity and biocatalytic sensors can be considered. With the former, the interaction of an analyte with a bio-receptor such as an antibody (immunosensors) or an oligonucleotide (DNA sensors) can be monitored whereas in the biocatalytic sensors an enzyme, a living cell, or a microorganism react catalytically with the analyte to yield a detectable product. With biosensors based on enzymatic inhibition, the interaction of the analyte with the enzyme can be evaluated as well. Biosensor(s) can also be classified based on the type of their transducer (Fig. 6.1).

It is worth noting that basically all the developed biosensors allow for the quantitative determination of an analyte as well as the determination of the binding affinity and binding kinetics of the biological element in the presence of an analyte. Biosensors can be miniaturized, inserted in a flow through device, need no or limited use of reagents, and may or may not be reused for multiple assays depending on the application. The close proximity of the biocomponent and the transducer allow for a high sensitivity to be achieved. The results can be exploited for the quantification of a given analyte in a complex matrix. Thanks to their unique configuration, they are particularly suitable for investigations devoted to the study of drug–biocomponent interaction.

High throughput screening (HTS) systems are needed especially in drug metabolism and in screening for enzyme inhibitors and for vaccine development (antibody screening). Several relatively recent review papers and book chapters have been described showing the interest in applying biosensors in drug discovery and analysis [2–6].

6.2 Immobilization Techniques

As mentioned before, a biosensor should have the biocomponent located in close contact with the transducer. For this propose, an immobilization method that forms a stable biomolecule layer and which provides a favorable environment for

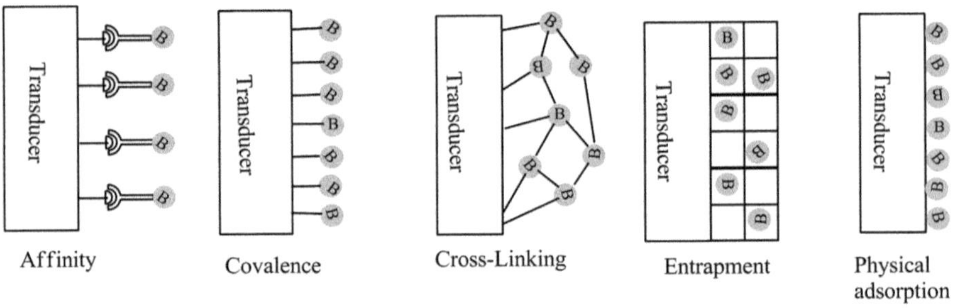

Fig. 6.2 Cartoon of different immobilization modes B = Biomolecule

maintaining the biological activity is required. Optimal immobilization of a biomolecule on the surface of a transducer is a major challenge in biosensor technology. Unfavorable orientation especially in affinity biosensors has negative impact on the recognition properties. This is also valid in enzyme immobilized biosensors since low sensitivity can be obtained due to poor access (1) of the substrate to the enzyme redox center and/or (2) the generated product to the transducer. Numerous techniques have been developed for the immobilization of various kinds of biomolecules. Many comprehensive reviews have been reported on methods of immobilization for biosensor development [7–14]. These methods include generally: physical adsorption or entrapment of the biomolecule onto the transducer surface or within the bulk of a composite matrix, retention behind a dialysis membrane or within a polymeric film or a sol gel network, and insolubilization by a cross-linking agent or covalently attachment onto the transducer surface (Fig. 6.2).

Electrostatic and van der Waals interactions of biomolecules with the surface of a transducer can be exploited for the preparation of a biosensor. Few microliters of a biomolecule solution are readily spiked onto the surface of a transducer and left to react during a defined period of time, and excess biomolecules that are not adsorbed are subsequently removed by gentle rinsing. The adsorbed biomolecules are generally retained by weak forces; thus, a change of pH, ionic strength, temperature, etc. may displace them from the transducer surface. As recent examples, ssDNA or dsDNA were physically adsorbed onto the surface of various carbon-based electrodes such as GCE, CPE, SPCE, and PGE for DNA–drug interaction studies [15–18]. The random immobilization of biomolecules, however, which occurs during physisorption cannot be regarded as optimal for efficient interaction studies.

The incorporation of biomolecules within the bulk material of a transducer such as a carbon composite matrix (graphite–paraffin, graphite–Teflon…) is another mild alternative method. Here, the biomolecule, generally in a dry state, is hand mixed with the graphite particles before blending the particles with the binding agent. The resulting paste can be manually poured into the cavity of an electrode or screen printed onto a miniaturized three-electrode strip [19]. These composites exhibit some advantages such as low background currents and ease of surface regeneration. The entrapped biocomponent is relatively stable in its hydrophobic environment, but it may leak out of the electrode into the bulk of the solution. The

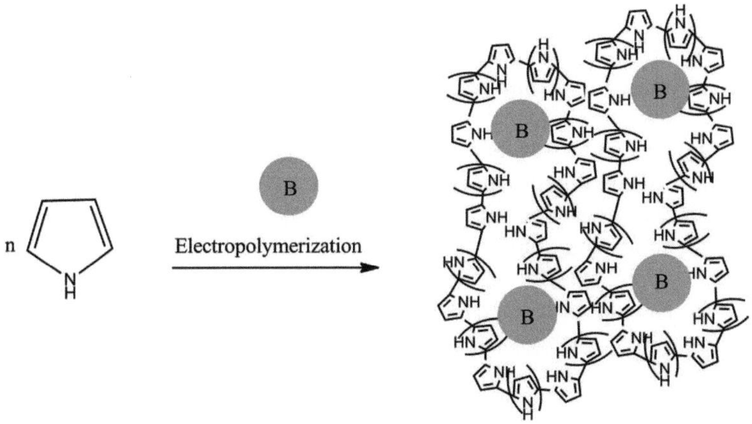

Fig. 6.3 Electropolymerization of pyrrole in the presence of a biomolecule (B)

random orientation of the biocomponent is a limiting factor of such biosensors. The stability of carbon paste-based biosensors was enhanced by replacing the soft binder (liquid paraffin) by solid paraffin. Tyrosinase (TYR) was gently blended, at low temperature, with solid paraffin and graphite particles. A TYR solid carbon paste electrode (SCPE) was used as amperometric sensor for the determination of dopamine. The SCPE proved to be more stable than the CPE under hydrodynamic conditions in flow injection analysis [20].

In order to reduce the leakage of entrapped biomolecules from the bulk matrices, some polymeric gels and matrices were casted onto the electrode surface in order to encapsulate the biomolecules. There are broad ranges of natural and synthetic polymers which have been used, namely chitosan, poly(allylamine), Nafion, polyaniline, polypyrrole, poly(ethylene-co-vinyl acetate) [21–29].

Pyrrole and its derivatives can be electrochemically polymerized in the presence of biomolecules. The polypyrrole layer can also serve for subsequent biomolecules attachment (Fig. 6.3) [21, 22].

By electrochemical polymerization, film thickness, permeation, and charge transport properties can be readily controlled by adjusting the electrochemical or chemical parameters for the polymer growth on the electrode surface. By this method, salicylate hydroxylase was immobilized using glutaraldehyde in a polypyrrole matrix electropolymerized on the surface of a GCE for determining salicylate in serum. The operational lifetime of the biosensor was at least 40 days [26].

Polyvinyl alcohol-bearing styrylpyridinium groups (PVA-SbQ), with soluble prepolymer-bearing photo-crosslinkable groups, have been employed to entrap enzymes [23]. AChE and ChOx were co-immobilized within a poly(carbamoyl) sulfonate (PCS) hydrogel on the surface of a PtE for inhibition studies. The AChE inhibition potency (IC_{50}) of tacrine hydrochloride and two structurally related reversible inhibitors (4-aminoquinaldine, heptylene-linked bis-9-amino-1,2,3,4-

tetrahydroacridine) were comparatively evaluated [24]. A polyacrylamide (PAA) gel was used for HRP encapsulation on the surface of a GCE and covered with a dialysis membrane for acetaminophen determination in three commercial pharmaceutical formulations [25].

Biomolecules can also be entrapped in a biohydrogel gel (alginate, chitosan, or agarose). Chitosan is a natural polyaminosaccharide obtained by N-deacetylation of chitin. An amperometric biosensor was developed for the highly sensitive determination of phenolic compounds such as dopamine and catechols by immobilizing tyrosinase in a chitosan film cross-linked with a dimethoxymethylsilane derivative onto a GCE. The chitosan matrix provided a large microscopic surface area and a porous morphology for high enzyme loading without affecting the activity of the enzyme. This biosensor, stored in PBS buffer pH 7.4 at 4 °C, retained 75 % of its activity for at least 70 days [27]. The enzyme CYP2B6 was entrapped in a chitosan matrix containing gold nanoparticle retained onto the surface of a GCE. This architecture provided a favorable environment for enzyme immobilization and allowed the direct electron transfer between the immobilized CYP2B6 enzyme and the electrode. The biosensor was used for studying bupropion and cyclophosphamide biotransformation by CYP2B6 [28]. Sol–gel technology can provide unique means to prepare three-dimensional networks suited for the encapsulation of biomolecules. Silica gels are highly porous and show excellent physical rigidity. AChE was encapsulated in a sol–gel chitosan composite-modified GCE. The hydroxyl groups in the sol–gel matrix formed strong hydrogen bonds with the immobilized enzyme and prevented it from leaking out of the film. This biosensor was used for the determination of galantamine and neostigmine, i.e., two inhibitors of AChE [29]. Collagen and gelatin matrices possess also an interesting three-dimensional network which can be beneficial to maintain the activity of entrapped biomolecules. Catalase and horseradish peroxidase were co-entrapped in a gelatin matrix to prepare a biosensor on the surface of a Teflon thin membrane covering a gold electrode for the determination of 5-aminosalicylic acid. The biosensor retained 91 % of its activity after 21 days of storage at 4 °C in a wet atmosphere [30].

Biomolecules bear several exposed functional groups (amine, thiol, carboxyl, and hydroxyl), and various methods can be selected for covalent binding onto the transducer surface. In most cases the electrode surface requires, however, activation in order to generate a substantial amount of reactive groups for subsequent attachment of the biomolecule. Chemical binding can occur by using multifunctional reagents such as glutaraldehyde (Glu) or carbodiimide (EDC) (Fig. 6.4) or by grafting thiol functionalities onto the biomolecule (e.g., in the case of oligonucleotides). Activation of the electrode surface can be performed electrochemically at high positive potentials, by chemical treatment, by radioactivation, or by self-assembling a linking arm onto the electrode surface (e.g., cysteamine onto gold). The GCE surface can be functionalized by electrochemical reduction of an aromatic diazonium salts, and further reactivation can be realized by a variety of chemical reactions with the amine functional group [31].

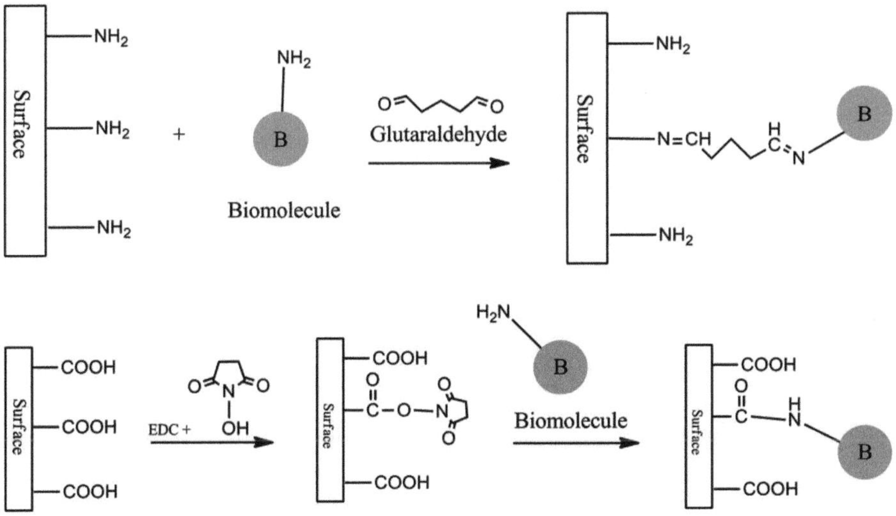

Fig. 6.4 Covalent coupling via two binding reagents

A covalent bond can be realized between an amine-terminated probe DNA and a carboxylic acid group from a functionalized conducting polymer (pTTBA) layer on a screen printed electrode. The latter was previously modified with electrodeposited gold nanoparticles and cardiolipin. This DNA–lipid-based biosensor was used for the detection of anticancer drugs in spiked urine [32].

The thiol mediated biomolecule immobilization onto a solid support, in particular on a gold surface, has been widely used in the construction of biosensors. This method allows a relatively well-controlled immobilization of biomolecules as a monolayer or multilayer configuration onto a gold electrode. The self-assembled monolayer can introduce reactive groups of various chemical diversities: amine, carboxyl, hydroxyl (Fig. 6.5). Glutaraldehyde or carbodiimide can be employed for further activation of these functional groups. Carbodiimide allows the binding between the carboxyl and amino functional groups. *N*-hydroxysuccinimide (NHS) is generally used in conjunction with carbodiimide in order to improve the immobilization efficiency.

Self-assembled monolayers with (−OH)- and ($-NH_2$)-terminated alkanethiols of different chain lengths were used for gold electrode modification and subsequent theophylline oxidase (TOX) immobilization. The direct electrochemistry of TOX at the electrode–solution interface and the bioelectrocatalytic oxidation of theophylline were investigated. Direct electron transfer of TOX or bioelectrocatalytic oxidation of theophylline was not observed at unmodified gold electrodes. However, direct electron transfer of TOX was strongly dependent on the nature of the alkanethiol head groups. SAMs of cysteamine or other amine-terminated alkanethiols provided orientation of TOX through the heme-domain which favored direct ET between TOX and the modified gold electrodes [33].

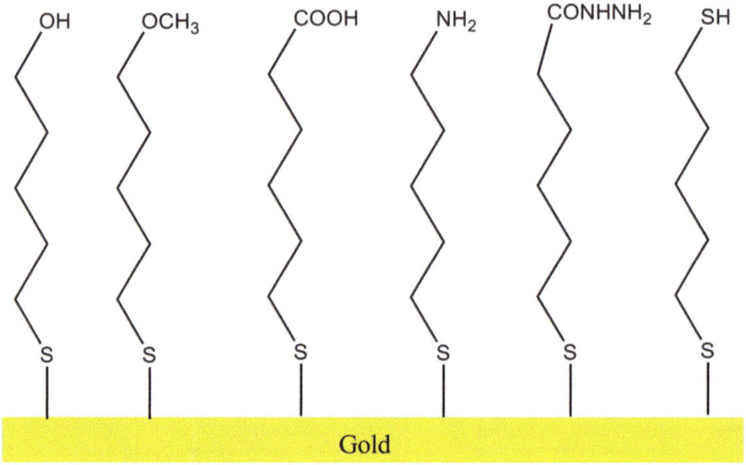

Fig. 6.5 Cartoon of several functional groups which can be self-assembled onto gold

Biomolecules can be linked to small particles (e.g., carbon, gold nanoparticles, silica, or clay material), natural polymers (e.g., cellulose or chitosan bead), or a synthetic polymer (e.g., Nylon and Nafion). Functionalization of carbon nanomaterials (CNT) and graphene sheets is usually preceded by an oxidation of the graphitic substrate with strong acids such as sulfuric acid [34]. The chemically created functional groups are eventually reduced and serve subsequently for the immobilization of desired biomolecules. The modification of microparticles allows for a high amount of biomolecules immobilization. When the particles are electroactive, improved electron transfer can be observed between the biomolecule (enzyme) and the electrode [35, 36].

The immobilization of a biocomponent onto nanosized superparamagnetic particles (MNP) has also been achieved for subsequent retention onto a magnetized transducer. High enzyme loading can be achieved with the resulting biomaterial readily trapped in any magnetized microenvironment. Glutaraldehyde activated streptavidin magnetic particles were used for tyrosinase immobilization, subsequently holding onto a magnetized carbon paste electrode for the amperometric study of tyrosinase inhibitors [37].

Affinity immobilization strategies can be advantageously exploited for biosensor development. Some methods use sugar–lectin, protein G-antibody at the electrode interface, or exploit the biotin–avidin (streptavidin) high binding strength. The aim of these methods is to create bonds between a modified transducer (e.g., by lectin or avidin) and a specific functional group on the surface of the biomolecule (e.g., carbohydrate residue, biotin, histidine).

6.3 Electrochemical-Based Biosensor

Among the different types of biosensors (Fig. 6.1), the electrochemical ones have been the most reported [38–42]. The electrode can convert the biological recognition event into a useful electrical signal: electron transfer, potential or impedance change at the electrode–solution interface. Electrodes are relatively readily modified, can be miniaturized for mass production, and arranged in different shapes. Their response is not affected by turbidity and light. High sensitivity can be achieved especially when electrochemical amplification is observed at the electrode–solution interface.

Different electrochemical modes can be employed in the use of biosensors for drug analysis: amperometry, voltammetry, potentiometry, and impedimetry are most commonly applied.

6.3.1 Amperometry and Voltammetry

Amperometric biosensors are based on the measurement of an oxidation or a reduction current of an electroactive species that is consumed or produced by a biological element or a redox tracer at the electrode surface at a constant applied potential. The steady state current (I) that flows is proportional to the analyte concentration following the reaction:

$$I = nFADC/d$$

where n is the number of electrons exchanged per molecule, F is Faraday's constant, A is the electrode area, D is the diffusion coefficient, C is the concentration of electroactive molecules, and d is the diffusional layer thickness.

The sensing design is relatively simple and consists of a three-electrode configuration comprising working (GCE, PtE, AuE, CPE...), reference, and auxiliary electrodes. Amperometric biosensors of different shapes and sizes have been developed since the development of the first amperometric biosensor for glucose in 1962 [43]. The commercial launch of a miniaturized three-electrode planar configuration with immobilized glucose oxidase can be considered as the starting point of a renewed interest in biosensor development [7].

6.3.1.1 Enzyme-Based Amperometric Biosensors

Among all the available biosensors, amperometric enzyme electrodes are the most described. In the presence of an analyte, three distinct operation modes can be considered (Fig. 6.6):

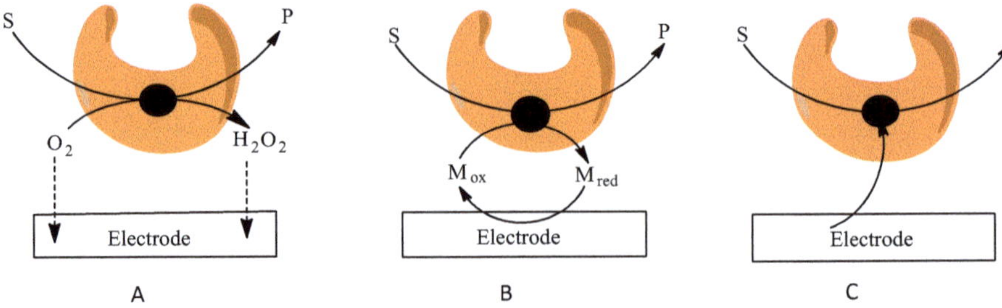

Fig. 6.6 Three generations of amperometric biosensors

1. Measuring the product of the enzymatic reaction or the coproduct (e.g., hydrogen peroxide, NADH) or a consumed co-substrate (oxygen). The signal, however, can be affected by fluctuations in the oxygen level or by interfering electroactive species such as ascorbic acid and uric acid (Fig. 6.6a).
2. Measuring the change in the oxidation–reduction pattern of a redox mediator coupled with the enzyme for shuttling electrons from its prosthetic group to the electrode (Fig. 6.6b). This alternative minimizes risks of co-substrate fluctuation; signals are amplified yet interferences can still occur due to reaction with the mediator.
3. Directly monitoring the electroactivity of the enzyme active center (Fig. 6.6c). This strategy attempts to eliminate the biosensor response dependency on the co-substrate concentration and to decrease the influence of interfering species. In practice, however, deep immersion of the prosthetic group within the protein structure of an enzyme introduces a kinetic barrier for electron transfer. The incorporation of nanoparticles (gold, carbon) or conducting organic salts, within the immobilized biocomponent layer, is an attractive strategy for solving the problems of poor electrically contact between the enzyme active center and the electrode material.

Enzyme activity is expressed as the amount of substrate converted to product per unit time under specific reaction conditions for pH and temperature. The latter parameters can affect the biosensor response but generally to a lesser extent as for the soluble enzyme. Enzyme activity can be affected by molecules such as inhibitors and activators. Indirect monitoring of such analytes can be achieved by measuring the change in the enzymatic product and/or electro-activity of the enzyme before and after interaction with the analyte.

6.3.1.1.1 Enzyme Biosensors for Metabolization Studies

Cytochrome P450 Family

CYP450 are a group of heme proteins characterized by an absorbance band at 450 nm and are important enzymes in drug discovery. They are versatile redox systems and known to be involved in most in vivo metabolism of fatty acids, steroid hormones, prostaglandins xenobiotics, and xenobiotics such as environmental toxins and drug compounds by catalyzing a variety of reactions such as oxidation and epoxidation group migration [44–46].

In the catalytical NADPH-dependent monooxygenation of a drug substrate (RH), the catalytic cycle entails electron supply to the heme iron in the presence of oxygen. The global reaction can be schematized with the following equation:

$$RH + O_2 + 2e^- + 2H^+ \overset{CYP450}{\rightarrow} ROH + H_2O$$

The electrons can be delivered by NAD(P)H, flavoproteins, a mediator, or an electrode. In the presence of O_2, an iron-hydroperoxo intermediate is produced. Then water is released from this peroxo structure to yield an active perferrylFeO species (such as FeO^{3+}), which leads to a single oxygen atom insertion into the bound substrate (RH), release of product (ROH), and regeneration of the CYP ferric form.

As shown in Fig. 6.7, the CYP450 catalytic cycle needs two reducing equivalents which are naturally supplied from NAD(P)H or a flavoprotein in order to generate the active form (FeO^{3+}). Thus, the determination of the consumption of NADPH or O_2 can be applicable approaches to determine CYP450 activity and for drug screening. These co-substrates, however, need to be present in excess in order to avoid any rate limitation, and it is not always possible to reliably determine small fluctuations in their concentration [47, 48]. Flavodoxin or cytochrome NADPH-reductase (CPR) linked to the CYP450 can shuttle electrons from the electrode to

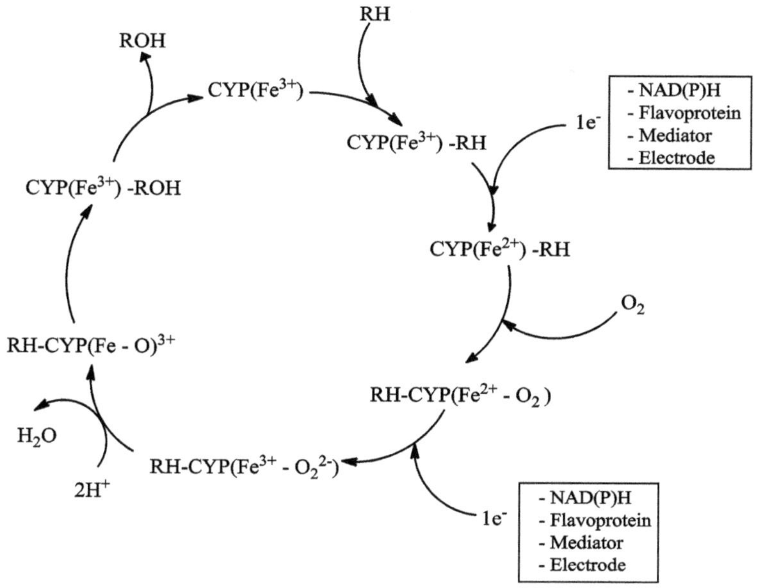

Fig. 6.7 CYP450 catalytic cycle

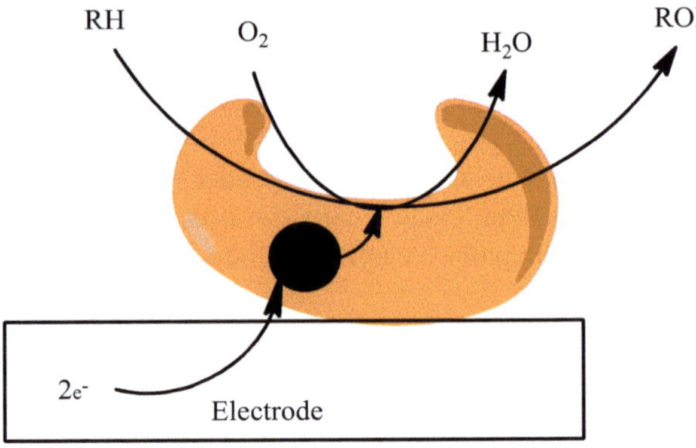

Fig. 6.8 Direct electron transfer of CYP450

the prosthetic group of the enzyme (Fig. 6.8). The latter approach permitted a closer mimic to the natural catalytic pathway and avoids the use of NADPH [49].

Electrochemical assay of CYP450 for drug metabolism studies was applied to testosterone and ketoconazole using the isoform CYP3A4 for substrate and inhibitor assays [49].

The requirement of NAD(P)H can also be avoided by conjugation of CYP450 with some mediators such as FMN, FAD, or riboflavins [50, 51].

Demethylation, p-hydroxylation, and o-dealkylation of typical substrates such as aminopyrine, aniline, 7-ethoxyresorufin, and 7-pentoxyresorufin were catalyzed by semi-synthetic flavocytochrome electrochemically reduced at a rhodium–graphite electrode polarized at −500 mV (vs. Ag/AgCl) [49].

Enhanced selectivity was achieved by direct immobilization of CYP450 on the electrode surface (third-generation biosensors). The electrode supplied electrons for the CYP cathodic reduction which is coupled with the substrate transformation in the presence of O_2. Figure 6.8 illustrates the active role of the electrode as electron source in the P450 catalytic cycle.

Since the electroactive center of CYP450 is deeply immersed in the protein structure, several original immobilization strategies of CYP450 have been implemented onto the electrode surface. A variety of immobilization methods were employed for direct electron transfer of CYP450 at the electrode surface [52]. Direct bioelectrocatalytic reaction of CYP450 was followed using colloids of clay or gold nanoparticles dispersed in a surfactant or chitosan onto the electrode surface [28, 53]. It was reported that in anaerobic solution, reversible oxidation and reduction currents were obtained at −0.305 and −0.450 V (vs. Ag/AgCl) for CYP2B4 and CYP2B6, respectively. In the presence of oxygen, an electrocatalytic reduction current was obtained with these biosensors by addition of substrates such as aminopyrine, benzphetamine, bupropion, lidocaine, and cyclophosphamide or an inhibitor such as methyrapone. C-hydroxylation and heteroatom release were identified as the main pathways for CYP2B6 mediated drug oxidation [28]. Direct electrochemistry of CYP450 was also achieved on a gold electrode negatively charged by a monolayer of 3-mercapto-1-propenesulfonic acid. The negatively charged CYP3A4 was alternated layer-by-layer in the presence of a polycation layer. Drug compounds such as midazolam, verapamil, quinidine, and progesterone showed catalytic activity at the biosensor depending on substrate concentration. This activity was inhibited by ketoconazole. A minor contribution of hydrogen peroxide to the catalytic cycle was observed [47].

Among the different available electrochemical techniques, cyclic voltammetry is well suited for studying the direct electron transfer of CYP450-based biosensor. The redox potential is affected by several factors such as the pH and the ionic strength of the buffer solution, the temperature, the immobilization technique, and the scan rate (Table 6.1) [54].

Table 6.1 CYP450 immobilized-based amperometric biosensors for drug analysis

Electrode	Analyte(s)	Amperometry E: V vs. Ag/AgCl	Comment	References
GCE/clay/Tween 80/CYP450 2B4	Aminopyrine Benzphetamine Metyrapon	−0.5	DL:0.4 mM amino-pyrine, 0.2 mM benzphetamine	[53]
GrE/ flavocytochrome RfP450 2B4	Aminopyrine	−0.5	–	[55]
Rh-GrE/ flavocytochrome RfP450 2B4	Aminopyrine	−0.5	–	[50, 51]
GCE/NAu–CHIT/ CYP450 2B6	Bupropion Cyclophosphamide	−0.45	–	[56]
GCE/DDAB/SG/ P450 cam	Camphor	–	–	[57]
PtE/EDA/Glu ({Co(Sep)$_3$ }/P450 cam)	Camphor Adamantanone Fenchone	−0.8	–	[58]
SPCE/CYP450 2B4	Cocaine	−0.25	CP: 23.05 ± 3.53 nM	[59]
PtE/PANI/ASA/ CYP450 3A4	Erythromycin	−0.538	DL: 0.076 μM	[60]
GCE/PANI/ CYP450 2D6	Fluoxetine	–	LR: 2–7 μM	[61]
PtE/DDAB–BSA/ GA/CYP450 3A4	Indinavir	−0.68	DL: 0.06 mg/L LR: up to 20 mM	[62]
GCE/ZrO$_2$/Pt-PLL/CYP450 2B6	Lidocaine	–	–	[56]
AuE/MSP/PDDA/ CYP4503A4	Midazolam Verapamil Progesterone Quinidine	–	–	[47]
SPAuE/Ppy/ CYP450 2B4	Phenobarbital	0.450	DL: 0.289 μM LR: 0.32–2.50 μM	[63]

(continued)

Table 6.1 (continued)

Electrode	Analyte(s)	Amperometry E: V vs. Ag/AgCl	Comment	References
SPCE/CYP450 2B4	Phenobarbital	+0.45	LR: 0.20–1.67 mM	[64]
SPAuE/MPA/AuNPs/CYP450 2B4	Phenobarbital	–	LR: 0.10–0.83 mM	[64]
AuE/thiolate/CPR-microsomes/CYP450 3A4	Testosterone (substrate) Ketoconazol (inhibitor)	CV	–	[49]
SPCE/DDAB/AuNPs/CYP450 2B4	Benzphetamine	CV and E_{app} :−0.25 V	–	[65]
SPE/CNT/CYP450 2B4	Benzphetamine	CV	–	[66]
SPE/CNT/CYP450 3A4	Cyclophosphamide	CV	DL: 12 μM	[66]
SPE/CNT/CYP450 2C9	Naproxen Flurbiprofen	CV	DL:82 μM (naproxen)	[66]

Peroxidase Enzymes

Horseradish peroxidase belongs to the heme-containing enzyme family which utilizes hydrogen peroxide to oxidize a wide variety of organic and inorganic compounds. The enzyme is often used as biocatalytic label in numerous immunoassays, yet it can also been used as a biocatalytic enzyme "mimicking" biotransformation processes. The most common HRP substrates are aromatic amines, indoles, and phenols. The primary products of oxidation by HRP are radicals leading subsequently to radical coupling and ultimately polymers [67]. HRP has been immobilized onto electrodes for drug peroxidation study in the presence of hydrogen peroxide. Drug compounds such as acetaminophen [25, 68], clozapine [69], levetiracetam [70], phenothiazine [71], promazine, promethazine [72], and rifampicine [22] were determined by amperometry at a HRP-modified electrode as shown in Fig. 6.9.

The simplified mechanism of such detection is as follows:

$$\text{Native HRP} + H_2O_2 + 2H^+ \rightarrow \text{Compound I} + 2H_2O$$
$$\text{Compound I} + AH \rightarrow \text{Compound II} + \dot{A}$$
$$\text{Compound II} + AH \rightarrow \text{Native HRP} + \dot{A}$$

The generated radical ($A^{\cdot}$) can dimerize or it can be further oxidized to other reaction products such as quinones. The biosensor signal corresponds to the reduction of the radical or of a subsequent reaction product. The intermediate radical or the reaction product can be scavenged in the presence of a substrate such as

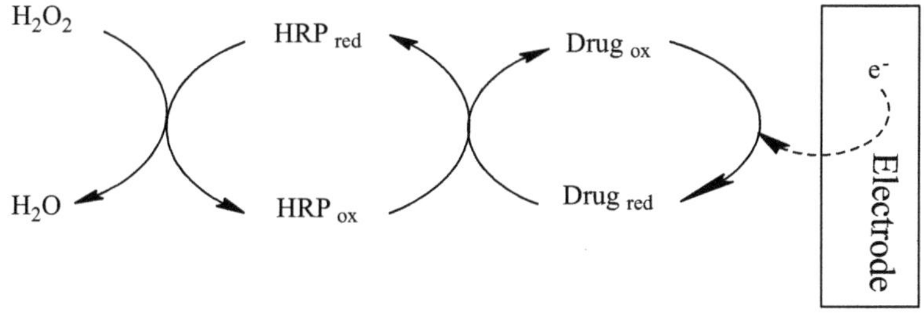

Fig. 6.9 Schematic drawing of a drug peroxidation at a horseradish peroxidase (HRP) immobilized electrode

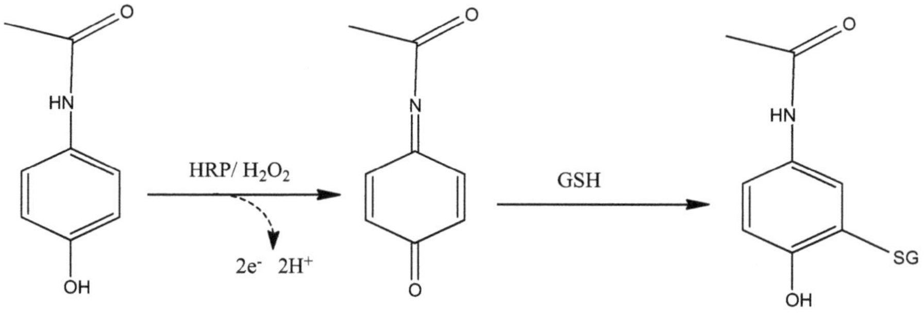

Fig. 6.10 Oxidation of APAP in the presence of H_2O_2 at the HRP-MMPs with formation of *N*-acetylbenzoquinoneimine (NAPQI) and subsequent glutathione conjugation [68]

ciprofloxacin or norfloxacin [73], or thiol derivatives such as glutathione. The latter was monitored as an inhibitor of the HRP biosensor response for acetaminophen (APAP) (Fig. 6.10). Thiols exhibit a fast Michael type reaction with the HRP oxidation product, *N*-acetyl-*p*-benzoquinoneimine (NAPQI), leading to a concentration-dependent decrease of the biosensor signal (Fig. 6.10). Inhibition parameters such as IC_{50} have been reported for different thiol compounds [69]. Such a biosensor configuration was found useful in screening for NAPQI scavengers since this metabolite exhibits substantial in vivo toxicity (Table 6.2).

Stereoselective Biosensors

Amino acid oxidase (AAOx) enzymes have been used for designing enantioselective biosensors for the assay of chiral drugs possessing an amino acid related fragment in their structure. The two enzymes L-amino acid (L-AAOx) and D-amino acid (D-AAOx) oxidases are FAD-dependent oxidoreductase enzymes, which use oxygen as co-substrate and convert it to hydrogen peroxide. Generally, the transduction system is based on monitoring hydrogen peroxide electrooxidation. Taking advantage of the high selectivity of these enzymes, simultaneous detection of *S*-and *R*-captopril [77] or *S*-and *R*-perindopril [78] has been possible using flow injection

Table 6.2 Examples of biosensors based on HRP for drug analysis

Electrode	Analyte(s)	Amperometry V vs. Ag/AgCl	Comment	References
GCE/PEI/ZrO/ HRP	Acetaminophen	−0.2	LR: 20 μM–0.26 mM DL: 0.1 μM	[74]
SCPE/MMP/HRP	Acetaminophen	−0.1	LR: 2–57 μM	[68]
SPE/ZrOPEI/HRP	Acetaminophen	−0.2	DL: 0.06 μM LR: 0.4 to 5 μM	[75]
GCE/HRP-PAA	Acetaminophen	−0.1	DL: 3 μM LR: 10–500 μM	[25]
SCPE/HRP/GA/BSA/ DM	Clozapine	0.0	LR: 1–10 μM DL: 0.2 μM	[76]
CPE/MMPs/HRP	Clozapine	0.0	–	[69]
SPE/HRP/PPY	Levetiracetam	−0.3	DL: 9.8 μM LR: 9.8–83 μM	[70]
SPCE/HPR	Promazine Promethazine	–	–	[72]
GCE/HRP/PPY	Rifampicin	+0.35	DL: 5.1 μM	[22]

analysis with sequential injection at a rate of more than 30 samples per hour. Application to pharmaceuticals has been realized.

Enzyme-Based Biosensors for Inhibition Studies

As stated above, various enzyme immobilized biosensors have been developed for the screening and determination of inhibitors such as heavy metals, toxins, pesticides, and drugs, which cause a reduction in the rate of an enzyme catalyzed reaction. Enzyme inhibitors can be classified in two groups: (1) reversible inhibitors which cause reversible inactivation of an enzyme thanks to a non-covalent binding with the inhibitor and (2) irreversible inhibitors which usually form a covalent bond at or near the enzyme active site. The determination of the $K_{m\ app}$ and V_{max} (I_{max}) parameters permit to specify reversible inhibitors criteria namely competitive, noncompetitive, uncompetitive, and mixed mode inhibitors. The strength and mode of binding of an inhibitor can be evaluated by studying enzyme inhibition kinetic parameters, such as inhibition constant (K_i) and half-maximal inhibition constant (IC_{50}) as well. True inhibitors act directly at the enzymatic site while other inhibitors can interfere with the reaction product formation. For these kinds of inhibitions, IC_{50} and K_i can be evaluated as well.

Since the depletion of the neurotransmitter acetylcholine by AChE has been demonstrated to be involved in the development of Alzheimer and other neurodegenerative diseases, reversible acetylcholinesterase inhibitors such as donepezil, neostigmine, galantamine, and eserine have been developed.

Acetylcholinesterase (AChE) and butyrylcholinesterase (BChE) are two types of relatively stable natural ChE enzymes which can be advantageously immobilized onto an electrode for inhibition studies. Since choline is not electrochemically active, the enzyme is co-immobilized with choline oxidase (ChOX). The second enzyme oxidizes the choline generated by the first enzyme to produce hydrogen peroxide which is detected as follows:

$$\text{Acetycholine} + H_2O \overset{\text{AChE}}{\rightarrow} \text{Choline} + \text{Acetate}$$
$$\text{Choline} + O_2 \overset{\text{ChOx}}{\rightarrow} \text{Betaine aldehyde} + H_2O_2$$

Reversible inhibition of the biosensor response by nicotine which acted on the catalytic activity of ChOX [79, 80] and by anti-Alzheimer drug candidates has been described [24, 29] In order to avoid the use of ChOX in a two enzyme configuration, the electrochemical monitoring of the catalytic cycle of AChE can be achieved thanks to the use of acetylthiocholine as substrate [29, 81]. The biosensor signal corresponds to the electrochemical oxidation of thiocholine on the surface of glassy carbon, carbon paste, or platinum electrode as follows:

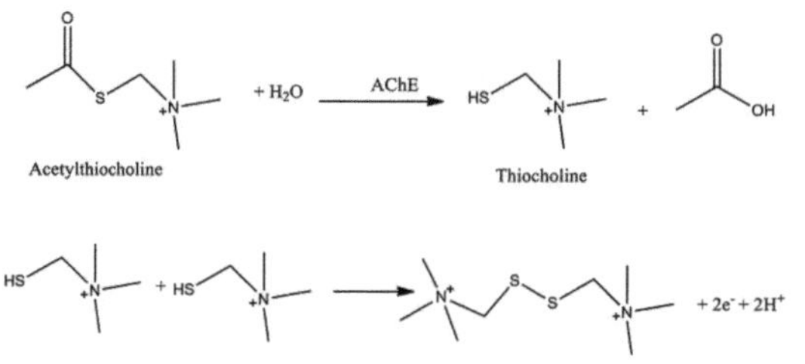

Oxidation reaction on the surface of these electrodes, however, is slow and requires high applied potentials with the added problem of electrode surface fouling. Different mediators such as metallic phthalocyanines (ferrophthalocyanine and cobalt phthalocyanine) [82], TCNQ (7,7,8,8-tetracyanoquinodimethane) [83], and hexacyanoferrate (Prussian Blue) [84] enabled the amperometric detection of thiocholine at low applied potentials without electrode passivation. A silver electrode poised at a potential close to 0.0 V vs. Ag/AgCl can also be implemented in a flow injection setup for amperometric detection of thiocholine based on the reaction of thiocholine with silver ions and/or silver chloride at the electrode–solution interface [85] (Fig. 6.11).

The inhibition of the AchE enzyme was studied in the presence of galantamine and neostigmine [29, 85].

Monoamine oxidase (MAO) is a copper-containing flavinoxidase enzyme and acting on the amine groups of each drug and producing hydrogen peroxide following:

$$RCH_2NH_2 + O_2 + H_2O \overset{\text{MAO}}{\rightarrow} RCHO + H_2O_2 + NH_3$$

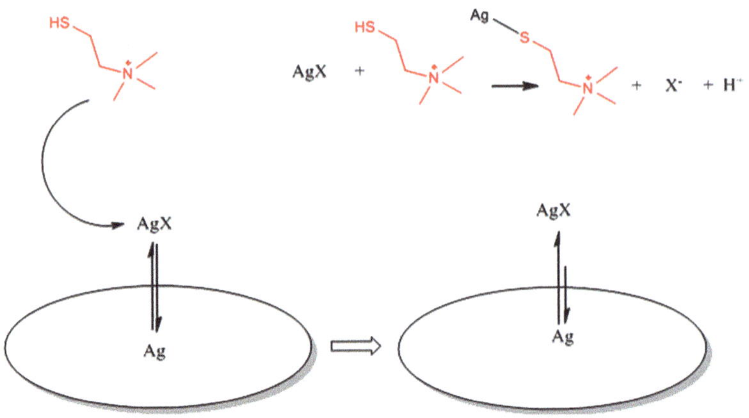

Fig. 6.11 Thiocholine amperometric detection on the surface of a silver electrode [85]

Dopamine as MAO substrate was determined using an enzyme monoamine oxidase immobilized biosensor. The enzyme was cross-linked on a glutaraldehyde-activated eggshell membrane that was deposited on a glassy carbon electrode [86].

Some antidepressants comprising amine groups in their structure can act as competitive inhibitors [86]. Enzyme inhibition was studied using amperometric biosensors for antidepressants such as desipramine and pyrazidol [88, 89] and nitrofuran drugs [89]. Detection of liberated hydrogen peroxide during enzymatic activity of MAOX permitted the determination of anti-inflammatory, antidepressant, and neurotransmitter drugs as well [90].

Lipoxygenases are a family of nonheme -containing enzymes that selectively catalyze the dioxygenation of polyunsaturated fatty acids like linoleic acid and arachidonic acid with the following reaction:

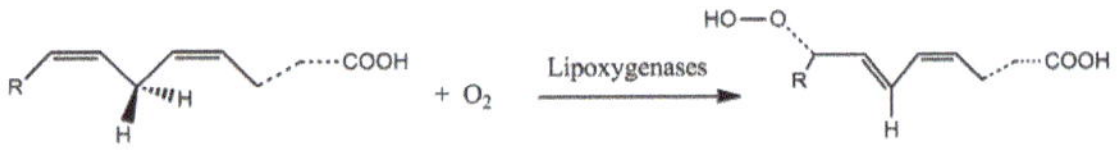

In mammals, lipoxygenases are involved in the synthesis of species which are in the development of inflammatory reactions and diseases, such as rheumatoid arthritis or psoriasis [91]. Therefore, drugs that inhibit the biosynthesis of these species have been considered as novel medication. Lipoxygenase (EC 1.13.11.12; 11.2 U/mg) was immobilized in a carbon paste electrode for the determination of inhibitors such as nordihydroguaiaretic acid, zileuton®, fenleuton®, and caffeic acid. The decrease in biosensor response for hydroperoxy linoleic acid can be correlated with the inhibitor concentration [92, 93].

Noninsulin-dependent diabetes mellitus (NIDDM) is a metabolic disorder that is characterized by high blood glucose level. One of the most beneficial therapy method for NIDDM is to control the blood glucose level by inhibition of an α-glucosidase enzyme. The enzyme α-glucosidase catalyzes the breakdown of starch and disaccharides to glucose as follows:

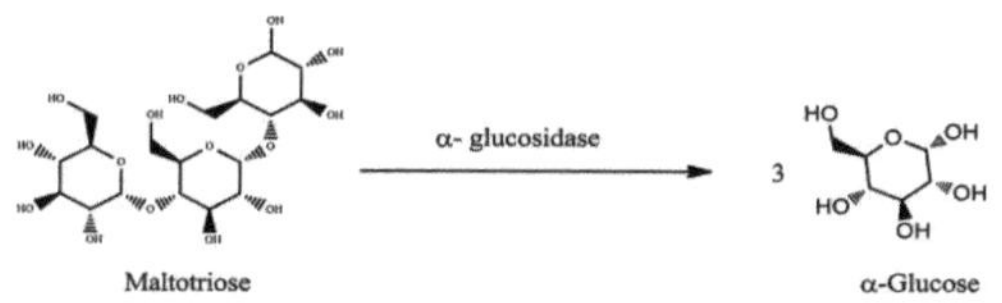

Inhibition of α-glucosidase by some drug compounds such as acarbose, miglitol, and voglibose permitted to control the blood glucose level in the NIDDM patients [94].A Clark oxygen sensor with co-immobilized glucose oxidase and mutarotase (MUT) was used in the downstream of a continuous-flow/stopped-flow system for monitoring the glucose liberated from maltose by the action of an immobilized α-glucosidase reactor which was also inserted in the flow line as follows:

- At the immobilized α-glucosidase reactor:

$$\text{Maltose} \overset{\alpha\text{-glucosidase}}{\rightarrow} 2\alpha - \text{D-Glucose}$$

- At the Clark oxygen sensor:

$$\alpha - \text{D-Glucose} \overset{\text{Mutarotase}}{\rightarrow} \beta - \text{D-Glucose}$$
$$\beta - \text{D-Glucose} + O_2 + H_2O \overset{\text{GOx}}{\rightarrow} \text{Gluconic acid} + H_2O_2$$

The inhibitory activity of acarbose and 1-deoxynojirimycin on α-glucosidase was determined by monitoring the decrease in the response of the Clark oxygen sensor for liberated glucose [95].

Xanthine oxidase (XOX) is a metalloprotein that catalyzes the purine catabolism, such as oxidation of hypoxanthine to xanthine and can further catalyze the oxidation of xanthine to uric acid as follows:

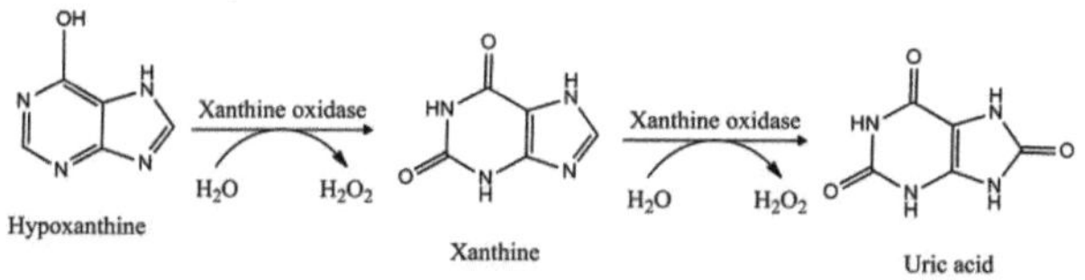

Default in purine metabolism results in an increase in the level of uric acid consequently leading to diseases such as gout, hyperuricaemia, and chronic heart failure [95]. An increase in the level of XOX also causes renal stone formation and reactive oxygen species (ROS) induced diseases [96]. Allopurinol is a potent inhibitor of xanthine oxidase (XOX) which, upon oxidation, yields the product oxypurinol. The latter compound inactivates the reduced form of XO, thus inhibiting the formation of uric acid [97].

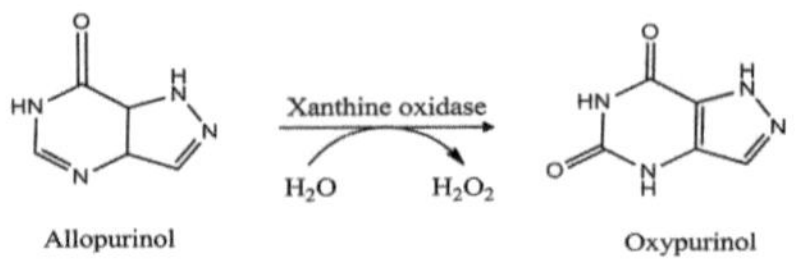

Xanthine oxidase was retained and cross-linked by GA into an anionic clay thin-layer deposited onto a Pt electrode for the determination of allopurinol by its inhibitory action of XOX product formation [98]. A non-exhaustive survey of different enzyme-based biosensors for inhibition studies is reported in Table 6.3 [95–108]. It is worth mentioning that already in 1990, a SPE was used for enzyme immobilization and inhibition studies namely for the assay of theophylline. Theophylline is an uncompetitive inhibitor of alkaline phosphatase. The enzyme was solubilized in a carbon suspension and adsorbed on the surface of a disposable screen printed test strip for determining theophylline concentration in whole blood based on the generation of p-aminophenol from *p*-aminophenyl phosphate by the action of bovine liver alkaline phosphatase [103].

Table 6.3 Enzyme biosensors based on amperometric determination of inhibitors

Electrode	Drug	Amperometry V vs. Ag/AgCl	Comment	References
GSDreactor + GOX/MUT reactor in FI system	Acarbose	−0.7	DL: 0.46 μM	[95]
GSDreactor + GOX/MUT reactor in FI system	1-Deoxynojirimycin	−0.7	DL: 0.23 μM	[95]
GCE/bismuth film/gelatin/GSD	Amaryl	−0.95	DL: 2 ppb	[99]
PtE/CHIT membranes/COD	Neostigmine (in the presence of AChE)	0.6	DL: 0.5 μM	[100]
PtE/PCS gel/AChE + COD	4-Aminoquinaldine	0.6	IC50: 29.6 nM	[24]
PtE/PCS gel/AChE + COD	Tacrine	0.6 V	IC50: 6.7 μM	[24]
PtE/PCS gel/AChE + COD	Heptylene-linked bis-9-amino-1,2,3,4-tetrahydroacridine	0.6 V	IC50: 280 nM	[24]
GCE/SiSG/AuNPs/CHIT/AChE	Galantamine	CV	–	[29]

(continued)

Table 6.3 (continued)

Electrode	Drug	Amperometry V vs. Ag/AgCl	Comment	References
GCE/SiSG/AuNPs/chit/AChE	Neostigmine	CV	–	[29]
PtE/PAAm/COD + AChE/DM	Choline, Nicotine	0.6 vs SCE	–	[80]
PtE/NYBM/ChOX + BChE	Physostigmine	0.6 V vs. SCE	LR:0.05-1.5 μM	[101]
PtE/NYBM/ChOX + BChE	Neostigmine	0.6 V vs. SCE	LR: 0.05–1.5 μM	[101]
PtE/NYBM/ChOX + BChE	Pyridostigmine	0.6 V vs. SCE	LR:0.05–0.8 μM	[101]
PtE/NYBM/ChOX + BChE	Edrophonium	0.6 V vs. SCE	LR:0.05–1.5 μM	[101]
PtE/NYBM/ChOX + BChE	Nicotine	0.6 V vs. SCE	LR: 3–80 μM	[101]
CPE/LOX	Fenleuton	DPV	20 μM	[92]
CPE/LOX	Nordihydroguaiaretic acid	DPV	20 μM	[92]
CPE/LOX	Zileuton	DPV	5 μM	[92]
CPE/LOX	Caffeic acid	0.9	30 % inhibition for 650 μM linoleic acid as substrate	[93]
Clark Electrode/Gel-k-cM/CYTOX	Naproxen	−0.65	DL: 0.05 μM	[102]
Clark Electrode/Gel-k-cM/CYTOX	Diclofenac	−0.65	DL: 0.05 μM	[102]
Clark Electrode/Gel-k-cM/CYTOX	Ibuprofen	−0.65	DL: 5 nM	[102]
Clark Electrode/Gel-k-cM/CYTOX	Tolmetin	−0.65	DL: 5 nM	[102]
CPE/MMP/TYR	Kojic acid	−0.1	IC50: 3.66 μM	[37]
CPE/MMP/TYR	Ascorbic acid	−0.1	IC50: 11.8 μM	[37]
CPE/MMP/TYR	Benzoic acid	−0.1	IC50: 71.6 μM	[37]
CBS/APO	Theophylline	0.15 and CV	DL: 0.3 mM	[103]
Clark electrode/DM/potato tissue (ALP) + GOX	Phosphate in drug formulations	–	–	[104]
PtE/PANI/PPO	Benzoic acid	0.1	DL:0.3 μM	[105]

(continued)

Table 6.3 (continued)

Electrode	Drug	Amperometry V vs. Ag/ AgCl	Comment	References
PtE/LDH/XOX	Allopurinol	0.55Vvs. SCE	IC50: 310 μM	[98]
CPE/XOX/DM	Allopurinol	0.0 V vs. SCE	LR: 0.20–50 μM	[106]
PGE/DNA/XOX	Quercetin	−0.73Vvs SCE	–	[107]
PtE/PPY/MAO	Fluoxetine Benzylamine	0.7	Fluoxetine: LR: 0.67–4.33 mM DL: 0.10 mM	[90]
PtSPE/MAO/ GA/BSA	Fluoxetine	0.75 V vs Ag	DL: 0.8 nM LR:0.1 mM–1 nM	[108]
PtSPE/MAO	Furazolidone	0.8 V vs Ag	DL: 8.3 nM	[89]
SPPtE/MAO	Furadonine	0.8 V vs Ag	DL: 85 nM	[89]
PtSPE/MAO	Furagin	0.8 V vs Ag	DL: 0.94 nM	[89]
PtSPE/MAO/ GA/BSA	Pyrazidol	0.75 V vs Ag	DL: 0.8 μM LR:0.1 mM–1 μM	[108]

Other Enzymes-Based Biosensors

Salicylate hydroxylase (SAH) can convert salicylates to catechols in the presence of NADH and oxygen as cosubstrate as in equation:

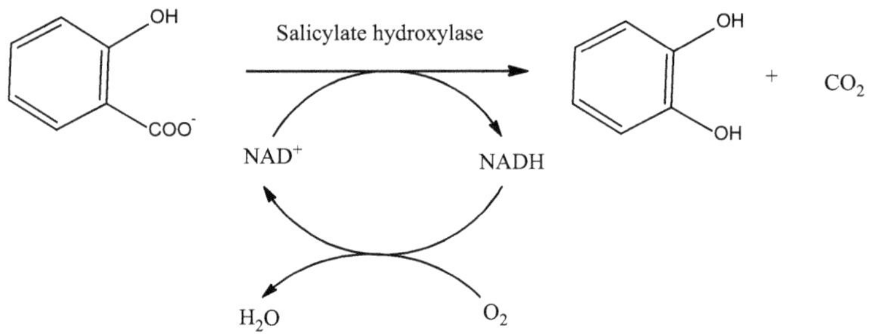

The coupling of this enzyme with tyrosinase (monophenolmonoxygenase) or polyphenol oxidase (PPO) to monitor oxygen consumption or quinone reduction has been proposed for salicylate detection [109–111]. Direct oxidation of catechol was also envisaged for the same aim [112, 113].

Theophylline oxidase (TOX), which is a heme-containing enzyme, oxidizes the bronchodilating drug theophylline in the presence of the naturally acceptor ferricytochrome *c* following equation:

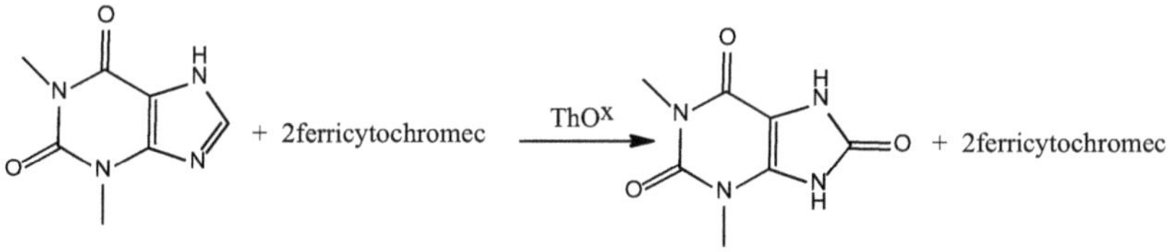

A redox mediator such as cyt. *c* within a Nafion layer on a platinum electrode was used for electronic coupling between the enzyme and the electrode [114]. A second redox mediator, $Fe(CN)_6^{4-}$, was used in order to facilitate shuttling the electrons between cyt. *c* and the electrode [115]. Cyt. *c* can be replaced by a non-physiological mediator in order to construct an enzyme electrode for the determination of theophylline in serum. Mediators such as ferrocene monocarboxylic acid and the phenazinemethosulfate-tetracyanoquinodimethane (NMP. TCNQ) organic salt [115] or an osmium-complex-based redox polymer [33] were described.

A third-generation biosensor based on direct electron transfer between the heme active center of theophylline oxidase at both graphite and at thiol-modified gold electrode was reported for theophylline. In a flow injection configuration, at an applied potential of 0.1 V vs. Ag/AgCl, a detection limit of 0.2 mM was obtained [116].

Microbial xanthine oxidase (XOX) was used for the construction of an amperometric enzyme electrode to detect theophylline. XOX catalyzes the oxidation of theophylline to 1–3 dimethyluric acid and hydrogen peroxide. The combination of this enzyme with peroxidase and ferrocene allowed the sensitive detection of liberated hydrogen peroxide [117].

Enzymes capable of oxidizing phenols, such as laccase, tyrosinase, and horseradish peroxidase were used for the sensing of direct oxidation of aminophenol drugs and for the assay of enzyme inhibitors. Although they react with the same family of compounds, catalytic cycles and consequently products and inhibitors of these enzymes are totally different.

Laccase is a copper enzyme which catalyzes the transformation of phenols and polyphenols to their corresponding radicals in the presence of oxygen [118].

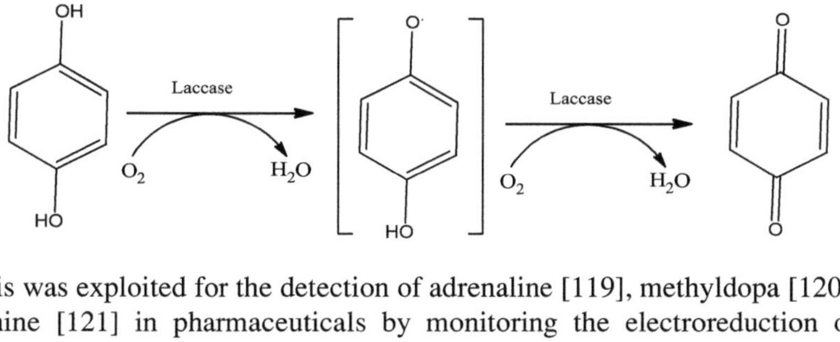

This was exploited for the detection of adrenaline [119], methyldopa [120], and morphine [121] in pharmaceuticals by monitoring the electroreduction of the product, or the oxygen consumption, the enzyme being co-immobilized with PQQ-glucose dehydrogenase [121].

L-cysteine was determined in a pharmaceutical formulation using a carbon paste electrode modified with laccase [122]. Hydroquinone was selected as enzyme substrate. The reaction of cysteine with the enzymatically generated *p*-quinone produced a decrease of the amperometric signal:

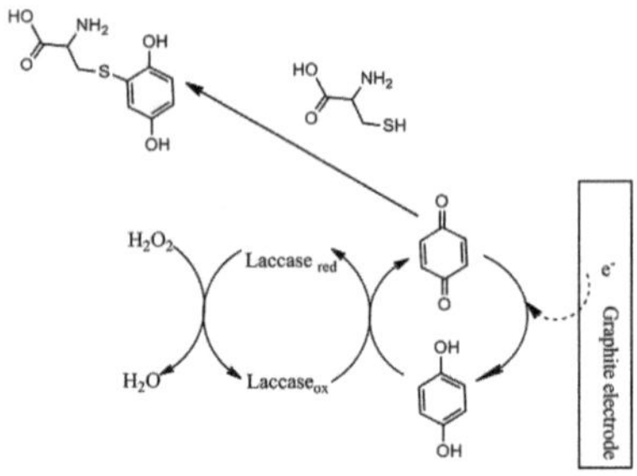

Tyrosinase contains two copper centers and catalyzes two different oxygen-dependent reactions: the o-hydroxylation of monophenols to o-diphenols (cresolase activity) and subsequent oxidation of *o*-diphenols to *o*-quinones (catecholase activity).

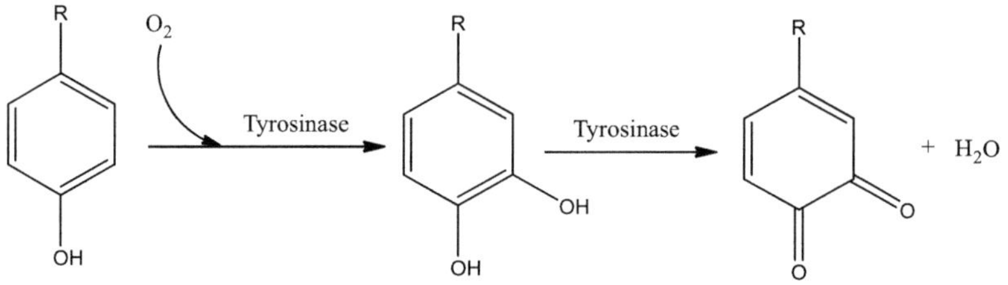

This enzyme cannot react with phenols having the ortho-position occupied, and it has many inhibitors, such as triazines, organophosphates, fluoride ions, benzoic acid, cinnamic acid, and sorbic acid which compete with the natural substrate for the binding to the enzyme site or which interrupt the catalytic cycle by reaction with intermediate products [123].

Mushroom tyrosinase is generally used as a biocatalyst in amperometric biosensors for catecholamine derivatives [124].The immobilization of this enzyme onto glutaraldehyde activated streptavidin paramagnetic microparticles retained onto a "magnetized" carbon paste electrode (CPE) was proposed for the amperometric assay of enzyme inhibitors [37]. The MNPs exerted little diffusion limitation towards the electrode surface, and the CPE exhibited a low background current at the applied potential, i.e., close to 0.0 V (vs. Ag/AgCl, 3 M KCl) with high sensitivity for the studied substrate tyrosine. The electrode signal corresponded to the reduction of the enzyme generated *o*-quinone (DOPAquinone) (Fig. 6.12). This biosensor was found of interest for the comparative study of inhibitors of skin melanogenesis (skin whitening agents). Kojic acid, benzoic acid, and azelaic acid

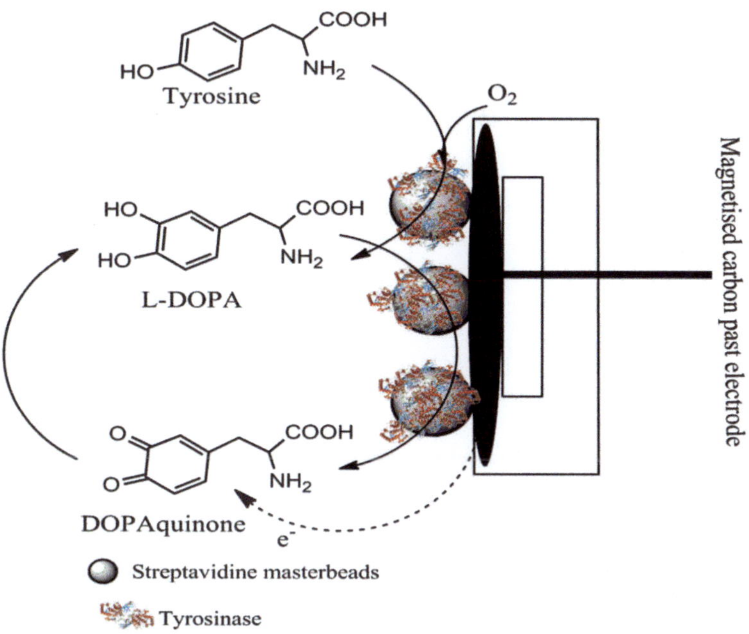

Fig. 6.12 Tyrosinase/NMPS/CPE-based biosensor amperometric response [37]

were characterized as melanin formation inhibitors by competing with tyrosine, i.e., the tyrosinase natural substrate, for binding to the enzyme active site. Ascorbic acid was also characterized as inhibitor of the melanogenesis process by interaction with the generated o-quinone [37]. The biosensor permitted the determination of the IC_{50} and the inhibition constant K_i for the studied compounds.

Penicillamine [125], pipemidic acid [126], and methimazole [127] were also determined via inhibition of the signal at a tyrosinase-based amperometric biosensor. The determination of the drug was accomplished by suppressing the substrate recycling process between tyrosinase and the electrode. It was observed that the affinity between the substrate and the drug affected severely the biosensor analytical performances. Catechol, 4-methylcatechol, and 4-tert-butylcatechol were studied as substrate. The former showed the greatest sensitivity for pipemidic acid [126] whereas 4-tert-butylcatechol was selected for penicillamine [125].

Antioxidant substances of synthetic and natural origins can be readily screened using amperometric biosensors with immobilized superoxide dismutase [128] or cytochrome *c* [129]. At the cytochrome *c* immobilized gold electrode, the enzyme specifically reacted with the superoxide anion in solution and the signal diminished in the presence of the antioxidant (Table 6.4).

The use of entire cells or tissues, as well as other components, is less common in biosensors applied to pharmacologically active compounds [4]. For economical source of enzymes, plan tissues can be used as biocatalysts in amperometric biosensors for drug assays. Dried plant tissue from avocado [139], soursop [140], and

Table 6.4 Other enzyme biosensors using amperometric detection

Electrode	Analyte(s)	Amperometry V vs Ag/AgCl	Comment	References
sCPE or GCE/TYR	Dopamine (DA) 3,4-dihydroxy phenylacetic acid (DOPAC) Norepinephrine (NE) Homovanillic acid	−200(CPE), +700 (GCE)	DL (μM)(GCE): (DA: 0.050, DOPAC & NE: 0.036), DL(μM) (CPE): (DA: 0.30, DOPAC: 0.84, NE: 0.25), LR: 0.09 μM−1 mM	[124]
AuE/APCPG/TYR	Pipemidic acid (in the presence of catechol)	0.0	LR: 0.02–70 μM, DL: 18 nM	[126]
GCE/APCPG/GA/TYR	Penicillamine (in the presence of catechol)	−0.15	LR: 0.02−80 μM	[125]
Clark Electrode/CTM/TYR +SAH	Salicylic acid	−0.8 V vs. Ag	DL: 4 μM−0.1 mM	[109]
CPE/SAH/TYR/GDH/NADH	Salicylic acid (in the presence of 10 mM glucose)	−0.1	DL: 3.5 μM	[111]
PtE/BSA/GA/LYOX	L-Lysine	+0.7 V vs. SCE	LR: up to 0.6 mM DL: 1 μM	[130]
GME/membranes/LYOX	Lysine	+1 V & −0.15	LR: up to 0.2 mM DL: 0.82 μM	[131]
AuE/aldrithiol/TOX	Theophylline	+0.1	LR: 0.2−2 mM	[116]
AuE/thiol SAM/TOX	Theophylline	+0.	LR: 2–5 mM DL: –	[33]
GrE/Os-polymer/TOX	Theophylline	+0.15	LR: 0.02−0.6 mM DL: 0.02 mM	[33]
GrE/DDAB/TOX	Theophylline	+0.15	LR: 0.2−8 mM DL: 0.2 mM	[33]

CPE/TOX/Cyt C/Nafion	Theophylline ($Fe(CN)_6^{3-}$ as a soluble mediator)	+0.4	LR:2 μM–0.1 mM DL: 2 μM	[114]
TCP/TOX/NMP · TCNQ	Theophylline	+0.1	LR: 0.03–0.67 mM DL: 0.03 mM	[115]
AuE/TOX	Theophylline (FMCA as a soluble mediator)	–	LR: 0.05–0.4 mM DL: 0.05 mM	[116]
CPE/microbial-XOx/HRP/Fc/DMem	Theophylline	0 V vs SCE	LR: up to 50 μM DL: 0.2 μM	[117]
Clark Electrode/MDH/SAH	Codeine & Morphine (in the presence $NADP^+$)	–	LR(codeine): 2–1,000 μM LR(morphine): 5–1,000 μM	[121]
GCE/GA/PPY/SAH	Salicylate	+0.17 V vs SCE	LR: 2.3–14 μM	[26]
PtE/3-APTS/GA/SAH	Salicylate	+0.55	LR: 7.25 μM–4.35 mM	[133]
GCE/PPy/SAH/GA/HCF	Salicylate (in the presence of NADH)	0.15 V vs. SCE	LR: 10 μM–0.1 mM	[112]
CFE/SAH	Salicylate (in the presence Of NADH, O_2)	–	LR: 0.1–1.93 μM and 1.96–12.0 μM DL: 64 nM	[113]
CFE/SAH	Salicylate	+0.3 V vs SCE	LR: 0.1–2 μM DL: 0.033 μM	[133]
GCE/SAH/pAPA	Salicylate	+0.4 V	DL: 3 nM	[110]
SPCE/SAH	Salicylate (in the presence of NADH and O_2)	+0.3 mV	–	[134]
GCE/PPO/SAH/pAPA	Salicylate	–0.2 V	LR: 0.056–3 μM	[110]
Elke Sen LLC/LOX	Chloramphenicol Penicillin	–	LR: 0.1–2.5 ppm for penicillin	[135]
Clark Electrode/LOX/PQQ-dependent GDH	Morphine	–	LR: 32 nM–100 μM	[121]
PtE/PVF/GA/Creatinase/SOX	Creatine	+0.7	LR: 20 μM–0.32 mM	[136]

(continued)

Table 6.4 (continued)

Electrode	Analyte(s)	Amperometry V vs Ag/AgCl	Comment	References
CRE/SOX/Creatine amidinohydrolase	Creatine (in the presence of $[Fe(CN)_6]^{3-}$/ $[Fe(CN)_6]^{4-}$)	+0.4	LR: up to 3.5 mM	[137]
CPE/(NPt-BMI.PF6)/LAC	Adrenaline	SWV	LR:1 μM–0.2 mM DL: 0.3 μM	[138]
CPE/CA/BMINTMS LAC	Methyldopa	SWV	LR: 35–370 μM DL: 5.5 μM	[120]
SPCE/CNT/TYR	Methimazole	CV	LR: 0.074–63.5 μM DL: 0.056 μM	[127]

Table 6.5 Biosensors based on cells or tissues

Electrode	Analyte(s)	Amperometry V vs Ag/AgCl	Comment	References
CPE/Corn(POX)/CHIT	Adrenaline (in the presence of H_2O_2)	−0.23	LR: 1–122 μM	[119]
CPE/TCNQ/ *Annonamuricata* L tissue (PPO)	Dopamine	+0.10	LR:0.2−20 mM	[140]
CPE/palm tissue (PPO)	Epinephrine	−0.10	LR:50–350 μM DL: 15 μM	[19]
Micro-chip/*E. coli*	Nalidixic acid	–	DL: 10 μg/mL	[142]
CPE/avocado tissue (PPO)	Acetaminophen	−0.12	LR:0.1–6 mM DL: 88 μM	[139]
Clark Electrode/ *P. alcaligenes*/CM/Gelatin/GA	Caffeine	–	LR:0.1−1 mg/mL	[141]
DOE/*S. cerevisiae*	Vitamin B1	–	LR: 5−100 nM	[142]
CPE/*D. bulbifera* (PPO)	Dopamine	CV, +0.15 V	LR: 2–8 mM DL: 0.75 mM	[143]
CPE/*D. bulbifera* (PPO)	Epinephrine	CV, −0.14 V	LR: 0.2−1 mM DL: 82 μM	[144]

palm tree fruits [19] for polyphenol oxidase sources or corn [119] for peroxidase enzymes can be readily mixed with graphite to obtain composite biosensors for acetaminophen and catecholamines sensing. Induced microbes of *pseudomonas alcaligenes* MTCC 5264 and *saccharomyces cerevisiae* NRRL-12632 cells have the capability to degrade L-lysine [130] and caffeine [141], respectively. They were covalently cross-linked in gelatin in a microbial biosensor sensitive to these drugs via O_2 consumption (Table 6.5).

6.3.1.2 Affinity Sensors

6.3.1.2.1 Immunosensors

Antibodies are proteins produced in animals as an immunological response to the presence of a foreign substance called antigen and have specific affinity for this antigen. Taking advantage on this specificity, antibodies or antigens can be immobilized directly onto an electrode surface to develop electrochemical immunobiosensors. Amperometric affinity sensors need the presence of an enzyme label and an electroactive substrate to generate a current related to the antibody–antigen interaction. Thus, and as in spectrophotometry, the ELISA technique for bioaffinity control can be adapted to amperometric transduction for electrochemical ELISA. Alkaline phosphatase, horseradish peroxidase, and glucose oxidase are the most popular enzyme labels for electrochemical ELISA.

An amperometric immunosensor for the detection of the antibiotic neomycin in meat samples was developed in a sandwich format. An antibody against neomycin was covalently linked to a conducting polypyrrole derivative onto the GCE. The antibiotic sample was allowed to react with the biosensor. Subsequently, the immunosensor was incubated with a conjugate comprising a secondary antibody linked to gold nanoparticles decorated carbon nanotubes labeled with hydrazine as catalyst for H_2O_2 detection [145].

Relatively few affinity sensors have been described for drug compounds assay. A general problem with immunosensors is the difficulty to regenerate the sensing surface. Single-use biosensors have been developed to solve such limitation. This consisted, for example, of a screen printed carbon electrode (SPCE) modified with an antibody and applied in ELISA strategies. An anti-progesterone monoclonal antibody was deposited onto a SPCE to fabricate a single-use biosensor for milk progesterone with admissible reproducibility. Using a continuous flow system and adapting competitive ELISA, three steps have been identified namely (1) competitive binding of sample/conjugate mixture (alkaline-phosphatase labeled progesterone; AP-prog), (2) establishment of a steady-state amperometric baseline current, and (3) measurement of an amperometric signal in the presence of the enzyme substrate 1-naphtylphosphate [146].

6.3.1.2.2 Oligonucleotide-Based Biosensors

The modification of an electrode surface with an oligonucleotide single or double stranded has gained considerable interest for drug assays and drug toxicity screening [147]. These have the aim to investigate drug–DNA interaction especially for "genotoxicity" or to determine drug compounds by selective interaction with synthesized oligonucleotides namely "aptasensor" [148–150]. Amperometric sensing at an oligonucleotide-modified electrode can be achieved by direct oxidation of guanine [18, 151, 152] or indirectly using a redox-active label which intercalates preferably at the double stranded configuration [153, 154]. There are several types of interactions associated with ligands that bind DNA. These include intercalation, noncovalent groove binding, covalent binding and cross linking, DNA cleaving, and nucleoside-analog incorporation.

The drug–DNA interaction can be studied using a double stranded-DNA (ds-DNA) sensor. A current decrease in guanine oxidation from ds-DNA can be used as indicator for the interaction between a drug compound and immobilized DNA. From these data, the binding constant *k* can be determined using the relation:

$$\log\left(\frac{1}{[\mathrm{Drug}]}\right) = \log K + \log\left(\frac{I_{\mathrm{DNA\text{-}Drug}}}{I_{\mathrm{DNA}} - I_{\mathrm{DNA\text{-}Drug}}}\right)$$

where [Drug] is the concentration of a drug and $I_{\mathrm{DNA\text{-}Drug}}$ and I_{DNA} are the oxidative DNA guanine oxidation current in the presence and the absence of the studied drug, respectively [151].Using this methodology, the effect of berberine on the structural stability of DNA from human cancer and noncancer cells was compared [155]. The guanine functionalities are electrooxidized at relatively high positive potentials at carbon-based electrodes, i.e., close to water oxidation, and current trends are oriented towards the use of boron-doped diamond electrodes for DNA biosensors since such electrodes offer an extended positive potential domain of investigation [156].

An assay based on a DNA–Cu(II) complex for electrocatalytic hydrogen peroxide reduction was used to study drug–dsDNA interactions based on an inhibitory effect of the antimalarial drug quinacrine on the hydrogen peroxide reduction. The signal was reversibly inhibited due to an electron blocking effect of the quinacrine–dsDNA interaction [157].

The advance in automatic synthesis of oligonucleotides specific to a target drug or a protein makes aptasensor an interesting strategy to develop biosensors for drug compounds. The three-dimensional shape of oligonucleotides allows the detection of a wide range of analytes such as ions, amino acids, organic dyes, antibiotics, vitamins, drugs, peptides, proteins, or even whole cells. Aptasensors are able to discriminate between closely related compounds on the basis of subtle structural differences [158]. A labeled aptasensor has an additional probe (enzyme, redox marker, or nanoparticle) covalently bound to the aptamer or to the target molecule.

Several detection mechanisms were reported by using labeled aptasensor (Fig. 6.13). In the aptasensor based on sandwich assays, small molecules such as toxins, drugs, or hormones can be detected with a primary aptamer binding the analyte and a second labeled aptamer binding the primary aptamer–analyte complex [159].

Alternatively, a competitive displacement by an unlabeled analyte of an analyte labeled with an electroactive tag can be monitored at an aptamer-modified electrode. The detection of this interaction can be studied by monitoring the labeled analyte electroactivity (Fig. 6.14). A review on this topic has recently been provided [160].

Electrochemical aptasensors have been developed taking advantage of the three-dimensional conformational change undergone by the aptamer upon target binding. The redox marker linked to the aptamer can move closer or away from the surface

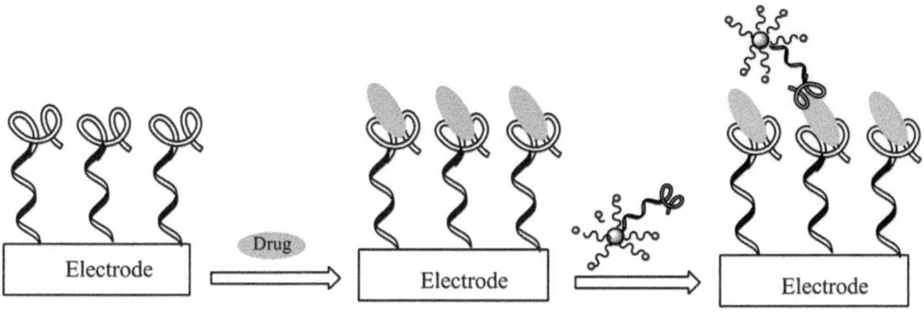

Fig. 6.13 Sandwich type aptasensor

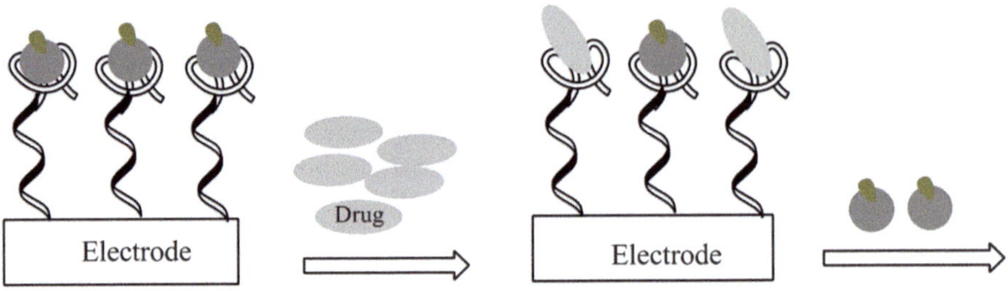

Fig. 6.14 Competitive displacement type aptasensor

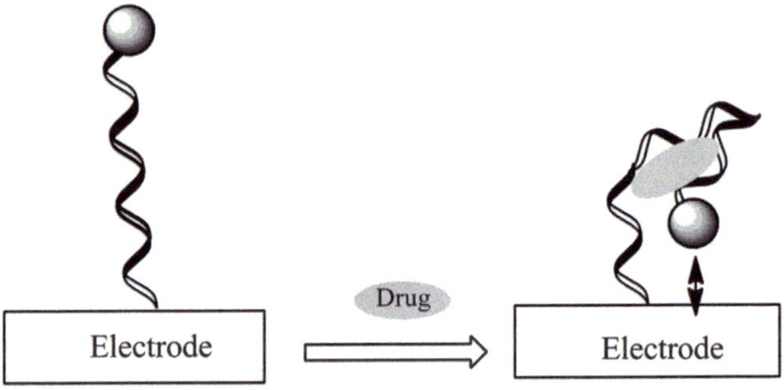

Fig. 6.15 Redox marker linked aptasensor

of electrode due to the selective analyte–aptamer interaction. Example of this development consisted in the immobilization of a sequence of RNA or DNA aptamers onto an electrode with covalent attachment of a redox tracer onto the terminal function of this sequence. When the interaction between the oligonucleotide and the drug occurred, a conformational change is triggered in the aptamer and the distance from the mediator to the electrode surface decreased with subsequent increase of the current due to mediator (Fig. 6.15). The conformational rearrangement is reversible allowing for multiple use of the aptamer-based sensor. Application of this strategy was illustrated for the assay of theophylline [154, 161]. A microfluidic setup with a miniaturized detection chamber (0.75 μL) based on the selective recognition between cocaine and a 32-base cocaine-binding aptamer with methylene blue (MB) as label was designed for the assay of micromolar concentrations of cocaine in undiluted blood serum. Microfluidic technology offers substantial advantages on mass transport and diffusion over conventional macroscale devices. The thiolated aptamer was self-assembled onto a 6-mercaptohexanol-modified gold electrode. The biosensor was operating in blood serum in real-time sensing for more than 1 h [162]. Based on a similar setup technology but with a particularly original parallel dual liquid flow design, the same group showed the possibility of this unique biosensor platform for in vivo continuous therapeutic drug monitoring. The in vivo concentration of doxorubicin and kanamycin in live rats could be determined during more than 4 h of continuous use. Square wave voltammetry was employed to detect MB reduction at the gold electrode [163].

The decrease of the oxidation current of a soluble mediator due to the blocking of the diffusion access to the electrode interface after drug–aptamer interaction was utilized, e.g., for a tetracycline antibiotic detection (Table 6.6) [153].

6.3.2 *Potentiometric Biosensors*

Catalytic and bioaffinity processes trigger a potential change at the electrode–solution interface. Potentiometric biosensors are for the vast majority based on ion-selective electrodes (ISE) in conjunction with immobilized enzymes [182]. Various ion selective membrane electrodes are commercially available and used for the selective detection of several ions such as K^+, Na^+, Ca^{2+}, NH_4^+, H^+, and CO_3^{2-}. The enzyme can be immobilized on the outer surface or captured inside the membrane. Penicillinase was immobilized on the surface of a glass pH electrode and the enzymatically formed penicilloic acid changed the pH in the interface [183–185].

The entrapment of lysine oxidase within a polyacrylamide gel around a pH electrode or as liquid layer trapped within a cellophane membrane has been used for biocatalytic recognition of lysine based on its degradation by lysine oxidase according to the following reaction:

$$\text{Lysine} + H_2O + O_2 \xrightarrow{\text{Lysine oxidase}} \text{2-Oxo-6-aminocaproic acid} + H_2O_2 + NH_3$$

This has been exploited for developing a potentiometric biosensor based on a solid state ammonium ion electrode for lysine assay in pharmaceuticals [186, 187]. The same ion selective electrode was also used for creatinine determination based on creatinine deiminase as biocatalytic recognition element [188]. Field-effect transistors (FET) such as Ion selective FETs (ISFETs) and LAPS (light addressable potentiometric sensor) have also been used for potentiometric biosensors. In biosensors based on ISFETs and LAPSs, the transducer monitors a local variation of ions concentration resulting from the enzymatic reaction. In the field effect transistors, the current flows along a semiconductor channel between two terminal electrodes usually the source and drain. An electric field generated by applying a voltage to the third electrode named gate can control the conductivity of this channel. A small change in gate voltage can cause a large variation in the current flow from the source to the drain. In the biosensor-based FET, changes in potential of gate during biochemical process can be monitored. Ion selective FETs (ISFETs) biosensors operate in a similar manner to FET, i.e., a local potential developed by generated ions associated with a biochemical process, and this potential can modulate the current flow along the semiconductor channel.

Table 6.6 Oligonucleotide-based biosensors

Electrode	Analyte (s)	Amperometry V vs. Ag/AgCl	Comment	References
CPE/calf thymus DNA	Actinomycin D	DPV	–	[164]
GCE/DNA	Benznidazole	DPV	–	[15]
SPE/(AuNPs–MWNT-SDS)/DNA	Berberine	DPV, CV	–	[155]
GCE/ssDNA	Carboplatin	DPV	LR: 8 μM–2.4 mM DL: 5.7 μM	[165]
GCE/dsDNA	Ciprofloxacin	DPV	LR: 40–80 μM DL: 9 μM	[151]
CPE/calf thymus dsDNA	Daunomycin	CV, +0.59	–	[166]
CPE/dsDNA or ssDNA	4,4′-dihydroxy chalcone	DPV	–	[16]
PGE/dsDNA	Efavirenz (EFV)	DPV,+1.02 and +1.40	LR: 2–24 ppm DL:0.6 ppm	[152]
GCE/dsDNA	Emodin	DPV	–	[167]
PGE/dsDNA	Fulvestrant	DPV	LR: 1–20 ppm DL: 0.41 ppm	[18]
PGE/dsDNA	Leuprolide	DPV and AdsDPV	LR:0.2–6.0 ppm DL: 0.06 ppm	[168]
CPE or PGE/dsDNA or ssDNA	Lycorine	DPV	DL(ds): 0.225 ppm(CPE) and 0.0302 ppm (PGE)	[169]
SPCE/DNA	1-methyl-2-piperidinoethylester of 2 hexoxyphenyl Carbamic acid	DPV	–	[170]
SPE/dsDNA	Cisplatin Carboplatin	SWV	–	[17]
GCE/DNA	Metronidazole	DPV	DL: 1.67 μM and without preconcentration: 3.25 μM	[171]

GCE/DNA	Metronidazole	DPV	DL: 1 μM	[172]
GCE/DNA	Metronidazole	DPV, −0.472 V vs. SCE	–	[172]
ITO/3-APTS/AuNPs/DNA	Mifepristone	DPV	LR: 0.4–6 μM DL: 0.2 μM	[173]
GCE/calf thymus DNA	Metronidazole	CV, −0.662 V vs. SCE	DL: 0.02 μM	[174]
PGE/dsDNA	Mitomycin C	DPV	–	[175]
GCE/DNA	Mitoxantrone	SWV	–	[176]
GCE/DNA	Nimorazole	DPV, −0.410 V vs. SCE	–	[172]
GCE/DNA	Secnidazol	DPV, −0.436 V vs. SCE	–	[172]
AuE/Cysteamine/DNA/SWCNTs	Taxol	SWV & CV	LR: 13.8 μM–0.13 mM DL: 0.9 mM	[177]
SPAuE/biotinylatedssD-NAaptamer/streptavidin	Tetracycline	SWV & CV	LR: 0.010–10 μM DL: 10 nM	[153]
PtE/Ppy/dsDNA	Salicylic acid (SA) AcetylSA	DPV	LD: 0.86 μM for SA 5.24 μM for acetylSA	[178]
GCE/DNA	Tinidazole	DPV, −0.432 V vs. SCE	–	[172]
SPE/AuNPs/pTTBA/dsDNA/cardiolipin	Mitoxantrone (MIT) Idarubicin (IDA) Doxorubicin (DOX) Daunomycin (DAN)	FI Eapp = −0.7 V	MIT(LR: 2–60 pM, DL: 1.2 ± 0.05 fM), IDA (LR: 5–55 pM DL: 2.2 ± 0.1 fM), DOX(LR:7.5–50 pM,DL:3.6 ± 0.2 fM), DAN(LR:8–50 pM, DL: 5.5 ± 0.3 fM)	[31]
AuE/alkanethiol/MB-tagged aptamer	Cocaine	DPV	DL: 500 μM	[179]
AuE/MB taged aptamer/6-mercapto-1-hexanol	Cocaine	DPV	–	[180]

(continued)

Table 6.6 (continued)

Electrode	Analyte (s)	Amperometry V vs. Ag/AgCl	Comment	References
AuE/1-mercapto hexanol/MB/RNA	Theophylline	DPV	LR: 20–100 μM	[154]
AuE/thiol group Fc/RNA aptamer	Theophylline	DPV	LR : 0.2–10 μM, DL: 0.2 μM	[161]
MAE/AuNPs/pTTCA/aptamer	Kanamycin	CV	Dissociation constant: 38 nM	[181]

A penicillin sensitive field-effect transistor with immobilized β-lactamase (penicillinase) was developed. A Ta_2O_5 layer acted as pH sensitive layer by monitoring the variation of H^+ concentration resulting from the hydrolysis of penicillin to penicilloic acid [189]. A similar approach based on H^+ sensing was developed using penicillin G acylase (benzylpenicillinamido hydrolase) to detect penicillin G [190].

An aptamer was assembled on an ISFET gate in order to develop a biosensor for cocaine in a sandwich format. The second thiolated aptamer fragment was bound to gold nanoparticles. A sandwich coupling of substrate with two aptamers caused a change in the interfacial electron transfer resistance and changed the potential of the ISFET device [191].

A light addressable potentiometric sensor principle is based on semiconductor as well. A multilayer structure of silicon/silicon oxide/silicon nitride is activated by a modulated light source in order to obtain a flow of photocurrent. The amplitude of this photocurrent is sensitive to the surface potential; therefore, this sensor is capable to measure the surface potential of the electrolyte–transducer interface. The surface potential itself depends on the biochemical process due to the enzyme activity. Penicillin was measured in the range 5–25 mM by utilizing this type of biosensor (Table 6.7) [192].

Table 6.7 Potentiometric-based biosensors

Electrode	Analyte(s)	Comment	References
SSAE/LYOX/NM	Lysine	LR: 25 μM–75 mM DL: 10.5 μM	[193]
AAAE/LYOX	Lysine	LR: up to 1 mM DL: 6.0 μM	[187]
GEE/PEVA/AO	L-ascorbic acid	LR: 8 μM–0.45 mM DL: 4.2 μM	[194]
AISM/penicillinase/EDAC	Penicillin V and G β-lactam antibiotics)	–	[195]
pHE/penicillinase/CM	Penicillin	LR: 1–10 mM	[184]
BB/penicillinase/PVA-PE pH-GE/penicillinase/PVA-PE	Penicillin-G	LR: 30 mM–0.1 M, DL: 20 mM	[183]
pH-GE/penicillinase/corning 12	Penicillin	–	[185]
pH-GE/penicillinase/fritted glass disk	Penicillin	LR: 3.5–1,100 ppm	[196]
AME/urease & creatinase/ PVC	Creatine	LR: 10 μM–1 mM	[136]
AISM/creatininedeiminase/ EDAC	Creatinine	LR: 0.02–20.0 mM DL: <0.015 mM	[188]
CO_2Elec/*var. calidolactis*/ silkF or β-lactamase	β-lactam residues	–	[197]
GlassElec/*P. aeruginosa*	Cephalosporins	LR: 0.1–11 μM	[198]
CO_2Elec/*B. stearo-thermophiles var. calidolactis*	Cephalosporins, Penicillins	–	[197]

(continued)

Table 6.7 (continued)

Electrode	Analyte(s)	Comment	References
CO_2Elec/silkF/*E. coli*	Quinolones Tetracyclines	–	[199]
LAPS, EIS, EnFETs/ penicillinase	Penicillin	LR: 0.05–10 mM	[200]
H^+ISFET/penicillin G acylase	Penicillin	LR: 0.5–8 mM	[190]
PenFET/penicillinase	Penicillin G	LR: 0.05–1 mM DL: 5 μM	[189]

6.3.3 Impedimetric Biosensors

The capacitance or the interfacial electron transfer resistance of a biosensor interface can be monitored by an impedimetric biosensor. Such biosensor monitors either change in the capacitance (C), impedance (Z), or its component resistance (R):

$$Z^2 = R^2 + \frac{1}{(2fC)^2}$$

Impedance transduction can be achieved by two ways: (1) faradaic mode using control of heterogeneous electron-transfer resistance in the presence of an electroactive species (usually noted F-EIS as faradaic-electrochemical impedance spectroscopy) or (2) non-faradaic mode using the capacitance of the equivalent circuit as monitored parameter [201]. The F-EIS was selected as transducer to detect neomycin B and tobramycin in a competitive mode at an aptamer sensor [202]. These were based on immobilization of an antibiotic to a mercaptopropionic acid-modified Au electrode, affinity binding of an aptamer, and incubation of the biosensor in the presence of the soluble antibiotic to extract the aptamer. The decrease in the charge resistance transfer Ret of a redox mediator $[Fe(CN)_6]^{3-/4-}$, deduced from the adoption of Randles equivalent circuit, was proportional to the antibiotic concentration.

Enzymatic reactions involving either consumption or production of charged species and consequently changes in solution conductivity can be detected with conductometric biosensors. Examples include the measurement of L-asparagine by L-asparaginase. The L-asparaginase catalyzes the hydrolysis of L-asparagine to L-aspartate and an ammonium [203].

$$\text{L-asparagine} + H_2O \xrightarrow{L-asparaginase} \text{L-aspartate}^- + {NH_4}^+$$

An impedimetric immunosensor with a monoclonal antibody-gold nanoparticle-sol–gel-modified electrode was realized for tetracyclines determination. The relative charge transfer resistance (Rct) increases linearly with doxorubicin concentration in the ranges of 0.1–1.0 and 2.5–50.0 pg m L^{-1}, with a detection limit of 0.09 pg mL^{-1}. Application to spiked human serum and urine samples was reported [204].

References

1. Thévenot DR, Toth K, Durst RA, Wilson GS (2001) Biosens Bioelectron 16:121–131
2. Comley J (2004) Drug Discov World 5:63–74
3. Keusgen M (2002) Naturwissenschaften 89:433–444
4. Yu D, Blankert B, Viré JC, Kauffmann JM (2005) Anal Lett 38:1687–1701
5. Parsajoo C, Kauffmann JM, ElKaoutit M (2012) Biosensors for drug testing and discovery. In: Higson S (ed) Biosensors for medical applications, 1st edn. Woodhead Publishing Series in Biomaterials, Cambridge
6. Sanvicens N, Mannelli I, Salvador JP, Valera E, Marco MP (2011) TrAC Trends Anal Chem 30:541–553
7. Turner APF, Karube I, Wilson GS (1987) Biosensors fundamentals and applications. Oxford University Press, Oxford
8. Guilbault GG (1984) Analytical uses of immobilized enzymes. Marcel Dekker, New York
9. Buell AK, White DA, Meier C, Welland ME, Knowles TPJ, Dobson CM (2010) J Phys Chem B 114:10925–10938
10. Blum LJ, Coulet PR (1991) Biosensor principles and applications. Marcel Dekker, New York
11. Mulchandani A, Rogers KR (1998) Enzyme and microbial biosensors: techniques and protocols. Humana, Totowa, NJ
12. Rusmini F, Zhong Z, Feijen J (2007) Biomacromolecules 8:1775–1789
13. Sassolas A, Blum LJ, Leca-Bouvier BD (2012) Biotechnol Adv 30:489–511
14. Guisan JM (2006) Immobilization of enzymes and cells. Humana, Totowa, NJ
15. La-Scalea MA, Serrano SHP, Ferreira EI, Oliveira Brett AM (2002) J Pharm Biomed Anal 29:561–568
16. Meric B, Kerman K, Ozkan D, Kara P, Erdem A, Kucukoglu O, Erciyas E, Ozsoz M (2002) J Pharm Biomed Anal 30:1339–1346
17. Mascini M, Bagni G, Pietro M, Ravera M, Baracco S, Osella D (2006) Biometals 19:409–418
18. Dogan-Topal B, Ozkan SA (2011) Electrochim Acta 56:4433–4438
19. Sapelnikova S, Dock E, Ruzgas T, Emnéus J (2003) Talanta 61:473–483
20. Petit C, Gonzalez-Cortes A, Kauffmann JM (1995) Talanta 42:1783–1789
21. Cosnier S (1999) Biosens Bioelectron 14:443–456
22. Alonso Lomillo MA, Kauffmann JM, Arcos Martinez MJ (2003) Biosens Bioelectron 18:1165–1171
23. Campás M, Marty JL (2006) Encapsulation of enzymes using polymers and sol–gel techniques. In: Guisan J (ed) Immobilization of enzymes and cells. Humana, Totowa, NJ
24. Lenigk R, Lam E, Lai A, Wang H, Han Y, Carlier P, Renneberg R (2000) Biosens Bioelectron 15:541–547
25. González-Sánchez MI, Rubio-Retama J, López-Cabarcos E, Valero E (2011) Biosens Bioelectron 26:1883–1889
26. Junior LR, Neto GDO, Fernandes JOR, Kubota LT (2000) Talanta 51:547–554
27. Wang G, Xu JJ, Ye LH, Zhu JJ, Chen HY (2002) Bioelectrochemistry 57:33–38
28. Liu S, Peng L, Yang X, Wu Y, He L (2008) Anal Biochem 375:209–216
29. Du D, Chen S, Cai J, Song D (2007) J Electroanal Chem 611:60–66
30. Akkaya A, Altug C, Pazarlioglu NK, Dinckaya E (2009) Electroanalysis 21:1805–1810
31. Allongue P, Delamar M, Desbat B, Fagebaume O, Hitmi R, Pinson J, Savéant JM (1997) J Am Chem Soc 119:201–207
32. Chandra P, Zaidi SA, Noh HB, Shim YB (2011) Biosens Bioelectron 28:326–332
33. Ferapontova EE, Shipovskov S, Gorton L (2007) Biosens Bioelectron 22:2508–2515
34. Loh KP, Bao Q, Ang PK, Yang J (2010) J Mater Chem 20:2277–2298
35. Jubete E, Loaiza OA, Ochoteco E, Pomposo JA, Grande H, Rodriguez J (2009) J Sens 78: 4–13
36. Rivas GA, Rubianes MD, Rodriguez MC, Ferreyra NF, Luque GL, Pedano ML, Miscoria SA, Parrado CN (2007) Talanta 74:291–307

37. Sima VH, Patris S, Aydogmus Z, Sarakbi A, Sandulescu R, Kauffmann JM (2011) Talanta 83:980–987
38. Guilbault GG, Schmid RD (1991) Biotechnol Appl Biochem 14:133–145
39. Wang J (1999) J Pharm Biomed Anal 19:47–53
40. Kauffmann JM, Guilbault GG (1995) Biosensors. In: Adamovics JA (ed) Biosensors in analysis of additive and misused drugs. Marcel Dekker, New York
41. D'Orazio P (2003) Clin Chim Acta 334:41–69
42. Sadik OA, Aluoch AO, Zhou A (2009) Biosens Bioelectron 24:2749–2765
43. Clark LC, Lyons C (1962) Ann N Y Acad Sci 102:29–45
44. Sono M, Roach MP, Coulter ED, Dawson JH (1996) Chem Rev 96:2841–2887
45. Bistolas N, Wollenberger U, Jung C, Scheller FW (2005) Biosens Bioelectron 20:2408–2423
46. Guengerich FP (2001) Chem Res Toxicol 14:611–650
47. Joseph S, Rusling JF, Lvov YM, Friedberg T, Fuhr U (2003) Biochem Pharmacol 65: 1817–1826
48. Krishnan S, Schenkman JB, Rusling JF (2011) J Phys Chem B 115:8371–8380
49. Mie Y, Suzuki M, Komatsu Y (2009) J Am Chem Soc 131:6646–6647
50. Shumyantseva VV, Bulko TV, Bachmann TT, Bilitewski U, Schmid RD, Archakov AI (2000) Arch Biochem Biophys 377:43–48
51. Shumyantseva VV, Bulko TV, Usanov SA, Schmid RD, Nicolini C, Archakov AI (2001) J Inorg Biochem 87:185–190
52. Yarman A, Wollenberger U, Scheller FW (2013) Electrochim Acta 110:63–72
53. Shumyantseva VV, Ivanov YD, Bistolas N, Scheller FW, Archakov AI, Wollenberger U (2004) Anal Chem 7:6046–6055
54. Johnson DL, Lewis BC, Elliot DJ, Miners JO, Martin LL (2005) Biochem Pharmacol 69: 1533–1541
55. Shumyantseva VV, Bulko TV, Alexandrova SA, Sokolov NN, Schmid RD, Bachmann T, Archakov AI (1999) Biochem Biophys Res Commun 263:678–680
56. Peng L, Yang X, Zhang Q, Liu S (2008) Electroanalysis 20:803–807
57. Iwuoha EI, Kane S, Ania CO, Smyth MR, Ortiz de Montellano PR, Fuhr U (2000) Electroanalysis 12:980–986
58. Iwuoha EI, Joseph S, Zhang Z, Smyth MR, Fuhr U, Ortiz de Montellano PR (1998) J Pharm Biomed Anal 17:1101–1110
59. Asturias-Arribas L, Alonso-Lomillo MA, Domanguez-Renedo O, Arcos-Martanez MJ (2011) Anal Chim Acta 685:15–20
60. Michira I, Akinyeye R, Somerset V, Klink MJ, Sekota M, Al-Ahmed A, Baker PGL, Iwuoha E (2007) Macromol Symp 255:57–69
61. Iwuoha EI, Wilson A, Howel M, Mathebe NGR, Jaime KM, Narinesingh D, Elied AG (2005) Anal Lett 37:929–941
62. Ignaszak A, Hendricks N, Waryo T, Songa E, Jahed N, Ngece R, Al-Ahmed A, Kgarebe B, Baker P, Iwuoha EI (2009) J Pharm Biomed Anal 49:498–501
63. Alonso-Lomillo MA, Gonzalo-Ruiz J, Dominguez-Renedo O, Muñoz FJ, Arcos-Martínez MJ (2008) Biosens Bioelectron 23:1733–1737
64. Alonso-Lomillo MA, Yardimci C, Dominguez-Renedo O, Arcos-Martanez MJ (2009) Anal Chim Acta 633:51–56
65. Shumyantseva VV, Bulko TV, Rudakov YO, Kuznetsova GP, Samenkova NF, Lisitsa AV, Karuzina II, Archakov AI (2007) J Inorg Biochem 101:859–865
66. Carrara S, Cavallini A, Erokhin V, De Micheli G (2011) Biosens Bioelectron 26:3914–3919
67. Veitch NC (2004) Phytochemistry 65:249–259
68. Yu D, Renedo OD, Blankert B, Sima V, Sandulescu R, Arcos J, Kauffmann JM (2006) Electroanalysis 18:1637–1642
69. Yu D, Blankert B, Kauffmann JM (2007) Biosens Bioelectron 22:2707–2711
70. Alonso-Lomillo MA, Dominguez-Renedo O, Matos P, Arcos-Martinez MJ (2009) Bioelectrochemistry 74:306–309

71. Petit C, Murakami K, Erdem A, Kilinc E, Borondo GO, Liegeois JF, Kauffmann JM (1998) Electroanalysis 10:1241–1248
72. Serradilla Razola S, Blankert B, Quarin G, Kauffmann JM (2003) Anal Lett 36:1819–1833
73. Torriero AAJ, Salinas E, Raba J, Silber JJ (2006) Biosens Bioelectron 22:109–115
74. Sima V, Cristea C, Lăpăduş F, Marian IO, Marian A, Săndulescu R (2008) J Pharm Biomed Anal 48:1195–1200
75. Sima V, Cristea C, Bodoki E, Duţu G, Sāndulescu R (2010) Centr Eur J Chem 8:1034–1038
76. Blankert B, Dominguez O, El Ayyas W, Arcos J, Kauffmann JM (2005) Anal Lett 37: 903–913
77. Stefan RI, van Staden JF, Aboul-Enein HY (2000) Biosens Bioelectron 15:1–5
78. Stefan RI, van Staden JF, Mulaudzi LV, Aboul-Enein HY (2002) Anal Chim Acta 467: 189–195
79. Campanella L, Favero G, Tomassetti M (2001) Anal Lett 34:855–866
80. Sánchez-Paniagua López M, Hervás Pérez JP, López-Cabarcos E, López-Ruiz B (2007) Electroanalysis 19:370–378
81. Pandey PC, Upadhyay S, Pathak HC, Pandey CMD, Tiwari I (2000) Sens Actuators B 62: 109–116
82. Shaidarova LG, Fomin AY, Ziganshina SA, Medyantseva EP, Budnikov GK (2002) J Anal Chem 57:150–156
83. Nunes GS, Montesinos T, Marques PBO, Fournier D, Marty JL (2001) Anal Chim Acta 434: 1–8
84. Ricci F, Arduini F, Amine A, Moscone D, Palleschi G (2004) J Electroanal Chem 563: 229–237
85. Parsajoo C, Kauffmann JM (2013) Talanta 109:116–120
86. Joshi P, Joshi H, Sanghi S, Kundu S (2010) Microchim Acta 169:383–388
87. Goodman LS, Gilman A (2001) The pharmacological basis of therapeutics, 10th edn. McGraw-Hill, New York
88. Medyantseva ÃP, Varlamova RM, Gimaletdinova DA, Fattakhova AN, Budnikov GK (2007) Pharm Chem J 41:341–344
89. Volotskaya DA, Medyantseva ÃP, Valieva ÃR, Varlamova RM, Fattakhova AN, Budnikov GK (2011) Pharm Chem J 44:624–627
90. Vela MH, de Jesus D, Couto CMCM, Araújo A, Montenegro MCBSM (2003) Electroanalysis 15:133–138
91. Lewis RA, Austen KF, Soberman RJ, Engl N (1990) J Med 323:645–655
92. Rittmannsberger A, Likussar W, Michelitsch A (2005) Biosens Bioelectron 21:655–660
93. Demirkol D, Gulsunoglu B, Ozdemir C, Dincer A, Zihnioglu F, Timur S (2012) Food Anal Method 5:244–249
94. Van de Laar FA, Lucassen PL, Akkermans RP, van de Lisdonk EH, Rutten GE, van Weel C (2005) Diabetes Care 28:154–163
95. Matsumoto K, Takayama K, Abesundara KJM, Matsui T (2003) Anal Chim Acta 479: 135–143
96. McCord JM, Engl N (1985) J Med 312:159–163
97. Nakamura M (1991) J Biochem 110:450–456
98. Shan D, Wang Y, Zhu M, Xue H, Cosnier S, Wang C (2009) Biosens Bioelectron 24: 1171–1176
99. Timur S, Anik Ü (2007) Anal Chim Acta 598:143–146
100. Hsieh BC, Matsumoto K, Cheng TJ, Yuu G, Chen RLC (2007) J Pharm Biomed Anal 45: 673–678
101. Campanella L, Cocco R, Tomassetti M (1992) J Pharm Biomed Anal 10:741–742
102. Campanella L, Persio GD, Pintore M, Tonnina D, Caretto N, Martini E, Lelo D (2009) Food Technol Biotechnol 47:172–177
103. Foulds NC, Wilshere JM, Green MJ (1990) Anal Chim Acta 229:57–62

104. Campanella L, Cordatore M, Mazzei F, Tomassetti M (1990) J Pharm Biomed Anal 8: 711–715
105. Li S, Tan Y, Wang P, Kan J (2010) Sens Actuators B 144:18–22
106. Martin GB, Rechnitz GA (1990) Anal Chim Acta 237:91–98
107. Zhao J, He X, Yang N, Sun L, Li G (2012) Int J Mol Sci 13:4873–4879
108. Medyantseva EP, Varlamova RM, Gimaletdinova DA, Fattakhova AN, Budnikov GK (2008) J Anal Chem 63:275–279
109. Campanella L, Gregori E, Tomassetti M (2006) J Pharm Biomed Anal 42:94–99
110. Cosnier S, Gondran C, Watelet JC (2001) Electroanalysis 13:906–910
111. Martin C, Dominguez E (1999) J Pharm Biomed Anal 19:107–113
112. Milagres BG, de Oliveira NG, Kubota LT, Yamanaka H (1997) Anal Chim Acta 347:35–41
113. De Carvalho RM, de Oliveira NG, Kubota LT (2001) Electroanalysis 13:131–136
114. Wang J, Dempsey E, Ozsoz M, Smyth MR (1991) Analyst 116:997–999
115. McNeil CJ, Cooper JM, Spoors JA (1992) Biosens Bioelectron 7:375–380
116. Christenson A, Dock E, Gorton L, Ruzgas T (2004) Biosens Bioelectron 20:176–183
117. Stredansky M, Pizzariello A, Miertus S, Svorc J (2000) Anal Biochem 285:225–229
118. Archibald FS, Bourbonnais R, Jurasek L, Paice MG, Reid ID (1997) J Biotechnol 53:215–236
119. Brondani D, Dupont J, Spinelli A, Vieira IC (2009) Sens Actuators B 138:236–243
120. Moccelini SK, Franzoi AC, Vieira IC, Dupont J, Scheeren CW (2011) Biosens Bioelectron 26:3549–3554
121. Bauer CG, Kühn A, Gajovic N, Skorobogatko O, Holt PJ, Bruce NC, Makower A, Lowe CR, Scheller FW (1999) Fresenius J Anal Chem 364:179–183
122. Santhiago M, Vieira IC (2007) Sens Actuators B 128:279–285
123. Burton SG (1994) Catal Today 22:459–487
124. Pravda M, Petit C, Michotte Y, Kauffmann JM, Vytřas K (1996) J Chromatogr A 727:47–54
125. Torriero AAJ, Salinas E, Marchevsky EJ, Raba J, Silber JJ (2006) Anal Chim Acta 580: 136–142
126. Bertolino FA, Vito IED, Messina GA, Fernández H, Raba J (2011) J Electroanal Chem 651: 204–210
127. Martinez NA, Messina GNA, Bertolino FA, Salinas E, Raba J (2008) Sens Actuators B 133: 256–262
128. Campanella L, Bonanni A, Bellantoni D, Favero G, Tomassetti M (2004) J Pharm Biomed Anal 36:91–99
129. Ignatov S, Shishniashvili D, Ge B, Scheller FW, Lisdat F (2002) Biosens Bioelectron 17: 191–199
130. Guerrieri A, Cataldi TRI, Ciriello R (2007) Sens Actuators B 126:424–430
131. Saurina J, Hernandez-Cassou S, Alegret S, Fabregas E (1999) Biosens Bioelectron 14: 211–220
132. Neumayr M, Sontag G, Pittner F (1995) Anal Chim Acta 305:26–31
133. de Carvalho RM, de Oliveira NG, Kubota LT (2000) Anal Lett 33:425–442
134. Frew JE, Bayliff SW, Gibbs PNB, Green MJ (1989) Anal Chim Acta 224:39–42
135. Rinken T, Riik H (2006) J Biochem Biophys Methods 66:13–21
136. Karakuş E, Erden PE, Pekyardımcı Ş, Kılıç E (2006) Artif Cells Blood Substit Immobil Biotechnol 34:337–347
137. Ramanavicius A (2007) Anal Bioanal Chem 387:1899–1906
138. Brondani D, Scheeren CW, Dupont J, Vieira IC (2009) Sens Actuators B 140:252–259
139. Fatibello-Filho O, Lupetti KO, Vieira IC (2001) Talanta 55:685–692
140. Bezerra VS, Filho JLDL, Montenegro MCBSM, Araújo AN, Silva VLD (2003) J Pharm Biomed Anal 33:1025–1031
141. Babu VRS, Patra S, Karanth NG, Kumar MA, Thakur MS (2007) Anal Chim Acta 582: 329–334
142. Ben-Yoav H, Biran A, Pedahzur R, Belkin S, Buchinger S, Reifferscheid G, Shacham-Diamand Y (2009) Electrochim Acta 54:6113–6118

143. Akyilmaz E, Yaoa Ã, Dinçkaya E (2006) Anal Biochem 354:78–84
144. Caruso ClS, da Cruz Vieira I, Fatibello Filho O (1999) Anal Lett 32:39–50
145. Zhu Y, Son JI, Shim YB (2010) Biosens Bioelectron 26:1002–1008
146. Pemberton RM, Hart JP, Mottram TT (2001) Biosens Bioelectron 16:715–723
147. Paleček E, Fojta M, Tomschik M, Wang J (1998) Biosens Bioelectron 13:621–628
148. Batchelor-McAuley C, Wildgoose GG, Compton RG (2009) Biosens Bioelectron 24:3183–3190
149. Lucarelli F, Marrazza G, Turner APF, Mascini M (2004) Biosens Bioelectron 19:515–530
150. Ozsoz MS (2012) Electrochemical DNA biosensors. Pan Stanford, Singapore
151. Nawaz H, Rauf S, Akhtar K, Khalid AM (2006) Anal Biochem 354:28–34
152. Dogan-Topal B, Uslu B, Ozkan SA (2009) Biosens Bioelectron 24:2358–2364
153. Kim YJ, Kim Y, Niazi J, Gu M (2010) Bioprocess Biosyst Eng 33:31–37
154. Ferapontova EE, Gothelf KV (2009) Electroanalysis 21:1261–1266
155. Ovádeková R, Jantová S, Letašiová S, Štepánek I, Labuda J (2006) Anal Bioanal Chem 386: 2055–2062
156. Yu Y, Zhou Y, Wu L, Zhi J (2012) Int J Electrochem Sci 2012:1–10
157. Gu T, Hasebe Y (2012) Biosens Bioelectron 33:222–227
158. Prieto-Simón B, Campàs M, Marty JL (2010) Bioanal Rev 1:141–157
159. Mairal T, Oezalp VC, Sanchez PL, Mir M, Katakis OSCI (2008) Anal Bioanal Chem 390: 989–1007
160. Radi AE (2011) Int J Electrochem Sci 2011:1–17
161. Ferapontova EE, Olsen EM, Gothelf KV (2008) J Am Chem Soc 130:4256–4258
162. Swensen JS, Xiao Y, Ferguson BS, Lubin AA, Lai RY, Heeger AJ, Plaxco KW, Soh HT (2009) J Am Chem Soc 131:4262–4266
163. Ferguson BS, Hoggarth DA, Maliniak D, Ploense K, White RJ, Woodward N, Hsieh K, Bonham AJ, Eisenstein M, Kippin TE, Plaxco KW, Soh HT (2013) Sci Transl Med 5: 213ra165
164. Gherghi IC, Girousi ST, Voulgaropoulos A, Tzimou-Tsitouridou R (2004) Int J Environ Anal Chem 84:865–874
165. Brett AMO, Serrano SHP, Macedo TA, Raimundo D, Helena Marques M, La-Scalea MA (1996) Electroanalysis 8:992–995
166. Wang J, Ozsoz M, Cai X, Rivas G, Shiraishi H, Grant DH, Chicharro M, Fernandes J, Palecek E (1998) Bioelectrochem Bioenerg 45:33–40
167. Wang L, Lin L, Ye B (2006) J Pharm Biomed Anal 42:625–629
168. Dogan-Topal B, Ozkan SA (2011) Talanta 83:780–788
169. Karadeniz H, Gulmez B, Sahinci F, Erdem A, Kaya GI, Unver N, Kivcak B, Ozsoz M (2003) J Pharm Biomed Anal 33:295–302
170. Vaníčková M, Lehotay J, Čižmárik J, Labuda J (2005) Bioelectrochemistry 66:125–127
171. Brett AMO, Serrano SHP, Gutz I, La-Scalea MA (1997) Bioelectrochem Bioenerg 42: 175–178
172. Brett AMO, Serrano SHP, Gutz I, La-Scalea MA, Cruz ML (1997) Electroanalysis 9: 1132–1137
173. Xu J, Zhu JJ, Zhu Y, Gu K, Chen HY (2001) Anal Lett 34:503–512
174. Jiang X, Lin X (2006) Bioelectrochemistry 68:206–212
175. Karadeniz H, Alparslan L, Erdem A, Karasulu E (2007) J Pharm Biomed Anal 45:322–326
176. Brett AMO, Macedo TRA, Raimundo D, Marques MH, Serrano SHP (1998) Biosens Bioelectron 13:861–867
177. Cheng W, Peng D, Lu C, Fang C (2008) Russ J Electrochem 44:1052–1057
178. Yousef Elahi M, Bathaie SZ, Kazemi SH, Mousavi MF (2011) Anal Biochem 411:176–184
179. Baker BR, Lai RY, Wood MS, Doctor EH, Heeger AJ, Plaxco KW (2006) J Am Chem Soc 128:3138–3139
180. White RJ, Phares N, Lubin AA, Xiao Y, Plaxco KW (2008) Langmuir 24:10513–10518
181. Zhu Y, Chandra P, Ban C, Shim YB (2012) Electroanalysis 24:1057–1064

182. Kauffmann JM, Guilbault GG, Blum LJ, Coulet P (eds) (1991) Potentiometric enzyme electrodes, handbook of biosensors. Marcel Dekker, New York, NY
183. Eppelsheim C, Aubeck R, Hampp N (1995) J Membr Sci 100:131–139
184. Nilsson H, Åkerlund AC, Mosbach K (1973) Biochim Biophys Acta 320:529–534
185. Papariello GJ, Mukherji AK, Shearer CM (1973) Anal Chem 45:790–792
186. Garcia-Villar N, Saurina J, Hernandez-Cassou S (2003) Anal Chim Acta 477:315–324
187. Saurina J, Hernandez-Cassou S, Alegret S, Fabregas E (1999) Biosens Bioelectron 14:67–75
188. Radomska A, Bodenszac E, Głab S, Koncki R (2004) Talanta 64:603–608
189. Poghossian A, Schöning MJ, Schroth P, Simonis A, Lüth H (2001) Sens Actuators B 76: 519–526
190. Liu J, Liang L, Li G, Han R, Chen K (1998) Biosens Bioelectron 13:1023–1028
191. Sharon E, Freeman R, Tel-Vered R, Willner I (2009) Electroanalysis 21:1291–1296
192. Seki A, Ikeda SI, Kubo I, Karube I (1998) Anal Chim Acta 373:9–13
193. García-Villar N, Saurina J, Hernández-Cassou S (2003) Anal Chim Acta 477:315–324
194. Fernandes JCB, Kubota LT, Neto GO (1999) Anal Chim Acta 385:3–12
195. Koncki R, Wałcerz I, Leszczyńska E (1999) J Pharm Biomed Anal 19:633–638
196. Cullen LF, Rusling JF, Schleifer A, Papariello GJ (1974) Anal Chem 46:1955–1961
197. Ferrini AM, Mannoni V, Carpico G, Pellegrini GE (2008) J Agric Food Chem 56:784–788
198. Kumar S, Kundu S, Pakshirajan K, Dasu VV (2008) Appl Biochem Biotechnol 151:653–664
199. Pellegrini GE, Carpico G, Coni E (2004) Anal Chim Acta 520:13–18
200. Poghossian A, Yoshinobu T, Simonis A, Ecken H, Lüth H, Schöning MJ (2001) Sens Actuators B 78:237–242
201. Katz E, Willner I (2003) Electroanalysis 15:913–947
202. González-Fernández E, de-los-Santos-Álvarez N, Lobo-Castañón MJ, Miranda-Ordieres AJ, Tuñón-Blanco P (2007) Biosens Bioelectron 26:2354–2360
203. Cullen DC, Sethi RS, Lowe CR (1990) Anal Chim Acta 231:33–40
204. Rezaei B, Askarpour N, Ensafi AA (2014) Talanta 119:164–169

Chapter 7
Electrochemical and Hyphenated Electrochemical Detectors in Liquid Chromatography and Flow Injection Systems for Drug Compound Analysis

Contents

Abbreviations

ACN	Acetonitrile
ACPG	Aminopropyl-controlled pore glass beads
AER	Anion exchange resin
amol	Attomole
BIA	Batch injection analysis
CFE	Porous carbon felt electrode
Chi	Chitosan
ChOx	Choline oxidase
CME	Chemically modified electrode
CNT	Carbon nanotube
CoHCF	Cobalt hexacyanoferrate
CSCNTs	Cup-stacked carbon nanotubes
ECL	Electrochemiluminescent
FIA	Flow injection analysis
GOx	Glucose oxidase
HMDE	Hanging mercury-drop electrode

S.A. Ozkan et al., *Electroanalysis in Biomedical and Pharmaceutical Sciences*, Monographs in Electrochemistry, DOI 10.1007/978-3-662-47138-8_7

HRP	Horseradish peroxidase
ITO	Indium tin oxide
LOX	Lysine oxidase
MBRS	Meldola's Blue-Reinecke Salt
MWCNT-COOH	Multiwalled carbon nanotube with carboxyl groups
NADH	Reduced form of nicotinamide adenine dinucleotide
NS	Not stated
OSA	1-Octanesulfonic acid
OSS	Sodium octyl sulfate
PB	Prussian blue
PDDA	Poly(diallyldimethylammoniumchloride)
PEDOT	Poly(3,4-ethylenedioxythiophene)
PGE	Pyrolytic graphite electrode
PNDGA	Poly(nordihydroguaiaretic acid)
PPy_{ox}	Overoxidized polypyrrole film
PVI	Poly(1-vinylimidazole)
Q	Quercetin
QDs	CdTe quantum dots
SDS	Sodium dodecyl sulfate
SIA	Sequential injection analysis
SPCE	Screen-printed carbon electrode
SWAdSV	Square-wave adsorptive-stripping voltammetry
SWCNT	Single-walled carbon nanotubes
SWV	Square-wave voltammetry
TBO	Toluidine blue O
TEA	Triethylamine
THF	Tetrahydrofuran

7.1 Introduction

Electrochemical detectors can be very well adapted to hydrodynamic conditions such as those encountered in liquid chromatography (LC), in capillary electrophoresis, or in flow injection analysis (FIA) applied to drug compound analysis [1–10]. Detectors with low cell volume based on conductivity [11–13], potentiometry [14, 15], amperometry (AD) [1–10], or pulsed amperometry (PAD) [16, 17] can be located online in close proximity, in a post-column configuration, to a liquid chromatographic setup or inserted as microelectrodes close to a capillary end [18, 19]. A miniaturized electrolysis cell can also be part of an online setup such as in a FIA system or of a pre- or post-LC column coupled to a mass spectrometer for electrochemical product generation and characterization [20–23]. Trends are directed to miniaturize the entire flow analytical system (e.g., lab on chip) and microfabricated electrochemical sensors as ideally suited for such configurations [24–26]. This chapter describes some basic concepts of most used and commercially available amperometric detectors and reactors coupled to LC or FIA. Examples are collected from authors' laboratory, and a nonexhaustive literature survey of applications to the assay of molecules of pharmacological interest is provided.

7.2 Basic Analytical Strategy for Implementing an Amperometric Detector Under Hydrodynamic Conditions

The decision for the selection of an electrochemical detector (ECD) in a flowing system should be based on the analytical performances achievable with the ECD compared to optical (UV–vis, fluorescence, refractive index, light scattering, etc.. . .) and mass spectrometry-based detectors. The final choice of instrumentation is generally depending on the available budget taking into account the performances of the detector in terms of ease of use, reliability, maintenance fees, and applicability to a broad spectrum of analytes or to very specific topics.

An amperometric detector allows for high signals to be obtained under hydrodynamic conditions: the total current is high, thanks to the Faraday constant (F), and is direct function of the concentration of the electroactive species expressed by the electrolysis efficiency (c_{in}–c_{out}) and the volume flow rate (f) eq. (7.1).

$$It = nFf(c_{\text{in}} - c_{\text{out}}) \tag{7.1}$$

c_{in} is the concentration of analyte entering the cell, c_{out} is the concentration of analyte leaving the cell, and n is the number of electrons transferred per analyte.

With respect to an ECD based on microelectrolysis (amperometry or coulometry), the fields of application, however, are limited to drug compounds which possess an electrophore, i.e., are electroactive at the selected working electrode in the selected mobile phase composition. Similarly to optical detectors, before implementing ECD in hydrodynamic conditions, preliminary experiments in a quiescent solution are required. This allows for the final selection of the detection wavelength or the potential applied to the working electrode (E_{app}) at the ECD. Alternatively the entire UV–vis spectrum can be recorded in-line by using a diode array detector (DAD) and the voltamperogram by using an array of detectors in series [27] or by rapid scanning voltammetry [28, 29]. Usually a voltammogram is recorded in order to check for the electroactivity of the analyte and to define the potential range at which the investigated compound is electroactive. In cyclic voltammetry, the capacity current can be high (depending on scan rate and electrode surface area) compared to amperometry; therefore, a relatively high concentration of analyte (0.05–1 mM) is required for well-defined faradaic currents to be obtained. It must be kept in mind that the higher the applied potential at an ECD in LC, the higher the background current and baseline noise. Thus, investigated molecules should preferably be electroactive in the potential range comprised between −0.4 and +0.9 V *vs* Ag/AgCl, 3 M KCl, for low limits of detection (LOD) and limits of quantification (LOQ) to be achieved. A wide electrochemical potential window, however, in addition to a very low background current, high resistance to deactivation via fouling, extreme electrochemical stability, and relative insensitivity to dissolved oxygen can be achieved using boron-doped diamond electrodes (BDDEs) under hydrodynamic conditions [30–34].

Yet, in order to precisely define the operational applied potential for electrochemical detection in LC, a hydrodynamic voltammogram needs to be recorded using the LC setup. Since the applied potential is constant, and the capacity current is low, the concentration of the injected analyte can be much lower compared to cyclic voltammetry (CV) and micromolar concentrations are accessible. A general rule of thumb for ECD in hydrodynamic experiments is the lower the analyte concentration, the less is the working electrode fouling, and the less electrode cleaning (renewing) is needed. The optimum applied potential is obtained by measuring the current at different potentials, under the final selected hydrodynamic experimental conditions, and tracing the resulting s-shaped (wave-shaped) curve (Fig. 7.1). The applied potential (E_{app}) is selected at the plateau of the curve corresponding to the diffusion-limiting current. At this potential, the signal is the highest and has slight electrode potential fluctuations, e.g., because of reference electrode stability problems will have negligible perturbation on the repeatability of the signal.

Implementation of an amperometric analytical strategy can be illustrated by studying D-cycloserine. D-cycloserine is a broad spectrum antibiotic used as a second line to treat tuberculosis. In complex media such as serum or in drug formulations, the drug compound must be analyzed by LC in order to achieve selectivity toward matrix components or impurities and degradation products such as cycloserine dimers (Fig. 7.2). D-cycloserine and its degradation products absorb UV light at 219 nm in phosphate buffer pH 2.8 and have been successfully analyzed in capsules by LC-UV [35]. The molecule, however, exhibits relatively low absorptivity for allowing sensitive assays to be performed in biological samples.

The voltamperogram of D-cycloserine at a carbon paste electrode (CPE) shows that the compound is oxidized with a peak potential close to 0.85 V (Fig. 7.3). The dimers are oxidized at potentials higher than 1 V (dimers I and II) or are not electroactive (dimer III). These data allow us to select two analytical options: (1) determination of only D-cycloserine by applying a potential lower than 1 V (approximately 150 mV more positive than D-cycloserine peak potential) and (2) determination of cycloserine and dimers I and II by applying a potential of +1.3 V.

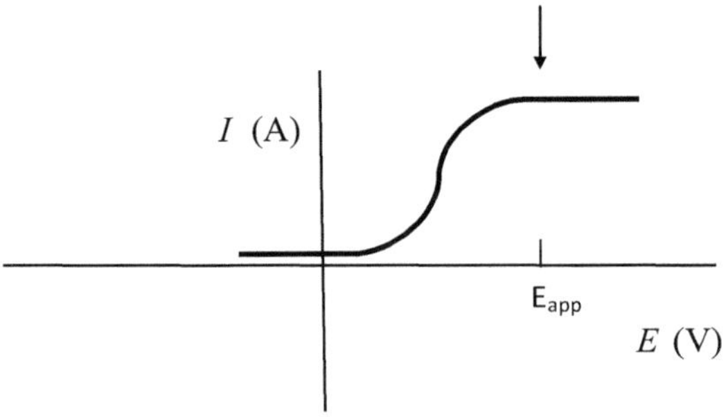

Fig. 7.1 Schematic s-type (wave-shaped) hydrodynamic voltammogram profile

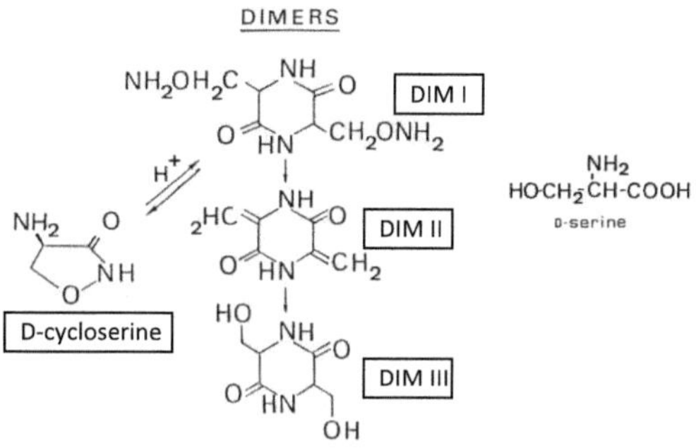

Fig. 7.2 Structure of D-cycloserine, some of its dimers, and D-serine

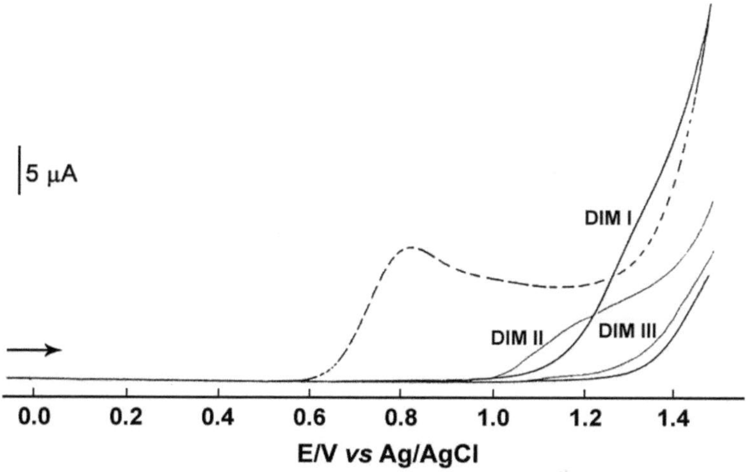

Fig. 7.3 Voltamperogram of D-cycloserine (*dotted lines*) and corresponding dimers (all molecules at 1 mM). Carbon paste electrode, phosphate buffer pH 6, scan rate 20 mV s^{-1} [unpublished data]

The use of a carbon paste-based EC detector in LC allows the selective detection assay of D-cycloserine in complex samples with a detection limit of 10 ng mL^{-1}. This compares favorably with LC-UV ($\lambda = 220$ nm) used under the same LC setup and same experimental conditions but with a detection limit of 500 ng mL^{-1} [data not published].

The are several basic rules to follow when using an amperometric detector in LC, namely selection of an electrolyte electrochemically inert in a large range of potentials, preferential use of an isocratic elution, highly stable hydrodynamic conditions, mobile phase preferably deoxygenated by nitrogen, and use of highly pure electrolytes. The mobile phase might contain traces of cations such as Fe(II), Fe(III), Ni(II), and Cr(III) which are electroactive and produce noisy backgrounds. A complexing agent such as EDTA (ethylenediaminetetraacetic acid) can be advantageously added in the flowing carrier provided that the applied potential is not too positive since EDTA is electroactive with an oxidation pattern being pH dependent; oxidation starts approx. at 0.6 V at pH 6 and at 0.9 V at pH 2 at a CPE [36]. Organic modifiers used in LC such as acetonitrile, methanol, and tetrahydrofuran are compatible with carbon-based electrodes such as the glassy carbon electrode (GCE) and the BDDE but are reactive at Pt and Au electrodes at positive potentials. The CPE contains a certain proportion of paraffin (liquid or preferably solid) and is compatible with a low amount (below 20 %, v:v) of a water-miscible organic solvent in the mobile phase [37, 38].

Similarly to an UV optical detector which exhibits wavelength exploration limits, the applied potential in amperometric detection is restricted to values where the mobile phase is not electroactive (Table 7.1). This depends on the nature of the working electrode and of the composition of the mobile phase. Most LC and FIA assays of pharmacologically active compounds are performed in aqueous-based solvents and positive potentials up to +1.2 V (*vs* Ag/AgCl) can be applied at the GCE and CPE and beyond +1.2 V at BDD electrodes [33]. In the negative potential range, provided dissolved oxygen has been removed in the mobile phase, potentials as low as –1.0 V can be achieved at carbon-based electrodes. The latter are usually preferred in electroanalysis of drug compounds [39], but for some applications such as for the detection of molecule containing glycans in their structure, gold or platinum electrodes are better suited provided an electrochemical surface cleaning (pulsed waveform) is repeatedly and automatically applied. Both the USP and European Pharmacopeia apply LC with (PAD) at a gold working electrode for the test of composition and related substances for antimicrobial compounds such as gentamicin, framycetin, amikacin, kanamycin, and tobramycin. A silver-based amperometric working electrode is the material of choice for halides, cyanide, thiocyanate, sulfide [40–42], and thiol-based compounds such as captopril, acetylcysteine, or aminothiols such as cysteine, glutathione, homocysteine, etc., with quantification limits in the submicromolar concentration range [6, 43, 44]. LC with carbon-based detectors such as those based on BDD, thanks to the high positive potentials applied, have been successful for the quantification of both oxidized and reduced forms of aminothiols [45]. Yet a silver-based amperometric detector with an applied potential as low as 0.08 V *vs* Ag/AgCl reacts selectively toward aminothiols and offers less complex chromatograms as those obtained with carbon-based detectors [44, 45]. The amperometric signal at the silver-based electrode corresponds to the reactivity of the analyte (halide, thiol ...) with silver

Table 7.1 Parallelism between voltammetry and molecular spectrophotometry

Method	Voltammetry	Spectrophotometry (UV–vis)
Signal generating unit of the molecule	Electrophore	Chromophore
Measured function	Voltammogram (I/E)	Spectrum (Abs/λ)
Experimentally controlled quantity	Applied potential (E)	Applied wavelength (λ)
Depending quantity	Current (I)	Absorption (Abs)
Quantity limiting the measurement range	Potential cutoff	Wavelength cutoff

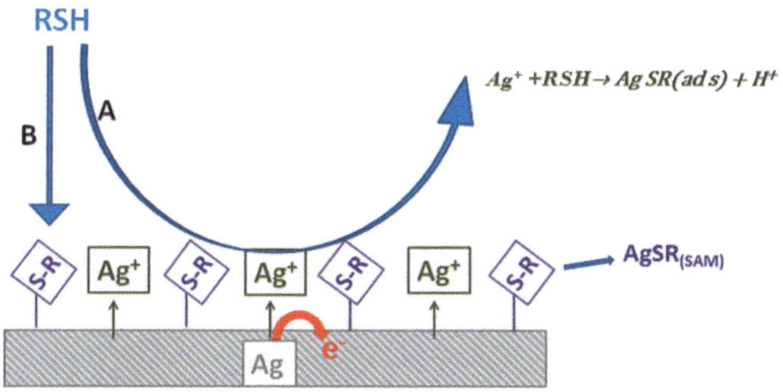

Fig. 7.4 Schematic illustration of the reaction of a small thiol-based species (RSH) at a silver electrode polarized at 0.08 V *vs* Ag/AgCl

ions at the electrode–solution interface which drives the electroxidative regeneration (i.e., the amperometric signal) of silver ions [45]. Chloride ions react with silver ions (Ag^{+}) with formation of a silver chloride deposit (AgCl) which is readily removed by the flowing mobile phase allowing for subsequent injections to be performed. Aminothiols react with both silver (Ag°) and silver ions: the former gives rise to an aminothiol self-assembled monolayer at the electrode ($AgSR_{SAM}$) while the latter gives a silver aminothiolate deposit ($AgSR_{ads}$) which is slowly removed under the LC flow (Fig. 7.4).

There is an extensive literature devoted to the development of modified electrodes and their application in hydrodynamic systems [9, 46–49], but no or few modified electrodes are commercially available for drug compound investigation. Limitations are essentially related to poor electrode storage and operational stability. The future, however, may benefit from the exponential interest and commercial availability of reliable screen-printed electrodes (SPEs). The latter can be modified by a great variety of (bio)catalyst and stability is not a critical factor since they are disposable. They are mass produced in a reproducible manner (RSD of geometric area below 5 %); there is no need for manual surface smoothing or renewing and no memory effects are encountered. All these criteria are of prime importance in routine analytical laboratories in the pharmaceutical industry. The suitability of a carbon-based SPE as amperometric detector for FIA [50] and LC [51] was illustrated for the assay of paracetamol in drug formulations.

7.3 Basic Theoretical Aspects [1, 52]

7.3.1 LC-Amperometric Detection (LC-AD)

Considering linear diffusion and a planar electrode surface, the limiting current for an electroactive species in a flowing system is given by the equation.

$$I_{\lim} = \frac{nFADC}{\delta}, \tag{7.2}$$

where D is the diffusion coefficient (cm^2/s), n is the number of electrons involved in the reaction, F is the Faraday constant, c is the concentration (mM), and A is the electrode surface (cm^2): the latter is changing in a quiescent solution, but it is almost stable under hydrodynamic conditions, i.e., when the diffusion of the analyte is controlled only by a change in the concentration of the analyte at the electrode–solution interface.

With respect to the inner geometry of cell, the limiting current is described for different cell designs, namely:

$$\underline{\text{Thin-layer cell}} \quad I_{\lim} = 1.47nFC\,(DA/b)^{2/3}\,f^{1/3}\,, \tag{7.3}$$

$$\underline{\text{Wall-jet cell}} \quad I_{\lim} = 0.89nFCD^{2/3}\nu^{-5/12}a^{-1/2}A^{3/8}\,f^{3/4}\,, \tag{7.4}$$

where f is the mean volume flow rate (cm^3/s), a the inlet diameter (cm), b the channel height, i.e., controlled by the gasket thickness (mm), A the electrode area, D the diffusion coefficient (cm^2/s), and ν the kinematic viscosity (cm^2/s).

Different models of EC cells are available for a flowing stream. The design of an electrochemical cell has to consider the requirements of low dead volume, high signal, low noise, and ease of construction and maintenance. Counter and reference electrodes have to be located as close as possible and preferably at the downstream side. Three major kinds of cell geometry are generally used: thin-layer, wall-jet, and "coulometric" cells.

7.3.2 *Wall-Jet*

The analyte jets perpendicularly to the working electrode surface and then flows radially over the surface (Fig. 7.5). The arrival of the analyte is governed by the radial laminar velocity profile. Approximately 30 % of the analyte can participate in

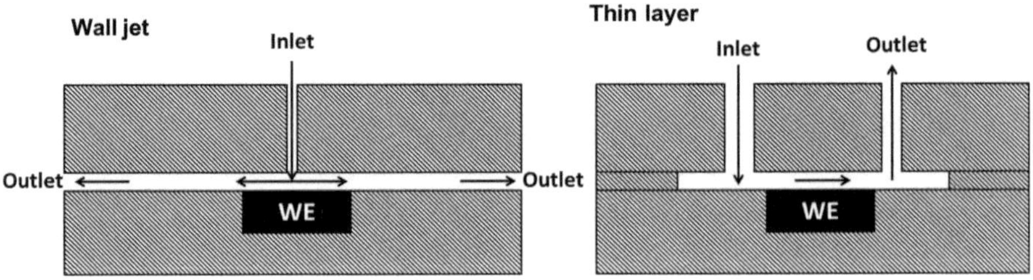

Fig. 7.5 Common detector configurations: wall-jet and thin-layer flow cells. *WE* working electrode

the response. The electrode surface has to be tenfold larger or more than the jet diameter. There is no major effect of the cell size except that it has to be big enough. The presence of an air bubble has no critical effect on the signal and this detector is less affected by surface adsorption compared to thin-layer cells due to the cleaning effect of the rapid incoming liquid jet.

7.3.3 *Thin-layer*

The solution flows parallel to the working electrode surface planar (Fig. 7.5). The detector body is formed from two solid blocks separated by a thin gasket; one of them contains the working electrode. To improve sensitivity, it is recommended to maximize electrode area and minimize cell volume (100 nL or less) by using the thinnest possible gasket and to use a high laminar flow rate. The percentage of the analyte which participates in the response is function of the flow velocity; it is generally lower than 20 % for mobile phase flow rates comprised between 0.7 and 1 mL $\min^{-1}$.

7.3.4 *"Coulometric"*

This configuration allows the laminar diffusion of the mobile phase through a porous graphitic working electrode structure. The electrode design permits to achieve electrolysis efficiency close to 100 % and the detector is often called "coulometric." The limiting current is expressed as in (7.1) with c_{out} being zero, i.e.,

$$I_{\text{lim}} = knFUc_{\text{in}}. \tag{7.5}$$

This detector is much more susceptible to fouling and there is usually no improvement in signal-to-noise ratio compared to the amperometric detectors described above. Of particular interest is that a porous guard electrode can be located online upfront and polarized at a lower potential compared to the ECD. This dual serial configuration allows, e.g., the oxidative removal of interfering species. The "coulometric" detector concept is very well suited when coupled online with a mass spectrometer for electrochemical products' identification (see below) [21, 27].

7.4 LC-Pulsed Amperometric Detection (LC-PAD) [53, 54]

PAD is an amperometric technique used essentially in an electrochemical detector operating under hydrodynamic conditions. It is based on the application of an automatic repetitive potential sequence on the working electrode for detection

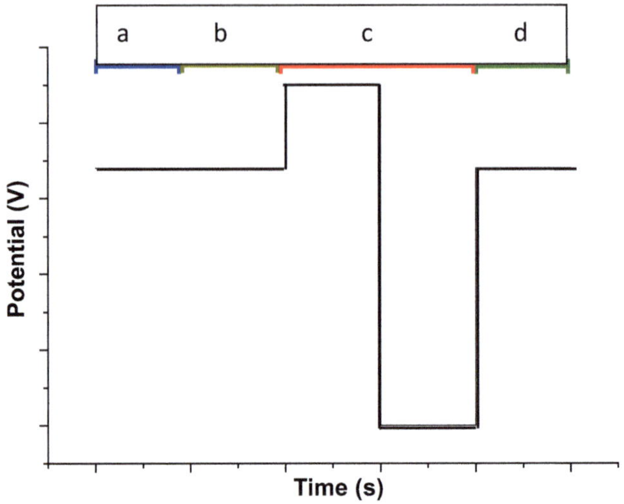

Fig. 7.6 Typical PAD wave form, relaxing time (**a**), measuring time (**b**), electrochemical cleaning time (**c**), and reconditioning time (**d**)

and surface cleaning. The potential sequence is called waveform. One of the applied potentials has to be capable of oxidizing the analyte to give the signal. After oxidation, an electrochemical cleaning step is applied in order to remove the formed layer on the surface (adsorbed analyte) and renew the electrode surface to its intial "naked" stage. The waveform consisted of three essential steps: measuring, cleaning, and relaxing (Fig. 7.6). The duration of one waveform is usually less than two seconds. PAD at Pt- or Au-based ECD is appropriate for the determination of amines, thiols, alcohols, and carbohydrates.

7.5 Hyphenated Techniques for LC and FIA

The possibility of simulating the oxidative metabolism of a drug compound is only one major reason for the renewed increasing attraction for the online coupling between electrochemistry and electrospray mass spectrometry (EC/MS) at both the industry and academia levels. Although this combination has been known from a few publications for a long time, it is now reaching an interesting investigation phase thanks to the continuous and important improvements in mass spectrometry coupled to liquid flow systems. Some very attractive new applications for EC/MS, in combination with (LC) and (FIA), have recently been reviewed [20–23, 55–57].

EC/MS with LC is useful for two different situations. On the one hand, the investigation of the electrochemical conversion of any single substance out of a complex mixture requires a separation. The favored procedure consists first of a

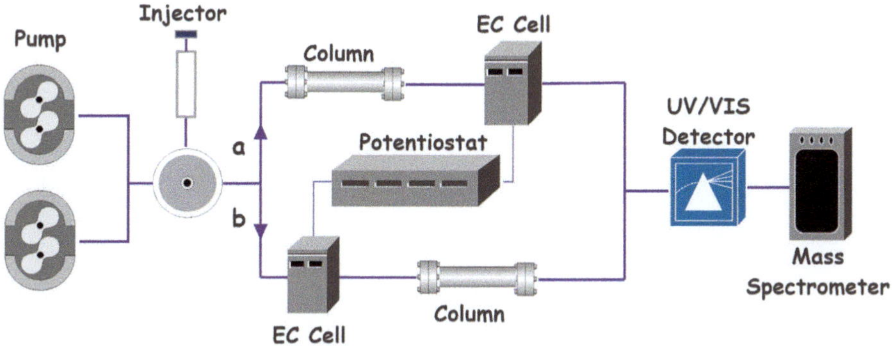

Fig. 7.7 Schematic setup of LC-EC/MS (route a) and EC/LC/MS (route b), including an optional UV/Vis detector prior to the mass spectrometer (adapted from [57])

reversed-phase LC separation of the individual constituents followed by electrochemical online post-column treatment and mass spectrometric detection (Fig. 7.7, route a). In this way, only the reaction products of one single compound are detected at one time. On the other hand, the separation of the oxidation products may provide useful information, for example, on the polarity of the individual products. In this case, a pure starting material is first oxidized electrochemically, and the products are analyzed by LC-MS (Fig. 7.7, route b). The integration of an additional UV/Vis detector prior to the mass spectrometer is possible in both cases. An interesting aspect is that LC-EC/MS detection is unrestrictedly compatible with gradient elution in LC, thus allowing for good LC separations to be obtained.

Porous graphite-based cells and flow through thin-layer cells with a large working area/cell volume ratio are suitable online in LC and FIA setups coupled with mass spectrometric detection of electrochemically generated products. The latter are formed by electrons' removal (in the case of an oxidation) from the investigated analyte (i.e., a faradaic process) at the working electrode–solution interface giving rise to, e.g., sulfoxidation, N-dealkylation, and/or aromatic ring hydroxylation [58, 59]. Production of hydrogen peroxide at the counter electrode can also generate interesting and metabolic relevant oxygenated species as recently demonstrated for lidocaine [60]. Instead of online flow system, an off-line EC-MS application can also be performed as illustrated by the voltammetric investigation of paracetamol at a carbon-based screen-printed electrode. The electrolysis of paracetamol in the absence or presence of glutathione in a few microliters of sample (50 μL) deposited on the SPE with subsequent injection into the LC-MS setup permitted to identify paracetamol phase I and phase II metabolites [61].

The coupling of an electrochemical cell with a nuclear magnetic resonance (NMR)-flow cell has also been recently reviewed [62] and can be regarded as an additional useful approach for the identification of electrochemically generated products as illustrated for paracetamol [63].

Tables 7.2 and 7.3 show the LC-EC and FIA-EC with their hyphenated EC-MS detection systems used in the methods, respectively.

Table 7.2 Application examples and their details of LC with EC and EC/MS examples on drug analysis

Analyte	Column type	Conditions	Applied potential (V)	Electrode type/ Detector	Linearity range	LOD/LOQ	Application media	References
Isoniazid Rifampicin	RP C18 column (150 × 4.6 mm, 5 μm)	Methanol: pH 6.0 phosphate buffer (65:35, v/v) and methanol: pH 6.0 phosphate buffer (7:93, v/v); (gradient elution); flow rate 1.0 mL min^{-1}	+0.9	GCE	0.01–100 μM	0.3 nM 0.5 nM	Pharmaceutical formulations, urine sample	[64]
Green tea catechins Gallic acid	RP C18 column (250 × 4.6 mm, 5 μm	ACN: pH 2.5 phosphate buffer (1:99, v/v) and ACN: pH 2.5 phosphate buffer (23:77, v/v) (gradient elution); flow rate 1.0 and 1.5 mL min^{-1}	E_1: −0.05 E_2: +0.2	Coulometric detector	10–1,000 ng mL^{-1}	0.4–3.1 ng mL^{-1}	Serum sample	[65]
Micronomicin	BDS C18 column (250 × 4.6 mm, 5 μm)	ACN: pH 2.6 aqueous solution (1:99, v/v); flow rate 0.3 mL min^{-1}	+0.05	Gold electrode	0.25–60 μg mL^{-1}	0.08 μg mL^{-1}	Pharmaceutical formulations	[66]
Serotonin Dopamine Norepinephrine	300SB C18 RP column (250 × 4.6 mm, 5 μm)	75 mM NaH_2PO_4: 1.7 mM OSA: 0.01 % triethylamine (v/v): 25 mM EDTA: 10 % ACN (v/v) (pH 3.5); flow rate 1.0 mL min^{-1}	+0.30 −0.15	Coulometric detector	50–1,000 ng mL^{-1} 5–1,000 ng mL^{-1} 5–1,000 ng mL^{-1}	20 ng mL^{-1} 2 ng mL^{-1} 2 ng mL^{-1}	Urine sample	[67]
Phenolic acids	$TC^{-}C18$ column (250 × 4.6 mm, 5 μm)	Methanol and water (gradient elution); flow rate 0.6 mL min^{-1}	+0.9	GCE	3.9–250 μg mL^{-1} 3.7–110.9 μg mL^{-1} 7.2–594.5 μg mL^{-1}	3.7–8.5 ng mL^{-1}	Plasma, urine and feces samples	[68]

Monoamine neurotransmitters	XDB‾C18 column (2.1 × 150 mm, 5 μm)	Methanol: pH 5.0 phosphate buffer (10:90, v/v); flow rate 0.3 mL min^{-1}	+0.60	Acetylene black nanoparticles-modified GCE	1.0×10^{-10}–1.0×10^{-5} M	5.0×10^{-11}–4.5×10^{-10} M	Rat striatum sample	[69]
D‾penicillamine	RP C18 column (150 × 4.6 mm, 5 μm)	Methanol: pH 2.5 phosphate buffer (11.5:88.5, v/v); flow rate 1.0 mL min^{-1}	+0.95	GCE	5–500 ng mL^{-1}	1.5 ng mL^{-1}	Skin specimens	[70]
Purine metabolites	Microsorb‾MV RP C18 column (150 × 4.6 mm, 5 μm)	ACN (10 %): pH 3.5 potassium dihydrogen phosphate and 0.52 mM sodium 1‾pentanesulfonate (gradient elution); flow rate 1.0 mL min^{-1}	+0.15 +0.45	Coulometric detector	0.01–10 mg L^{-1} 0.001–50 mg L^{-1} 0.01–50 mg L^{-1} 0.005–50 mg L^{-1} 0.05–10 mg L^{-1}	0.001–0.01 mg L^{-1}	Brain tissue and serum samples	[71]
Clarithromycin	C8 column (100 × 4.6 mm, 5 μm)	ACN: 0.045 M phosphoric acid (37:63, v/v, pH 6.7); flow rate 1.2 mL min^{-1}	+0.85	GCE	0.05–5.0 μg mL^{-1}	0.05 μg mL^{-1}	Plasma sample	[72]
Berberine Jatrorrhizine Coptisine Palmatine	C18‾WR column (250 × 4.6 mm, 5 μm)	ACN: pH 7.0 phosphate buffer (60:40, v/v); flow rate 0.8 mL min^{-1}	+1.3	GCE	0.04–80 μM 0.06–80 μM 0.08–80 μM 0.07–80 μM	0.01 μM 0.02 μM 0.03 μM 0.03 μM	Rat plasma sample	[73]
Sildenafil Vardenafil *N*-desmethyl sildenafil *N*-desethyl vardenafil	C18 column (100 × 2.1 mm, 3 μm)	ACN: 20 mM sodium dihydrogen phosphate (30:70, v/v, pH 3.5); flow rate 0.2 mL min^{-1}	+1.52	BDDE	10–400 ng mL^{-1}	3 ng mL^{-1} 3 ng mL^{-1} 4 ng mL^{-1} 2 ng mL^{-1}	Plasma sample	[74]

(continued)

Table 7.2 (continued)

Analyte	Column type	Conditions	Applied potential (V)	Electrode type/ Detector	Linearity range	LOD/LOQ	Application media	References
Ascorbic acid Aminothiols Methionine	HS C18 column (250 × 4.6 mm, 5 μm)	Methanol: 0.05 % trifluoroacetic acid solution (5:95, v/v); flow rate 1.5 mL min^{-1}	+0.85	Gold electrode	0.2–10,000 ng mL^{-1}	60 pg mL^{-1} 55–80 pg mL^{-1} 65 pg mL^{-1}	Biological matrices	[75]
Lipoic acid Dihydrolipoic acid	HS C18 RP column (250 × 4.6 mm, 5 μm)	ACN: 0.05 M phosphate buffer (52:48, v/v, pH 2.4); flow rate 1.5 mL min^{-1}	+1.0	GCE	0.05–1,000 ng mL^{-1} 0.5–10,000 ng mL^{-1}	–	Plasma sample	[76]
L-dopa Carbidopa 3-O-methyldopa Entacapone	C18 column (250 × 4.6 mm, 5 μm) RP C8 column (150 × 4.6 mm, 5 μm)	Methanol: pH 2.88 phosphate buffer (8:92, v/v); flow rate 1.5 mL min^{-1}; methanol: ACN: pH 1.90 phosphate buffer (17.5:22.5:60, v/v/v); flow rate 1.3 mL min^{-1}	+0.80	GCE	100–4,000 ng mL^{-1} 25–4000 ng mL^{-1} 200–10,000 ng mL^{-1} 20–4,000 ng mL^{-1}	4 ng mL^{-1} 4 ng mL^{-1} 4 ng mL^{-1} 10 ng mL^{-1}	Plasma sample	[77]
Captopril	C18 column (15 cm × 4.1 mm, 5 μm)	ACN: pH 3.0 phosphate buffer (30:70, v/v); flow rate 1.0 mL min^{-1}	E_1: +0.60 E_2: +0.95	Coulometric detector	2–70 μg mL^{-1}	0.6 μg mL^{-1}	Pharmaceutical formulations	[78]
Tryptamines Phenethylamines Piperazines	TSK-gel ODS-100 V column (4.6 × 250 mm, 3 μm)	Methanol/ACN: phosphate buffer (gradient elution); flow rate 1.0; 0.5 and 0.8 mL min^{-1}	+1.0 +1.1 +1.2	Porous graphite electrode	1–25 μg mL^{-1} 3–40 μg mL^{-1} 5–60 μg mL^{-1} 3–200 μg mL^{-1} 0.1–2 μg mL^{-1} 1–10 μg mL^{-1}	17.1 ng–11.7 μg mL^{-1}	Legal substances in Japan	[79]

Lipoic acid	HS C18 RP column (250 × 4.6 mm, 5 μm)	ACN: pH 2.5 phosphate buffer (50:50, v/v); flow rate 1.5 mL min^{-1}	+1.0	GCE	0.001–10 μg mL^{-1}	200 pg mL^{-1}	Plasma sample	[80]
Nicotine	C18 column (250 × 4.6 mm, 5 μm)	0.15 M SDS: 6 % (v/v) Pentanol: pH 6.0 sodium dihydrogen phosphate: 0.001 M KCl; flow rate 1.0 mL min^{-1}	+0.8	–	0.03–2 μg mL^{-1}	0.17 ng mL^{-1}	Pharmaceutical formulations and biological fluid	[81]
6-,8-, and 10-Gingerols 6-Shogaol	RP C18 column (4.6 × 250 mm, S-5 μm)	ACN: water: ammonium acetate (59:39:2, v/v/v) and ACN: pH 4.5 ammonium acetate (98:2, v/v) (gradient elution); flow rate 0.8 mL min^{-1}	+0.60 +0.55 +0.50	Coulometric detector	0.1–5.0 μg mL^{-1}	0.037–0.075 μg mL^{-1}	Plasma sample	[82]
Dopamine Serotonin Their metabolites	MD column (150 × 3.2 mm)	ACN:aqueous mobile phase (75 mM monobasic sodium phosphate buffer, 0.5 mM EDTA, 0.81 mM sodium octylsulfonate, 5 % THF) (5:95, v/v); flow rate 0.37 mL min^{-1}	E_1: −0.1 E_2: +0.3	Coulometric detector	1–200 ng mL^{-1}	1 ng mL^{-1}	Cerebrospinal fluid	[83]

(continued)

Table 7.2 (continued)

Analyte	Column type	Conditions	Applied potential (V)	Electrode type/ Detector	Linearity range	LOD/LOQ	Application media	References
Kanamycin	Platinum EPS column (150 × 4.6 mm, 5 μm)	Sodium sulfate: sodium octanesulfonate: pH 3.0 phosphate buffer (gradient elution); flow rate 1.0 and 0.3 mL min^{-1}	E_{oxd}: +0.75 E_{red}: −0.15	Gold electrode	5–600 ng	1.7 ng	Pharmaceutical formulations	[17]
Danofloxacin Difloxacin Sarafloxacin (for veterinary use)	C18 column (3.9 × 150 mm, 4 μm)	ACN: 0.05 M $NaClO_4$ (20:80, v/v); flow rate 0.6 mL min^{-1}	+0.8	Coulometric detector and amperometric detector	10–150 $ng.g^{-1}$ 50–315 $ng.g^{-1}$ 25–100 $ng.g^{-1}$	<10 $ng.g^{-1}$	Chicken tissue samples	[84]
Risperidone 9-hydroxyrisperidone	Chiral-AGP column (100 × 4.0 mm, 5 μm)	Methanol: pH 6.2 phosphate buffer (15:85, v/v); flow rate: 0.8 mL min^{-1}	+0.50 +0.65 +0.95 +0.95	Coulometric detector	1.0–100 ng mL^{-1} 1.0–100 ng mL^{-1}	0.5 ng mL^{-1} 0.5 ng mL^{-1}	Plasma sample	[85]
Pyridoxamine -5′-phosphate Pyridoxamine Pyridoxal 5′-phosphate Pyridoxal Pyridoxine 4-pyridoxic acid	C18 column (250 × 4.6 mm, 5 μm)	Methanol (12 %, v/v): pH 3.2 sodium phosphate; flow rate: 1.0 mL min^{-1}	+0.35 +0.80	Coulometric detector	12.4–227.3 nM	0.8 nM 1.1 nM 1.5 nM 1.3 nM 1.1 nM 2.1 nM	Plasma and serum samples	[86]
Abacavir Didanosine Lamivudine Zidovudine	C18 UG120 column (250 × 4.6 mm, 5 μm)	ACN: 25 mM phosphate buffer (1:9, v/v, pH 8.0); flow rate: 0.6 mL min^{-1}	+2.0	BDDE	–	100 ng mL^{-1} 5 ng mL^{-1} 2 ng mL^{-1} 1 ng mL^{-1}	Plasma sample	[87]
Lincomycin	C8 column (100 × 2.1 mm, 5 μm)	ACN:methanol: pH 6.0 phosphate buffer (5:20:75, v/v/v); flow rate: 0.4 mL min^{-1}	+2.0	SPCE	1–50 μM	0.08 μM	Urine, feeds, honey, milk samples	[88]

Artesunate Dihydroartemisinin Amodiaquine Desethylamodiaquine	C4 column (4.6 × 100 mm, 5 μm)	ACN: pH 4.0 potassium dihydrogen phosphate (11:89, v/v); flow rate 1.0 mL min^{-1}	+0.8	Thin-layer dual GCE	20–1,600 ng mL^{-1}	8 ng mL^{-1}	Plasma sample	[89]
8-hydroxy-2′deoxyguanosine	C18 column (150 × 4.6 mm 5 μm)	Methanol (1 %): acetonitrile (2.5 %): 50 mM potassium dihydrogen phosphate: 2 mM KCl (pH 4.45); flow rate 0.5 mL min^{-1}	+0.6	GCE	50–400 nM	3 nM	Urine sample	[90]
Coenzyme Q10	Microsorb-MV RP C18 column (150 × 4.6 mm; 5 μm)	-	+0.6	Coulometric detector	116–2,317 nM	116 nM	Blood platelets	[91]
Zearalenone and its metabolites	RP C18 column (150 × 4.6 mm, 2.7 μm)	Methanol:water (60/40, v/v); flow rate 0.7 mL min^{-1}	+0.85	CNT-GCE	5–50 ng mL^{-1}	1.3–1.4 ng mL^{-1}	Human, bovine, and swine urine samples	[92]
Azithromycin	XDB-CN column (150 × 4.6 mm, 5 μm)	ACN: methanol: pH 6.0 phosphate buffer (20:20:60, v/v/v); flow rate of 1.0 mL min^{-1}	+1.2	Amperometric detector	10–400 ng mL^{-1}	10 ng mL^{-1}	Plasma sample	[93]
Pipecuronium bromide	CN column (250 × 4.6 mm; 5 μm)	ACN: pH 6.4 tetramethylammonium hydroxide (4:6, v/v); flow rate 1.5 mL min^{-1}	+0.8	Porous graphite electrode	50–1,000 ng mL^{-1}	8 ng mL^{-1}	Pharmaceutical formulations	[94]
Morphine	CN-RP column, (4.6 × 250 mm, 5 μm)	Methanol (10 %): 20 mM potassium dihydrogen phosphate (pH 3.0); flow rate 1.0 mL min^{-1}	+0.97	–	0.781–250 ng mL^{-1}	–	Serum sample	[95]

(continued)

Table 7.2 (continued)

Analyte	Column type	Conditions	Applied potential (V)	Electrode type/ Detector	Linearity range	LOD/LOQ	Application media	References
Artesunate Dihydroartemisinin	Eclipse XDB-C18 column (4.6 × 150 mm, 5 μm)	ACN: ESA Acid Metabolite A (40:60, v/v)	−0.4	Coulometric detector	2–1,000 ng mL^{-1} 2–1,000 ng mL^{-1}	1.17 ng mL^{-1} 0.69 ng mL^{-1}	Dog plasma sample	[96]
Amikacin	Hypurity Elite column (150 × 4.6 mm, 5 μm)	Sodium-1-octanesulfonate: anhydrous sodium sulfate: THF: pH 3.0 phosphate buffer (0.15: 2: 1.5; w:w:v), 1.0 mL min^{-1}	+0.05 +0.70 −0.15	Gold electrode	0.06–4.00 μg mL^{-1}	0.06 μg mL^{-1}	Cerebrospinal fluid	[97]
Serotonin 5-hydroxytryptophan 5-hydroxytryptophol 5-hydroxyindole-acetic acid	LC-18 DB column (15 cm × 4.0 mm, 3 μm)	Methanol (3 %): citric acid: sodium phosphate dibasic: EDTA (pH 3.18); flow rate 1.0 mL min^{-1}	+0.65	–	10–200 pg mL^{-1}	3.7 pg mL^{-1} 2.3 pg mL^{-1} 2.8 pg mL^{-1} 4.6 pg mL^{-1}	Plasma sample	[98]
Biotin	60 RP-select B column (250 × 4 mm; 5 μm)	ACN: pH 3.0 aqueous acetic acid (15:85; v/v); flow rate 1.0 mL min^{-1}	+0.6 +0.9	Coulometric detector	15–3,600 ng mL^{-1}	5 ng mL^{-1}	Pharmaceutical formulations	[99]
p-nitrophenol	C18 column (250 × 4.6 mm, 5 μm)	ACN: water (40:60, v/v, pH 2.0); flow rate 0.5 mL min^{-1}	−0.1	GCE	7–500 ng on column	1 ng on column	Urine sample	[100]
Rocuronium bromide Eight impurities	Silica column (250 × 4.6 mm, 5 μm)	ACN: pH 7.4 tetramethyl-ammonium hydroxide (1:9, v/v); flow rate 1.5 mL min^{-1}	+0.9	GCE	0.05–125 μg mL^{-1} 0.025–10 μg mL^{-1}	15 ng mL^{-1} 7–225 ng mL^{-1}	Pharmaceutical formulations	[101]

Serotonin Sertonin metabolite	C18 column (150×0.5 mm, 5 μm)	0.15 M sodium phosphate; 0.5 mM EDTA, 0.14 mM OSA, 8 % methanol; flow rate 8 μL.min^{-1}	+0.5	GCE	3×10^{-11}–10^{-8} M 3×10^{-10}–10^{-6} M	-	Rodent brain and human spinal cord samples	[102]
p-aminobenzoic acid Its metabolites	C18 column (250×4.6 mm, 5 mm)	Methanol: water (20:80, v/v, pH 4.41); flow rate 0.4 mL min^{-1}	+1.4	CPE	125–1,800 ng mL^{-1}	2 ng mL^{-1}	Urine sample	[103]
Teicoplanin component	C8 column (150×4.6 mm, 5 μm)	ACN: pH 4.4 phosphate buffer (15:85, v/v); flow rate 0.6 mL min^{-1}	E_1: +0.3 E_2: +0.8	Coulometric detector	0.025–10 μg mL^{-1}	1.0 ng mL^{-1}	Plasma and cerebrospinal fluid samples	[104]
Catechins	ODS-3 Capillary EX-Nano column (150×0.2 mm, 3 μm)	Methanol: water: phosphoric acid (27.5:72.5:0.5, v/v/v); flow rate 4.0 μL. min^{-1}	+0.85	GCE	–	54–89 amol	Plasma sample	[105]
Sildenafil *N*-desmethyl sildenafil	C8 column (150×4.6 mm, 5 μm)	ACN: methanol: 0.05 M phosphate buffer (18.5:34.5:47.0, v/v/v, pH 7.68); flow rate 1.3 mL min^{-1}	+1.2	GCE	7.858–449.032 ng mL^{-1} 8.675–130.122 ng mL^{-1}	7.858 ng mL^{-1} 8.675 ng mL^{-1}	Plasma sample	[106]
Rotigonite	C18 silica microbore column (150×1 mm, 3 μm)	ACN (35 %, v/v): 50 mM sodium dihydrogen phosphate dihydrate: pH 4.5 sodium octyl sulfonate; flow rate 30 μL.min^{-1}	+0.85	GCE	1–100 nM	1 nM	Rat brain sample	[107]

(continued)

Table 7.2 (continued)

Analyte	Column type	Conditions	Applied potential (V)	Electrode type/ Detector	Linearity range	LOD/LOQ	Application media	References
Vertilmicin sulfate Micronomycin sulfate Etimicin Sulfate Sisomicin sulfate	Poly(styrene-divinylbenzene) PLRP-S column (250 × 4.6 mm, 8 μm)	38 g.L^{-1} sodium sulfate: 0.4 g.L^{-1} sodium 1-octanesulfonate: 13 mL.L^{-1} THF: pH 3.0 phosphate buffer; flow rate 1.0 mL min^{-1}	E_1: +0.05 E_2: +0.75 E_3: −0.15	Gold electrode	0.005–0.025 mg mL^{-1} 0.005–0.020 mg mL^{-1} 0.02–0.15 mg mL^{-1} 0.002–0.012 mg mL^{-1}	1.0 μg mL^{-1} 1.1 μg mL^{-1} 1.0 μg mL^{-1} 0.5 μg mL^{-1}	–	[108]
Norepinephrine Epinephrine Dopamine	RP column (150 × 4.6 mm, 5 μm)	ACN: 50 mM potassium dihydrogen phosphate, 2.5 mM sodium octyl sulfonate, 0.1 g.L^{-1} EDTA (3.5:96.5, v/v, pH 3.5); flow rate 1.2 mL min^{-1}	+0.6	–	15–500 μg L^{-1} 5–500 μg L^{-1} 50–500 μg L^{-1}	15 μg L^{-1} 5 μg L^{-1} 50 μg L^{-1}	Urine sample	[109]
Levodopa methyl ester Levodopa 3-methoxytyrosine L-tyrosine	RP C18 column (150 × 4.6 mm, 5 μm)	Methanol:ACN: 50 mM potassium dihydrogen phosphate (8:2:90, v/v/v, pH 3.2); flow rate 0.8 mL min^{-1}	+0.9	–	0.0206–10.34 μg mL^{-1} 0.0108–0.657 μg mL^{-1} 0.00864–1.80 μg mL^{-1} 0.00612–1.28 μg mL^{-1}	6 ng mL^{-1} 3.15 ng mL^{-1} 2.04 ng mL^{-1} 2.88 ng mL^{-1}	Pharmaceutical formulations	[110]
Ethyl glucuronide	RP C16 column (250 × 4.6 mm, 5 μm)	Acetic acid: t-butanol (99.5:0.5, v/v); flow rate 1.0 mL min^{-1}	+0.1	Gold electrode	5.9–49.9 μg mL^{-1}	0.01 μg mL^{-1}	Urine sample	[111]

Olanzapine	RP C8 column (150 × 4.6 mm, 5 μm)	Methanol: pH 3.5 phosphate buffer (21:79, v/v); flow rate 1.2 mL min^{-1}	+0.35	Porous graphite electrode	0.2–100.0 ng mL^{-1}	0.2 ng mL^{-1}	Rat brain tissue sample	[112]
Metoclopramide Hydrochlorothiazide Imipramine Diclofenac	C18 column (125 × 4 mm, 5 μm)	Methanol: pH 3.0 phosphate buffer (30:70, v/v) and 0.01 M acetic acid and methanol: pH 3.0 phosphate buffer (60:40, v/v) methanol: 0.008 M phosphoric acid (68:32, v/v); flow rates 1.0 mL min^{-1} and 1.2 mL min^{-1}	+1.10 +1.30 +1.10 +1.15	GCE	5–120 ng mL^{-1} 10–400 ng mL^{-1} 2–100 ng mL^{-1} 5–2,000 ng mL^{-1}	0.8 ng mL^{-1} 5.0 ng mL^{-1} 1.0 ng mL^{-1} 2.0 ng mL^{-1}	Serum or plasma samples	[113]
Clioquinol	C18 column (300 × 3.9 mm, 7 μm)	Methanol: ACN (1:1, v/v); flow rate 1.0 mL min^{-1}	E_1: +0.3 E_2: +0.7	Amperometric detector	5–2,000 ng mL^{-1} in plasma 10–1,000 ng g^{-1} in tissues	2 ng mL^{-1} in plasma 5 ng mL^{-1} in tissues	Hamster plasma and tissue samples	[114]
Levodopa 3-O-methyldopa Dopamine Carbidopa 3,4-dihydroxyphenyl acetic acid	C18 HR-80 column (8 cm × 4.6 mm, 3 μm)	Methanol: phosphate buffer: patented ion pairing agent; flow rate 0.7 mL min^{-1}	E_1: +0.05 E_2: −0.40	Coulometric detector	–	4 ng mL^{-1} 5 ng mL^{-1} 5 ng mL^{-1} 40 ng mL^{-1} 5 ng mL^{-1}	Plasma sample	[115]
Dopamine	HP ODS column (4.6 × 200 mm, 5 μm)	pH 5.4 phosphate buffer; flow rate 1.0 mL min^{-1}	+0.4	MWNT-COOH CME	5.0×10^{-9}–5.0×10^{-5} M	2.5×10^{-9} M	Rat striatal microdialysate	[116]
Ethyl glucuronide	RP C18 column (250 × 4.6 mm, 5 μm)	ACN: 1 % acetic acid (2:98, v/v); flow rate 1.0 mL min^{-1}	+0.2	Gold electrode	–	0.03 μg mL^{-1}	Urine sample	[117]
Nitrazepam	C18 column (250 × 4.6 mm, 5 μm)	Methanol: pH 4.1 acetate buffer (60:40, v/v); flow rate 1.0 mL min^{-1}	+0.5	GCE	0.070–11.24 μg	5 ng	Bovine and human serum samples	[118]

(continued)

Table 7.2 (continued)

Analyte	Column type	Conditions	Applied potential (V)	Electrode type/ Detector	Linearity range	LOD/LOQ	Application media	References
Diclofenac	C18 column (125 × 4 mm, 5 μm)	Methanol: water (68:32, v/v, pH 3.2); flow rate 1.0 mL min^{-1}	+1.15	GCE	5–2,000 ng mL^{-1}	2 ng mL^{-1}	Plasma sample	[119]
Epinephrine Norepinephrine Metanephrine Nor-metanephrine	C18 column (120 × 4.6 mm, 5 μm)	0.075 M SDS: 1.6 % butanol: pH 7 phosphate buffer; flow rate 1.0 mL min^{-1}	+0.7	Carbon electrode	0.5–50 ng mL^{-1}	2.7–17 pg mL^{-1}	Serum sample	[120]
Amitriptyline Nortriptyline	C18 column (250 × 4.6 mm, 5 μm)	0.15 M SDS: 6 % (v/v) pentanol (pH 7.0); flow rate 1.5 mL min^{-1}	+0.65	–	120–250 ng mL^{-1} 50–150 ng mL^{-1}	0.25 ng mL^{-1} 0.31 ng mL^{-1} in serum	Serum sample	[121]
Gliclazide	C18 column (150 × 4.6 mm, 5 μm)	70 mM: pH 7.5 disodium tetraborate; flow rate 0.8 mL min^{-1}	+0.8	GCE	50 nM–4.00 μM	10 nM	Plasma sample	[122]
Flavonoids Phenolic acids	RP-C18 column (4.6 × 150 mm, 5 μm) and XDB-C18 column (3.0 × 150 mm, 3.5 μm)	Methanol: pH 2.1 water (gradient elution); flow rate 0.4 mL min^{-1}	+1.1	Amperometric detector	–	1.45–22.27 nM	Plasma sample	[123]
Nalbuphine Butorphanol Morphine	C18 column (4.6 × 150 mm, 5 μm) for blood and C18 column (2 × 150 mm) for brain microdialysate	Methanol: pH 5.5 phosphate buffer (35:65, v/v); flow rate 1.0 mL min^{-1} and for morphine methanol: pH 4.0 phosphate buffer (25:75,v/v); flow rate 0.2 mL min^{-1}	+0.85 +0.85 +0.75 For morphine in brain +0.80	Amperometric detector	25–1,000 ng mL^{-1}	25 ng mL^{-1} 50 ng mL^{-1} 25 ng mL^{-1} 0.5 ng mL^{-1} for morphine in microdialysate samples	Blood and brain microdialysate samples	[124]

Homocysteine Cysteine Methionine	C18 column (25 cm × 4.6 mm, 5 μm)	Methanol: pH 3.0 phosphate buffer (5:95, v/v); flow rate 0.8 mL min^{-1}	+0.6 +1.2 +0.1	Platinum/poly (methyl violet)-modified electrode	2×10^{-7}–1×10^{-4} M 1×10^{-7}–1×10^{-4} M 4×10^{-6}–5×10^{-3} M	1.0×10^{-7} M 7.5×10^{-8} M 5.0×10^{-7} M	Hypothyroid patients' plasma sample	[125]
Glutathione	C18 column (25 cm × 4.6 mm, 5 μm)	Methanol: pH 3.0 sodium dihydrogen phosphate (2:98, v/v); flow rate 1.0 mL min^{-1}	+0.5	Porous graphite electrode	1–300 ng mL^{-1}	0.30 pmol	Human colon carcinoma cells and epidermoid human larynx carcinoma cells	[126]
Acetaminophen	RP C18 column (250 × 4.6 mm, 5 μm)	pH 7.0 sodium dihydrogen phosphate; flow rate 1.0 mL min^{-1}	+0.5	–	0.05–5 μg mL^{-1} in micellar solution, serum, and urine	0.56 ng mL^{-1} in micellar solution 0.83 ng mL^{-1} in serum 0.74 ng mL^{-1} in urine	Serum and urine samples	[127]
Minoxidil	C8 column (150 × 2 mm, 5 μm)	Methanol: pH 8.0 phosphate buffer (27:73, v/v); flow rate 0.4 mL min^{-1}	+0.8	GCE	30–500 ng mL^{-1}	1 ng mL^{-1}	Hamster skin follicles	[128]
Thyroxine Triiodothyronine Reverse triiodothyronine	YMC-Pack ODS-AM column (15 cm × 4.6 mm)	ACN: pH 6.0 phosphate buffer: (27.5:72.5, v/v); flow rate 1.0 mL min^{-1}	E_1: +0.3 E_2: +0.5	Coulometric detector	7.4–148.2 pM 1.5–74.1 pM 1.5–74.1 pM	0.10 pM 0.020 pM 0.025 pM	Plasma sample	[129]
Quercetin Kaempferol	C18 column (250 × 4.6 mm, 5 μm)	Methanol: ACN: $NaClO_4$ (30:30:40, v/v/v); flow rate 1.0 mL min^{-1}	+0.50	GCE	10–270 μg mL^{-1}	0.5 μg mL^{-1} 0.1 μg mL^{-1}	Phyto-pharmaceutical formulations	[130]

(continued)

Table 7.2 (continued)

Analyte	Column type	Conditions	Applied potential (V)	Electrode type/ Detector	Linearity range	LOD/LOQ	Application media	References
Imipramine Desipramine	C18 column (250 × 4.6 mm, 5 μm)	0.15 M SDS: 6 % (v/v) pentanol buffered pH 7.0; flow rate 1.5 mL min^{-1}	+0.65	–	50–1,000 ng mL^{-1}	0.34 ng mL^{-1} 0.24 ng mL^{-1}	Serum sample	[131]
Troglitazone stereoisomers	Semi-micro chiral column (1.0 × 150 mm, 5 μm)	Methanol:acetic acid (1000:1, v/v); flow rate 20 μL.min^{-1}	+0.7	–	0.1–50 ng mL^{-1}	1.3 fmol on column	Rat plasma sample	[132]
5-aminosalicylic acid Acetyl-5-aminosalicylic acid	C18 column (150 × 4.6 mm, 5 μm)	Methanol: 10 mM Na_2HPO_4 (10:90, v/v); flow rate 0.8 mL min^{-1}	E_1: −0.05 E_2: +0.45	Coulometric detector	50–3,200 ng mL^{-1}	5 ng mL^{-1} 10 ng mL^{-1}	Plasma sample	[133]
Serotonin Catecholamines	ODS column (150 × 1.0 mm, 3 μm)	ACN: pH 3.2 citrate buffer: (17.5:82.5, v/v); flow rate 100 μL.min^{-1}	+0.75	GCE	10–100 ng mL^{-1} –	– –	Biological tissue sample	[134]
Oxycodone Hydrocotarnine	RP18 column (4.6 × 50 mm, 5 μm)	Methanol: ACN: pH 8.0 phosphate buffer (2:1:7, v/v/v)	+0.8	Coulometric detector	10–100 ng mL^{-1}	0.5 ng mL^{-1} 0.6 ng mL^{-1}	Cancer patient serum sample	[135]
Pralidoxime methylsulfate	ODS column (3,250 × 4.6 mm)	Methanol: pH 2.6 phosphate buffer (8:92, v/v); flow rate 1.0 mL min^{-1}	E_1: +0.30 E_2: +0.95	Coulometric detector	0.25–50 μg mL^{-1}	0.2 μg mL^{-1}	Serum sample	[136]
Levodopa Dopamine Dihydroxyphenyl acetic acid Homovanillic acid	RP C18 column (4.6 × 250 mm, 5 μm)	Methanol (8 %, v/v): 0.05 M citric acid: 50 μM EDTA: pH 2.9 sodium octylsulfonate; flow rate 1.0 mL min^{-1}	E_1: +0.30 E_2: −0.20	Coulometric detector	10–100 μM	125 fmol 250 fmol 50 fmol 150 fmol	Striatum of freely moving rats	[137]

Glutathione L-Cysteine	C18 Varian column (4.6 × 250 mm, 5 μm)	pH 3.0 phosphate buffer; flow rate 1.0 mL min^{-1}	−1.15	Nanosized CoHCF chemically modified GCE	2×10^{-7}–2×10^{-4} M	1.2×10^{-7} M 1×10^{-7} M	Rat striatum	[138]
Dopamine Serotonin Homovanilic acid Hydroxyindoleacetic acid Tyramine	C18 column (120 × 4.6 mm, 5 μm)	0.15 M SDS buffered pH 3.0; flow rate 1.0 mL min^{-1}	+0.7	GCE	0.025–5 μg mL^{-1}	0.9 ng mL^{-1} 0.42 ng mL^{-1} 1.8 ng mL^{-1} 0.17 ng mL^{-1} 0.8 ng mL^{-1}	Serum sample	[139]
Gentamicin	C18 octadecylsilane (100 × 4.6 mm)	Methanol: water: acetic acid (67:28:5; v/v/v); flow rate 1.0 mL min^{-1}	+0.7	GCE	1–50 μg mL^{-1}	1 μg mL^{-1}	–	[140]
Sinistrin	RCX-10 anion exchange column, (250 × 4.1 mm)	50 mM NaOH; flow rate 1.0 mL min^{-1}	+1.0	Gold electrode	50–500 mg L^{-1}	2 mg L^{-1}	Serum sample	[141]
Daidzein Genistein	dC18 column (150–2.1 mm, 3.5 μm)	ACN: pH 5.5 acetate buffer (gradient elution); flow rate 0.4 mL min^{-1}	+0.45	Porous graphite electrode	0–10 μg mL^{-1} 0–600 pg mL^{-1}	1.8 nM 1.5 nM	Urine sample, soybean products	[142]
Hyperforin	RP-18 column (5 μm)	Methanol: ACN: pH 3.7 ammonium acetate buffer: (40:50:10, v/v/v); flow rate 0.8 mL min^{-1}	+1.1	GCE	0.054–5.4 μg mL^{-1}	0.050 ng on column	Pharmaceutical formulations (herbal)	[143]
Salbutamol	ODS column (150 × 4 mm, 5 μm)	Methanol: pH 6.0 sodium dihydroxy phosphate (10:90, v/v); flow rate 1.0 mL min^{-1}	+0.30 +0.40 +0.55 +0.65	Porous graphite electrode	0.84–84.0 ng mL^{-1} for plasma 4.2–420.0 ng mL^{-1} for urine	0.5 ng mL^{-1} for plasma 2.5 ng mL^{-1} for urine	Plasma and urine samples	[144]

(continued)

Table 7.2 (continued)

Analyte	Column type	Conditions	Applied potential (V)	Electrode type/ Detector	Linearity range	LOD/LOQ	Application media	References
Levodopa 3-O-methyldopa	HR80T column (4.6 × 80 mm, 3 μm)	7.8 g sodium dihydrogen phosphate dihydrate: 50 mg 1-heptanesulfonic acid: 50 μL reagent MB (pH 2.6); flow rate 1.5 mL min^{-1} and 2.0 mL min^{-1}	+0.10 +0.40 +0.05 -0.35	Coulometric detector	19–2,000 ng mL^{-1} 19–2,000 ng mL^{-1}	– –	Human plasma	[145]
Amlodipine Nitrendipine Felodipine Lacidipine Lercanidipine	ABZ+ Plus column (25 cm × 4.6 mm, 5 μm)	ACN: 10 mM acetate buffer (72:28, v/v, pH 5.0); flow rate 1.0 mL min^{-1}	+1.10	GCE	4.5–15 μg mL^{-1}	90 ng mL^{-1}–1.55 μg mL^{-1}	Pharmaceutical formulations	[146]
Norepinephrine 3-methoxy-4-hydroxyphenylglycol 3,4-dihydroxyphenylglycol	ODS C18 column (25 cm × 4.6 mm, 5 μm)	0.1 M sodium acetate: 0.02 M citric acid: 0.4 mM OSS: 0.2 mM EDTA (pH 4.85); flow rate 0.8 mL min^{-1}	+0.78	GCE	20–60 ng mL^{-1} 7–23 ng mL^{-1} 6–20 ng mL^{-1}	0.6 ng mL^{-1} 0.5 ng mL^{-1} 0.2 ng mL^{-1}	Rat cerebral cortex and plasma samples	[147]
Triclocarban	ProntoSIL 1203-C18-ace-EPS column 50 × 2 mm, particle size 3 μ, pore size 12 nm	–EC: 100 μM Triclocarban in ACN:1 mM ammonium acetate (50:50, (v/v); flow rate:500 μL. min^{-1} –LC/MS: methanol: water (90:10, v/v) (eluent A) and methanol (eluent B), gradient conditions	+2.5	BDDE	NS	NS	Human and rat liver microsome samples	[148]

Clozapine	C8 15 cm–2.1 mm, 5 μm particle size	–EC: ACN: pH 7.0 ammonium acetate (30:70, v/v) –LC/MS: 30 psi nebulizer gas, 8 L dry gas.min^{-1} of 365 °C and –5,000 V on capillary inlet	+0.4	CPE	NS	NS	Mechanism identification	[59]
Phenothiazine	RP-18 150 × 2.1 mm	–EC: ACN:pH 3.0 formate; flow rate: 0.3 mL min^{-1} gradient, –LC/MS: nebulizer gas flow rate 2.5 L. min^{-1}, probe temperature 450 °C, probe voltage 3 kV	+0.8 V	GCE	NS	5×10^{-8} M	Mechanism identification	[149]
Paracetamol	ProntoSIL phenyl column, 250 × 2.0 mm; 5 μm particle size	–EC: pH 7.0 ammonium acetate –LC/MS: 30 psi nebulizer gas, 50 psi dry gas/heating gas with a temperature of 400 °C, and an ionspray voltage of –4,500 V; flow rate 0.3 mL min^{-1}. MS/MS collision energy: –25 V, cell exit potential of –35 V, and collision-activated dissociation (CAD) gas pressure of 2 psi.	+0.6 V	GCE	NS	NS	Mechanism identification	[150]

(continued)

Table 7.2 (continued)

Analyte	Column type	Conditions	Applied potential (V)	Electrode type/ Detector	Linearity range	LOD/LOQ	Application media	References
Amodiaquine Amsacrine Mitoxantrone	Zorbax Eclipse XDB-C8 150 mm × 4.6 mm, 5 μm particle size	−EC-LC/MS: ACN: pH 7.4 ammonium formate (50:50, v/v); flow rate:50 μL. min^{-1} end plate offset, −200 V; capillary, 4,500 V; nebulizer gas (N_2), 1.6 bar; dry gas (N_2), 8.0 L. min^{-1}; dry temperature, 200 °C; capillary exit, 90.0 V; skimmer 1, 30.0 V; skimmer 2, 23.0 V; hexapole 1, 20.0 V; hexapole 2, 20.6 V; hexapole rf, 150 V; transfer time, 49.0 μs; prepulse storage, 5.0 μs; detector, −1,000 V	+0.7	GCE	NS	NS	Mechanism identification	[151]

Na-thyroxine and its degradation products	Merck LiChrospher 60 - RP-select B Column (5-μm packing, 125 × 4.6 mm i.d.).	−EC: ACN:water: pH 3.12 formic acid (40:60:0.5 v/v/v); flow rate: 0.7 mL min^{-1} −LC/MS: 0.3 % acetic acid in 40 % ACN, pH 3.27, full scan mode from mass 150–850 amu in 2 s. Vaporizer of the APCI and the heated capillary: 420 and 180 °C, CID (source) voltage of 12 eV	+0.8	Amperometric detector	0.1–100 ng mL^{-1}	0.1 ng μl^{-1}	Pharmaceutical formulations	[152]
Sodium phenylbutyrate and metabolites	4.6 × 250-mm Shiseido C18 5-μm column	−EC-LC/MS: Mobile phase A (0.01 M ammonium acetate) flow rate: 0.3 mL min^{-1}. Mobile phase B (60 % ACN/0.01 M ammonium acetate) flow rate: 0.2 mL min^{-1}, positive and negative ion scan modes (*m/z* 100–2,000, ion spray voltage 4.5–5.5 kV)	channels 1 to 12: 0–840 mV in 70-mV increments	Coulometric electrode	NS	NS	Mechanism identification	[153]

Table 7.3 The application examples and their details of FIA with EC examples on drug analysis

Analyte	Method type	Conditions	Applied Potential/Current	Electrode Type	Linearity Range	LOD/LOQ	Application media	References
S-Enalapril *S*-Ramipril *S*-Pentopril	SIA/Amperometry	0.1 M NaCl Flow rate:3.61 mL min^{-1}	+0.65 V	CPE	0.08–1.50 μM 0.12–0.60 μM 0.2–6.0 μM	4.33 nM 0.01 μM 0.1 μM	NS	[154]
2-nitrophenol 4-nitrophenol 2,4-dinitrophenol	FIA/Amperometry	Methanol:pH 4.0 Britton–Robinson buffer (4:1 v/v) Flow rate:0.5 mL min^{-1} Injection volume:20 μL	−1.0 V	Bismuth film-coated GCE	1.0–10.0 μg L^{-1} 1.0–10.0 μg L^{-1} 1.0–20.0 μg L^{-1}	0.3 μg L^{-1} 0.6 μg L^{-1} 0.7 μg L^{-1}	NS	[155]
Uric acid	FIA/Amperometry	pH 7.0 phosphate buffer Flow rate:0.1 mL min^{-1} Injection volume:5 μL	+0.5 V	poly(*N,N*-dimethylaniline) film-coated GCE	2–21 μM	2.0 μM	NS	[156]
Iodide	FIA/Amperometry	pH 5.0 phosphate buffer	+1.0 V	BDD thin film	0.8–200 μM	0.01 μM	Nuclear Emergency Pharmaceutical formulations	[157]
Urea	FIA/Conductimetry	pH 8.8 Glycine–NaOH Flow rate: 1.00 mL min^{-1} without dialyzer, 0.30 mL min^{-1} sample line with dialyzer 0.80 mL min^{-1} buffer line with dialyzer Injection volume:500 μL	-	Pt conductimetry	0.5–150 mM	0.5 mM	Serum sample	[158]

Insulin	FIA/Amperometry	pH 7.4 phosphate buffer (0.02 % Tween-80) Flow rate:1.0 mL min^{-1} Injection volume:200 µL	+0.8 V	CNT-modified GCE	100–1,000 nM	14 nM	NS	[159]
α-fetoprotein	FIA/Amperometry	pH 7.4 Tris–HCl buffer-0.1 % Tween-20 Flow rate:0.2 mL min^{-1}	0.0 V	SPCE	5–500 ng mL^{-1}	NS	Serum sample	[160]
Dopamine	FIA/Amperometry	pH 6.0 phosphate buffer Flow rate:1.2 mL. s^{-1} Injection volume:95 µL	+0.20	5,5-ditetradecyl-2-(2-trimethyl-ammonioethyl)-1,3-dioxane bromide self-assembled bilayer membrane-modified GCE	5×10^{-10}–5×10^{-4} M	5×10^{-11} M	NS	[161]
ß-D-glucose	FIA/Amperometry	pH 7.2 phosphate buffer -0.1 M NaCl Flow rate:841 µL. min^{-1} Injection volume:77 µL	+1.15 V	Tubular biosensor (graphite:resin (1:1) containing 5 % enzyme)	7.8×10^{-3}–0.5 M	5.6×10^{-3} M	Commercial medical glucose-containing solutions	[162]
Acetaminophen	FIA/Amperometry	1.2 mM EDTA-0.1 M KNO_3 (pH 6.5) Flow rate:3.89 mL min^{-1} Injection volume:100 µL	+0.6 V	PGE	2.0×10^{-5}–10.0×10^{-5} M	1.4×10^{-8} M	Pharmaceutical formulations	[163]

(continued)

Table 7.3 (continued)

Analyte	Method type	Conditions	Applied Potential/ Current	Electrode Type	Linearity Range	LOD/LOQ	Application media	References
Galactose	FIA/Amperometry	0.1 M NaOH Flow rate:1.0 mL min^{-1}	+0.45 V	GCE modified with SWCNT	100–600 μM	10 μM	NS	[164]
Hydrazine	FIA/Amperometry	pH 7.4 phosphate buffer Flow rate:500 μL. min^{-1} Injection volume:20 μM	+0.2 V	Copper–palladium alloy nanoparticle plated SPCE	2–100 μM	270 nM	Cigarette tobacco sample	[165]
Metam	FIA/Amperometry	pH 5.6 acetate buffer Flow rate:2.4 mL min^{-1} Injection volume:207 μL	+1.30 V	GCE, HMDE	1.3×10^{-2}–1.3 mg L^{-1}	1.3 μg.L^{-1}	Pharmaceutical formulations	[166]
Sertraline	FIA-SWAdSV	pH 8.2 borate buffer Flow rate:1.1 mL min^{-1} Injection volume:500 μL	-1.2 V	Multimode mercury electrode	0.20×10^{-6}–1.20×10^{-6} M	1.5×10^{-7} M	Pharmaceutical formulations	[167]
Fluvoxamine	FIA-SWAdSV	pH 2 H_3PO_4/H_2PO_4 buffer Flow rate:1.1 mL min^{-1} Injection volume:500 μL	−0.6 V	Mercury-drop electrode	5.00×10^{-7}–1.00×10^{-5} M	4.4×10^{-7} M	Pharmaceutical formulations serum samples	[168]
Paroxetine	FIA-SWAdSV	pH 8.8 borate buffer Flow rate:1.5 mL min^{-1}	−1.55 V	Mercury-drop electrode	3.00×10^{-6}–2.10×10^{-5} M	4.8×10^{-7} M	Pharmaceutical formulations	[169]

Lincomycin	FIA/Amperometry	pH 7 phosphate buffer, flow rate: 1 mL min^{-1} Injection volume:20 μL	+1.2 V	BDD thin film	0.5–125 μM	0.02 μM	Pharmaceutical formulations	[170]
Vitellogenin	SIA/Amperometry	pH 8.0 HCl–Tris buffer Flow rate:2 μL.s^{-1} Injection volume:300 μL	+0.2 V	NS	0–500 ppb	NS	NS	[171]
Minoxidil	FIA/Amperometry	pH 7.25 phosphate buffer Flow rate:0.74 mL min^{-1} Injection volume:20 μL	+0.80 V	GCE	1×10^{-7}–1×10^{-4} M	2.2×10^{-6} M	Pharmaceutical formulations	[172]
Ascorbic Acid	FIA/Amperometry	0.1 M H_2SO_4 Flow rate:4.6 mL min^{-1}	+0.35 V +0.75 V	CPE modified with a polyaniline film containing electrodeposited palladium	2.5 μmol–2.5 nmol 2.5 μmol–25 pmol	0.025 nmol 0.25 pmol	Pharmaceutical formulations	[173]
Cysteine	FIA/Pulsed	pH 10.0 phosphate buffer Flow rate:0.40 mL min^{-1} Injection volume:200 μL	−0.6 V *vs* saturated calomel electrode	Polycrystalline gold rod	1.0×10^{-6}–6.0×10^{-6} M	5.0×10^{-7} M	Pharmaceutical formulations	[174]
Epinephrine	FIA/Amperometry	pH 6.0 phosphate buffer Flow rate:2.0 mL min^{-1} Injection volume:100 μL	−0.10 V *vs* Ag/AgCl	CPE	5.0×10^{-5}–3.5×10^{-4} M	1.5×10^{-5} M	Pharmaceutical formulations	[175]

(continued)

Table 7.3 (continued)

Analyte	Method type	Conditions	Applied Potential/ Current	Electrode Type	Linearity Range	LOD/LOQ	Application media	References
Insulin	FIA/Amperometry	pH 7 phosphate buffer Flow rate:0.8 mL min^{-1} Injection volume:75 μL	+0.65 V	Three-dimensional chemically modified carbon ceramic electrode	100–500 pM	40 pM	NS	[176]
Isoniazid	BIA/Amperometry	0.1 M NaOH Flow rate:75.3 μL. s^{-1} Injection volume:100 μL	+0.15 V	GCE	2.5×10^{-8}–1.0×10^{-3} M	4.1×10^{-9} M	Pharmaceutical formulations	[177]
Vitamins B1 Vitamins B2 Vitamins B6	FIA/Amperometry	0.01 M H_2SO_4 Flow rate:4.25 mL min^{-1} Injection volume:600 μL	+0.90 V +1.00 V +0.80 V	Graphite Electrode-Ruthenium(III) Hexacyanoruthenate (II) film	0.5–50 μmol 1.0–50 μmol 0.2–50 μmol	0.2 μmol 0.5 μmol 0.1 μmol	Pharmaceutical formulations	[178]
p-nitrophenol	FIA/Amperometry	0.1 M Tris (pH 9.8, 1 M $MgCl_2$)–0.2 M, pH 7.2 phosphate buffer (2:1) Flow rate:4.2 mL min^{-1} Injection volume:50 μL	+1.1 V	SPCE	5×10^{-8}–5×10^{-6} M	2×10^{-8} M	Urine sample	[179]
Acetaminophen	FIA/Amperometry	pH 8.0 phosphate buffer Flow rate:2.0 mL min^{-1} Injection volume:150 μL	+0.6 V	Carbon film resistor electrodes	8.0×10^{-7}–5.0×10^{-4} M	1.4×10^{-7} M	Pharmaceutical formulation	[180]

Diclofenac sodium	FIA/Amperometry	pH 4.0 acetate buffer Flow rate:2.0 mL min^{-1}	−0.8 V	Bismuth film electrode	6.0–50.0 μM	4.3 μM	Pharmaceutical formulation	[181]
Acethylcholine	FIA/Amperometry	pH 7.0 phosphate buffer, pH 8.5 borate buffer Flow rate:0.1 mL min^{-1} Injection volume:20 μL	+0.7 V	Dual platinum electrode/ overoxidized poly(pyrrole) and poly(2-naphthol) films	100 nM–200 μM	100 nM	Brain tissue homogenates and cerebrospinal fluid samples	[182]
l-dopa	FIA/Amperometry	pH 7.0 phosphate buffer Flow rate:1 mL min^{-1}	+0.25 V	Gold nanoelectrode	5–300 ng mL^{-1}	3.0 ng mL^{-1}	Urine sample	[183]
L-lactate	FIA/Amperometry	pH 6.9 phosphate buffer Flow rate:0.85 mL min^{-1} Injection volume: 50 μL	-0.1 V	Prussian Blue-modified GCE	4.4–2240 μM	0.84 μM	Blood sample	[184]
Procaine	FIA/Amperometry	pH 6.0 sodium acetate Flow rate:3.8 mL min^{-1} Injection volume:100 μL	+0.80 V	SPCE	9.0×10^{-6}–1.0×10^{-4} M	6.0×10^{-6} M	Pharmaceutical formulation	[185]
Carcinoembryonic antigen α-fetoprotein β-human Choriogonadotropin Carcinoma antigen 125	FIA-Amperometry	pH 6.5 phosphate buffer Flow rate:3.6 mL min^{-1} Injection volume:120 μL	-0.5 V	Chi/TBO/SPCE	0.5–31.3 μg L^{-1} 0.4–41.7 μg L^{-1} 0.5–55.6 kIU.L^{-1} 0.4–44.4 IU.L^{-1}	0.19 μg L^{-1} 0.28 μg L^{-1} 0.28 kIU.L^{-1} 0.20 IU.L^{-1}	Serum samples	[186]

(continued)

Table 7.3 (continued)

Analyte	Method type	Conditions	Applied Potential/ Current	Electrode Type	Linearity Range	LOD/LOQ	Application media	References
Chloramphenicol	FIA/Amperometry	pH 6.0 phosphate buffer-1 % ethanol Flow rate:1 mL min^{-1} Injection volume:20 μL	−0.7 V	BDDE	0.1–50 μM	0.03 μM	Sterile eye drops and milk samples	[187]
Chlorine	FIA/Amperometry	pH 7.0 phosphate buffer Flow rate:1 mL min^{-1} Injection volume:20 μL	+1.1 V	BDDE	0.1–2 mg L^{-1}	8.3 μg L^{-1}	NS	[188]
Choline	FIA/Amperometry	pH 8.0 phosphate buffer Flow rate:1.5 mL min^{-1} Injection volume:20 μL	+0.6 V	Platinum electrode/ chitinous mem-brane/ choline oxidase	0.05–5.0 mM	1.0 μM	Serum sample	[189]
Dopamine	FIA/Amperometry	pH 7.0 phosphate buffer Flow rate:7.4 μL. min^{-1} Injection volume:0.15 μL	+0.2 V	Mercaptopropionic acid-modified elec-troless gold microelectrode	0.1–10 μM	74 nM	Pharmaceutical formulation	[190]
Interleukin-6	FIA/Amperometry	pH 7.2 phosphate buffer Flow rate:5 μL. min^{-1} Injection volume:25 μL	+0.10 V	Gold electrode	0–400 pg mL^{-1}	0.41 pg mL^{-1}	Serum sample	[191]

Bovine serum albumin	FIA/Amperometry	pH 7.4 phosphate buffer Flow rate:1 mL min^{-1}	+1.3 V	BDDE	50–400 μg mL^{-1}	190 ng mL^{-1}	NS	[192]
Dopamine	FIA/Amperometry	10 mM phosphate buffer Flow rate:1 mL min^{-1} Injection volume:5 μL	+0.15 V	GC/MWCNTs/Q/ Nafion electrode	1.4–300 μM	1.4 μM	Serum sample	[193]
Epinephrine Norepinephrine	FIA/Amperometry	Phosphate buffer Flow rate:3.5 mL min^{-1} Injection volume:10 μL	+0.172 V +0.244 V	MWCNTs–PNDGAChi SPE	$1x10^{-4}$–$1x10^{-1}$ M	NS	NS	[194]
Glucose	FIA/Amperometry	pH 7.4 phosphate buffer Flow rate:1 mL min^{-1} Injection volume:150 μL	−0.1	PEDOT[GOx]/PB	100 μM–1 M	100 nM	Serum sample	[195]
Glucose	FIA/Amperometry	pH 6.98 phosphate buffer Flow rate:0.4 mL min^{-1}	+0.6 V	PDDA/GOx/ AER	1 μM–2 mM	0.8 μM	Serum sample	[196]
Salbutamol	FIA/Amperometry	pH 8.3 Tris buffer Flow rate:40 μL. min^{-1} Injection volume:15 μL	+0.6 V	CNT/Si	5–100 μM	1 μM	Pharmaceutical formulation	[197]
Tamoxifen	FIA-SWV	0.05 M H_3PO_4 Flow rate:0.5 mL min^{-1} Injection volume:50 μL	-0.1 V	Gold wire	1.0×10^{-11}–3.0×10^{-6} M	3.0×10^{-12} M	Urine and plasma samples	[198]

(continued)

Table 7.3 (continued)

Analyte	Method type	Conditions	Applied Potential/ Current	Electrode Type	Linearity Range	LOD/LOQ	Application media	References
Choline	FIA/ electrochemiluminescence	pH 8.5 phosphate buffer-50 μM luminal Flow rate:1.0 mL min^{-1} Injection volume: 10 μL	+0.8	GCE/ChOx/ ACPG	0.05–2.0 μM	0.05 μM	Urine sample	[199]
Lactate	FIA/Amperometry	pH 10 phosphate buffer, 0.1 M KCl Flow rate:0.8 mL min^{-1}	+0.05 V	MBRS-SPCE	0.55–10 mM	0.55 mM	Serum sample	[200]
L-glutamate	FIA/Amperometry	pH 7.0 sodium phosphate buffer Flow rate:200 μL. min^{-1} Injection volume:20 μL	0.0 V	HRP/CSCNTs/GCE	3.0 nM–10 μM	1.2 nM	NS	[201]
N-acetylcysteine	FIA/Amperometry	0.1 M NaOH Flow rate:2.0 mL min^{-1} Injection volume:75 μL	0.0 V	Pyrolytic graphite rod	5.0×10^{-5}–1.0×10^{-3} M	9.0× 0.1 μM	Pharmaceutical formulation	[202]
Dopamine	FIA/ECL	pH 10.5 phosphate buffer –40 mM TEA–0.1 M KNO_3	pulse voltage with 0 V of threshold +1.1 V of limit for ECL excitation	QDs/CNTs/Chi/ ITO	10 pM–4 nM	3.6 pM	cerebro-spinal fluid sample	[203]

Gallic Acid	SIA/Amperometry	pH 7.0 phosphate buffer Flow rate:18 mL min^{-1} Injection volume:37.5 μL	−0.10 V	GCE	1–100 ppm	7.50 ppm	Ginger powder samples	[204]
Glucose	FIA/Amperometry	pH 7.4 phosphate buffer- 0.15 M NaCl Flow rate:0.8 mL min^{-1} Injection volume: 50 μL	+0.3 V	GOx-SWCNT conjugate/PVI-Os bilayer	0.3–10 mM	0.3 mM	NS	[205]
NADH	FIA/Amperometry	pH 7.0 phosphate buffer Flow rate:4.1 mL min^{-1} Injection volume:200 μL		Poly(TB) film-modified CFE	1–30 μM	0.3 μM	NS	[206]
Ascorbate Dopamine Norepinephrine	FIA/Potentiometry	pH 8.6 sodium citrate Flow rate:0.44 mL min^{-1} Injection volume:100 μL	1.0 μA	Polymeric membrane permanganate ion-selective electrode	1.0×10^{-5}–2.5×10^{-7} M	7.8×10^{-8} M 1.0×10^{-7} M 1.0×10^{-7} M	Vegetable and Pharmaceutical formulation	[207]
NADH	FIA/Amperometry	pH 7.0 phosphate buffer Flow rate:0.7 mL min^{-1} Injection volume:20 μL	+0.2 V	Poly (thionine)-modified disk electrode	20 μM–1 mM	1.7 μM	NS	[208]
Glucose Galactose Xylose Fucose	FIA/Amperometry	pH 4.5 KH_2PO_4 Flow rate:0.5 mL min^{-1}	+1.95 V	BDDE	5–50 nmol	105.1 pmol 101.2 pmol 96.1 pmol 52.9 pmol	NS	[209]

(continued)

Table 7.3 (continued)

Analyte	Method type	Conditions	Applied Potential/ Current	Electrode Type	Linearity Range	LOD/LOQ	Application media	References
Fructose Sorbitol Xylitol Glycerol Ethylene glycol Maltose Sucrose Lactose Raffinose Maltohexaose Maltoheptaose Cyclomaltohexaose Cyclomaltoheptaose		Injection volume:20 μL				25.8 pmol 10.2 pmol 11.7 pmol 12.7 pmol 10.8 pmol 11.2 pmol 10.0 pmol 10.7 pmol 9.9 pmol 10.0 pmol 7.5 pmol 9.8 pmol 5.0 pmol		
Glucose	FIA/Amperometry	pH 5.0 acetate buffer Flow rate:0.9 mL min^{-1} Injection volume:100 μL	+0.25 V	Pt/MoO_3/GCE/ GOx	0.05–0.5 mM	0.048 mM	NS	[210]
l-lysine	FIA/Amperometry	pH 5.0 acetate buffer Flow rate:0.6 mL min^{-1} Injection volume:20 μL	+0.7 V	Pt/PPy_{ox}/LOX	0.02–2 mM	4 μM	Pharmaceutical formulation	[211]
Codeine Promethazine	BIA/Amperometry	0.1 M acetate buffer Flow rate:4.5 mL min^{-1} Injection volume:150 μL	+1.30 V/ 50 ms	BDDE	8–40 mg L^{-1} 5–25 mg L^{-1}	0.135 mg L^{-1} 0.065 mg L^{-1}	Pharmaceutical formulation	[212]

7.6 Conclusion

Electrochemistry applied in the form of a detector combined with an analytical setup based on a flowing liquid such as in LC and FIA is facing continuous technological progress and appears to be complementary to optical detectors with respect to drug compounds' quantification. Electrochemistry provides unique mechanistic information when used as an electrolysis cells coupled online with mass spectrometry in terms of possibility to generate drug metabolites and to mimic the drug metabolism phase I and II patterns. Progress is expected in the use of specifically designed ECDs for microfluidic chip-based systems and for nano-LC applications.

References

1. Štulík K, Pacáková V (1987) Electroanalytical measurements in flowing liquids. Ellis Horwood, Chichester
2. Wang J (ed) (1988) Electroanalytical techniques in clinical chemistry and laboratory medicine. VCH, New York
3. Smyth MR, Vos JG (eds) (1992) Analytical voltammetry, vol 27. Elsevier, Amsterdam
4. Kissinger PT, Heineman WR (eds) (1996) Laboratory techniques in electroanalytical chemistry, 2nd edn. Marcel Dekker, New York
5. Christian GD, Dasgupta PK, Schug KA (eds) (2014) Analytical chemistry, 7th edn. Wiley, NJ
6. Flanagan RJ, Perrett D, Whelpton R (eds) (2005) Electrochemical detection in HPLC, Analysis of drugs and poisons. RSC Chromatography Monographs, Cambridge
7. Trojanowicz M, Szewczynska M, Wcislo M (2003) Electroanalysis 15:347–365
8. Wang C, Xu J, Zhou G, Qu Q, Yang G, Hu X (2007) Comb Chem High Throughput Screen 10:547–554
9. Siangproh W, Leesutthipornchai W, Dungchai W, Chailapakul O (2009) J Flow Inject Anal 26:5–25
10. Ozkan SA (2007) Chromatographia 66:S3–S13
11. Tsai E, Ip DP, Brooks MA (1992) J Chromatogr A 596:217–224
12. Ouyang J, Gao X, Baeyens WRG, Delanghe JR (2005) Biomed Chromatogr 19:266–271
13. DeBorba BM, Rohrer JS, Bhattacharyya L (2004) J Pharm Biomed Anal 36:517–524
14. Bazylak G, Nagels LJ (2002) J Chromatogr A 973:85–96
15. Daems D, Van Camp G, Fernandez Guisez Y, Prinsen E, Nagels LJ (2013) Anal Chim Acta 777:25–31
16. Hanko VP, Roher JS (2007) J Pharm Biomed Anal 43:131–141
17. Manyanga V, Dhulipalla RL, Hoogmartens J, Adams E (2010) J Chromatogr A 1217: 3748–3753
18. Polesello S, Valsecchi SM (1999) J Chromatogr A 834:103–116
19. Cvacka J, Quaiserová V, Park J, Show Y, Muck A Jr, Swain GM (2003) Anal Chem 75: 2678–2687
20. Jurva U, Wikström HV, Weidolf L, Bruins AP (2003) Rapid Commun Mass Spectrom 17: 800–810
21. Gamache PH, Meyer DF, Granger MC, Acworth IN (2008) Online electrochemical–LC–MS techniques for profiling and characterizing metabolites and degradants. In: Ramanathan R (ed) Mass spectrometry in drug metabolism and pharmacokinetics. Wiley, Hoboken, NJ
22. Faber H, Vogel M, Karst U (2014) Anal Chim Acta 834:9–21

23. Blankert B, Kauffmann J-M (2010) Electroanalytical methods as tools for predictive drug metabolism studies. In: Zacharis CK, Tzanavaras PD (eds) Reviews in pharmaceutical and biomedical analysis. Bentham Science Publishers, Sharjah, pp 76–83
24. Baldwin RP, Roussel TJ, Crain MM, Bathlagunda V, Jackson DJ, Gullapalli J, Conklin JA, Pai R, Naber JF, Walsh KM, Keynton RS (2002) Anal Chem 74:3690–3697
25. Lillehoj PB, Huang M-C, Truong N, Ho C-M (2013) Lab Chip 13:2950–2955
26. Krejcova L, Nejdl L, Merlos Rodrigo MA, Zurek M, Matousek M (2014) Biosens Bioelectron 54:421–427
27. Acworth IN, Naoi M, Parvez H, Parvez S (1997) Coulometric electrode array detectors for HPLC, vol 6. VSP Publishing, Utrecht, The Netherlands
28. Nagels LJ, Mush G, Massart DL (1989) J Pharm Biomed Anal 7:1479–1483
29. Nagels LJ, Kauffmann J-M, Dewaele C, Parmentier F (1990) Anal Chim Acta 234:75–81
30. Ghen G (2007) Talanta 74:326–332
31. Ivandini TA, Honda K, Rao TN, Fujishima A, Einaga Y (2007) Talanta 15:648–655
32. Martins I, de Souza CL, Doretto KM, Rath S (2010) Electroanalysis 22:455–462
33. Brocenschi RF, Rocha-Fihlo RC, Duran B, Swain GM (2014) Talanta 126:12–19
34. Bailey B, Waraska, Acworth I (2010) Advanced protocols in oxidative stress II. Method Mol Biol 594:327–339
35. Pendela M, Dragovic S, Bockx L, Hoogmartens J, Van Schepdael A, Adams E (2008) J Pharm Biomed Anal 47:807–811
36. Verbiese-Genard N, Kauffmann J-M, Hanocq M, Molle L (1984) J Electroanal Chem 170:243–254
37. Petit C, Gonzales-Cortes A, Kauffmann J-M (1995) Talanta 42:1783–1789
38. Pravda M, Petit C, Michotte, Kauffmann J-M, Vytras K (1996) J Chromatogr A 727:47–54
39. Gomes Cavalheiro ET, Brett CMA, Oliveira-Brett AM, Fatibello-Filho O (2012) Bioanal Rev 4:31–53
40. Jedral W, Hulanicki A (1988) J Electroanal Chem 251:359–367
41. Cheng J, Jandik P, Avdalovic N (2005) Anal Chim Acta 536:267–274
42. Kimbeni Malongo T, Patris S, Macours P, Cotton F, Nsangu, Kauffmann J-M (2008) Talanta 76:540–547
43. Sarakbi A, Aydogmus Z, Dago A, Mertens D, Dewert J-Y, Kauffmann J-M (2013) Anal Chim Acta 786:22–28
44. Sarakbi A, Kauffmann J-M (2014) Food Chem 153:321–326
45. Park HJ, Mah E, Bruno RS (2010) Anal Biochem 407:151–159
46. Wang E, Ji H, Hou W (1991) Electroanalysis 3:1–11
47. Baldwin RP, Thomsen KN (1991) Talanta 38:1–16
48. Trojanowicz M (2009) Anal Chim Acta 653:36–58
49. Trojanowicz M (2011) Anal Chim Acta 688:8–35
50. Stefano JS, Montes RHO, Richter EM, Munoz RAA (2014) J Brazil Chem Soc 25:484–491
51. Sarakbi A, Aydogmus Z, Sidali T, Gokce G, Kauffmann J-M (2011) Electroanalysis 23:29–36
52. Wang J (ed) (2006) Analytical electrochemistry. Wiley, Hoboken, NJ
53. LaCourse WR (1997) Pulsed electrochemical detection in high-performance liquid chromatography. Wiley, New York
54. Austin-Harrison DS, Johnson DC (2005) Electroanalysis 1:189–197
55. Gamache PH, Meyer DF, Granger MC, Acworth IN (2004) J Am Soc Mass Spec 15: 1717–1726
56. Álvarez-Lueje A, Pérez M, Zapata C (2012) Electrochemical methods for the in vitro assessment of drug metabolism. In: J Paxton (ed) Topics on drug metabolism, Dr. ISBN: 978-953-51-0099-7, InTech, Available from: http://www.intechopen.com/books/topics-on-drugmetabolism/electrochemical-methods-for-the-in-vitro-assessment-of-drug-metabolism
57. Karst U (2004) Angew Chem Internat Ed 43:2476–2478

58. Blankert B, Hayen H, van Leeuwen SM, Karst U, Bodoki E, Lotrean S, Sandulescu R, Dominguez O, Arcos J, Kauffmann J-M (2005) Electroanalysis 17:1501–1510
59. Van Leeuwen SM, Blankert B, Kauffmann J-M, Karst U (2005) Anal Bioanal Chem 382: 742–750
60. Nouri-Nigjeh E, Permentier HP, Bischoff R, Bruins AP (2010) Anal Chem 82:7625–7633
61. Kauffmann J-M, Van Antwerpen, Sarakbi A, Feier B, Sidali T, Aydogmus Z (2011) Electroanalysis 23:2643–2650
62. Bussy U, Boujtita M (2015) Talanta 136:155–160
63. Simon H, Melles D, Jacquoilleot S, Sanderson P, Zazzeroni R, Karst U (2012) Anal Chem 84:8777–8782
64. Yan H, Zhou Y, Xie Q, Zhang Y, Zhang P, Xiao H, Wang W, Yao S (2014) Anal Methods 6:1530–1537
65. Narumi K, Sonoda J-I, Shiotani K, Shigeru M, Shibata M, Kawachi A, Tomishige E, Sato K, Motoyaa T (2014) J Chromatogr B 945–946:147–153
66. Yuana Y, Chopra S, Deng X, Zhang M, Fan X, Hu C, Jin S, Schepdael AV, Adams E (2013) J Chromatogr A 1295:90–98
67. Oppolzer D, Moreno I, da Fonseca B, Passarinha L, Barroso M, Costa S, Queiroz JA, Gallardo E (2013) Biomed Chromatogr 27:608–614
68. Jia P, Wang S, Meng X, Lan W, Luo J, Liao S, Xiao C, Zheng X, Li L, Liu Q, Zheng J, Zhou Y, Zheng X (2013) Talanta 107:103–110
69. Lin L, Yang J, Lin R, Yu L, Gao H, Yang S, Li X (2013) J Pharm Biomed Anal 72:74–79
70. Saracino MA, Cannistraci C, Bugamelli F, Morganti E, Neri I, Balestri R, Patrizi A, Raggi MA (2013) Talanta 103:355–360
71. Burdett TC, Desjardins CA, Logan R, McFarland NR, Chen X, Schwarzschild MA (2013) Biomed Chromatogr 27:122–129
72. Zaater MF, Tahboub YR, Ghanem E (2012) J Chromatogr Sci 50:763–768
73. Liu L, Chen Z (2012) Anal Chim Acta 737:99–104
74. Bartosová Z, Jirovsky D, Horna A (2011) J Chromatogr A 1218:7996–8001
75. Khan MI, Iqbal Z (2011) J Chromatogr B 879:2567–2575
76. Khan A, Iqbal Z, Watson DG, Khan A, Khan I, Muhammad N, Muhammad S, Nasib HA (2011) Iqbal N, ur-rehman F, Kashif M. J Chromatogr B 879:1725–1731
77. Bugamelli F, Marcheselli C, Barba E, Raggi MA (2011) J Pharm Biomed Anal 54:562–567
78. Khamanga SM, Walker RB (2011) Talanta 83:1037–1049
79. Min JZ, Yamashita K, Toyo'oka T, Inagaki S, Higashi T, Kikura-Hanajiri R, Goda Y (2010) Biomed Chromatogr 24:1287–1299
80. Khan A, Khan MI, Iqbal Z, Ahmad L, Shah Y, Watson DG (2010) J Chromatogr B 878: 2782–2788
81. Chin-Chen M-L, Rambla-Alegre M, Durgavanshi A, Bosec D, Esteve-Romero J (2010) J Chromatogr B 878:2397–2402
82. Zick SM, Ruffin MT, Djuric Z, Normolle D, Brenner DE (2010) Int J Biomed Sci 6(3): 233–240
83. Hubbard KE, Wells A, Owens TS, Tagen M, Fraga CH, Stewart CF (2010) Biomed Chromatogr 24:626–631
84. Cáceres MIR, Cabanillas AG, Díaz TG, Canas MAM (2010) J Chromatogr B 878:398–402
85. Locatelli I, Mrhar A, Grabnar I (2009) J Pharm Biomed Anal 50:905–910
86. Marszałł ML, Lebiedzinska A, Czarnowski W, Makarowski R, Kłos M, Szefer P (2009) J Chromatogr B 877:3151–3158
87. Suzuki K, Katayama M, Takamatsu K, Kaneko S, Miyaji K, Ishikawa H, Matsuda Y (2009) J Chromatogr A 1216:3117–3121
88. Chiu M-H, Yang H-H, Liu C-H, Zen J-M (2009) J Chromatogr B 877:991–994
89. Lai C-S, Nair NK, Muniandy A, Mansor SM, Olliaro PL, Navaratnam V (2009) J Chromatogr B 877:558–562
90. Domijan A-M, Peraica M (2008) Arh Hig Rada Toksikol 59:277–282

91. Miles MV, Tang PH, Miles L, Steele PE, Moye MJ, Horn PS (2008) Biomed Chromatogr 22:1403–1408
92. Andrés F, Zougagh M, Castaneda G, Ríos A (2008) J Chromatogr A 1212:54–60
93. Supattanapong S, Konsil J (2008) Southeast Asian J Trop Med Public Health 39:978–987
94. Błazewicz A, Fijałeka Z, Samsel K (2008) J Chromatogr A 1201:191–195
95. Mashayekhi SO, Ghandforoush-Sattari M, Hain RDW (2008) J Clin Pharm Ther 33:419–427
96. Gu Y, Li Q, Melendez V, Weina P (2008) J Chromatogr B 867:213–218
97. Brajanoski G, Hoogmartens J, Allegaert K, Adams E (2008) J Chromatogr B 867:149–152
98. Tekes K (2008) J Chromatogr Sci 46:169–173
99. Zerzanova A, Zizkovsky V, Kucera R, Klimes J, Jesensky I, Dohnal J, Barron D (2007) J Pharm Biomed Anal 45:730–735
100. Honeychurch KC, Hart JP (2007) Electroanalysis 19:2176–2184
101. Błazewicz A, Fijałeka Z, Warowna-Grzeskiewicz M, Boruta M (2007) J Chromatogr A 1149: 66–72
102. Parrot S, Lambas-Senas L, Sentenab S, Denoroy L, Renaud B (2007) J Chromatogr B 850: 303–309
103. Wang L-H, Huang W-S, Tai H-M (2007) J Pharm Biomed Anal 43:1430–1436
104. Mochizuki N, Ohno K, Shimamura T, Furukawa H, Todo S, Kishino S (2007) J Chromatogr B 847:78–81
105. Kotani A, Takahashi K, Hakamata H, Kojima S, Kusu F (2007) Anal Sci 23:157–163
106. Al-Ghazawi M, Tutunji M, AbuRuz S (2007) J Pharm Biomed Anal 43:613–618
107. Kehr J, Hub X-J, Yoshitake T, Scheller D (2007) J Chromatogr B 845:109–113
108. Wang J, Wang D, Ni K, Hu X (2007) J Liq Chromatogr Rel Technol 30:1001–1013
109. Rozet E, Morello R, Lecomte F, Martin GB, Chiap P, Crommen J, Boos KS, Hubert P (2006) J Chromatogr B 844:251–260
110. Wang J, Fang Y (2006) Biomed Chromatogr 20:904–910
111. Shah R, LaCourse WR (2006) Anal Chim Acta 576:239–245
112. Saracino MA, Gandolfi O, Dall'Olio R, Albers L, Kenndler E, Raggi MA (2006) J Chromatogr A 1122:21–27
113. Chmielewska A, Konieczna L, Plenis A, Lamparczyk H (2006) J Chromatogr B 839:102–111
114. Bondiolotti GP, Pollera C, Pirola R, Bareggi SR (2006) J Chromatogr B 837:87–91
115. Karimi M, Carl JL, Loftin S, Perlmutter JS (2006) J Chromatogr B 836:120–123
116. Lin L, Qiu P, Yang L, Cao X, Jin L (2006) Anal Bioanal Chem 384:1308–1313
117. Kaushik R, LaCourse WR, Levine B (2006) Anal Chim Acta 556:267–274
118. Honeychurch KC, Smith GC, Hart JP (2006) Anal Chem 78:416–423
119. Chmielewska A, Konieczna L, Plenis A, Bieniecki M, Lamparczyk H (2006) Biomed Chromatogr 20:119–124
120. Bose D, Durgbanshi A, Carda-Broch S, Gil-Agusti M, Capella-Peiro M-E, Esteve-Romero J (2005) J Liq Chromatogr Rel Technol 28:3265–3281
121. Bose D, Durgbanshi A, Martinavarro-Dominguez A, Capella-Peiro M-E, Carda-Broch S, Esteve-Romero J, Gil-Agusti M (2005) J Pharmacol Toxicol Methods 52:323–329
122. Kuo C-Y, Wu S-M (2005) J Chromatogr A 1088:131–135
123. Bolarinwa A, Linseisen J (2005) J Chromatogr B 823:143–151
124. Groenendaal D, Blom-Roosemalen MCM, Danhof M, de Lange ECM (2005) J Chromatogr B 822:230–237
125. Xua H, Zhang W, Zhu W, Wang D, Ye J, Yamamoto K, Jin L (2005) Anal Chim Acta 545: 182–188
126. Spadaro A, Bousquet E, Santagati NA, Vittorio F, Ronsisvalle G (2005) Chromatographia 62:11–15
127. Bose D, Durgbanshi A, Martinavarro-Domínguez A, Capella-Peiró ME, Carda-Broch S, Esteve-Romero JS, Gil-Agustí MT (2005) J Chromatogr Sci 43:313–318
128. Huang T, Garceau ME, Ramstad T, Stehle RG (2005) J Pharm Biomed Anal 38:532–536

129. Kimura T, Nakanishi K, Nakagawa T, Shibukawa A, Matsuzaki K (2005) J Pharm Biomed Anal 38:204–209
130. Aguilar-Sanchez R, Ahuatl-Garcia F, Davila-Jimenez MM, Elizalde-Gonzalez MP, Guevara-Villa MRG (2005) J Pharm Biomed Anal 38:239–249
131. Bose D, Martinavarro-Domínguez A, Gil-Agustí M, Carda-Broch S, Durgbanshi A, Capella-Peiró M-E, Esteve-Romero J (2005) Biomed Chromatogr 19:343–349
132. Suzuki N, Miyashita N, Kotani A, Kusu F, Kawasaki T (2005) J Pharm Biomed Anal 38: 155–161
133. Palumbo G, Bacchi S, Primavera L, Palumbo P, Carlucci G (2005) Biomed Chromatogr 19: 350–354
134. Patel BA, Arundell M, Parker KH, Yeoman MS, O'Hare D (2005) J Chromatogr B 818: 269–276
135. Kokubun H, Ouki M, Matoba M, Kubo H, Hoka S, Yago K (2005) Anal Sci 21:337–339
136. Houze P, Borron SW, Scherninski F, Bousquet B, Gourmel B, Baud F (2005) J Chromatogr B 814:149–154
137. Cannazza G, Stefano AD, Mosciatti B, Braghiroli D, Baraldi M, Pinnen F, Sozio P, Benatti C, Parenti C (2005) J Pharm Biomed Anal 36:1079–1084
138. Liu M, Li P, Cheng Y, Xian Y, Zhang C, Jin L (2004) Anal Bioanal Chem 380:742–750
139. Bose D, Durgbanshi A, Capella-Peiro ME, Gil-Agusti M, Esteve-Romero J, Carda-Broch S (2004) J Pharm Biomed Anal 36:357–363
140. Somé IT, Semde R, Moustapha O, Amighi K, Guissou PI, Duez P, Dubois J (2004) C R Chimie 7:1087–1093
141. Oettl K, Payerl D, Zitta S, Müller T, Estelberger W, Reibnegger G (2004) Anal Biochem 331: 183–188
142. Klejdus B, Vacek J, Adam V, Zehnálek J, Kizek R, Trnková L, Kubá V (2004) J Chromatogr B 806:101–111
143. Rückert U, Eggenreich K, Wintersteiger R, Wurglics M, Likussar W, Michelitsch A (2004) J Chromatogr A 1041:181–185
144. Zhang XZ, Gan YR, Zhao FN (2004) J Chromatogr Sci 42:263–267
145. Saxer C, Niina M, Nakashima A, Nagae Y, Masuda N (2004) J Chromatogr B 802:299–305
146. Baranda AB, Jiménez RM, Alonso RM (2004) J Chromatogr A 1031:275–280
147. Sastre E, Nicolay A, Bruguerolle B, Portugal H (2004) J Chromatogr B 801:205–211
148. Baumann A, Lohmann W, Rose T, Ahn KC, Hammock BD, Karst U, Schebb NH (2010) Drug Metab Dis 38:2130–2138
149. Hayen H, Karst U (2003) Anal Chem 75:4833–4840
150. Lohmann W, Karst U (2006) Anal Bioanal Chem 386:1701–1708
151. Lohmann W, Karst U (2007) Anal Chem 79:6831–6839
152. Kazemifard GA, Moore DE, Aghazadeh A (2001) J Pharm Biomed Anal 25:697–711
153. Ebbel EN, Leymarie N, Schiavo S, Sharma S, Gevorkian S, Hersch, Matson WR, Costello CE (2010) Anal Biochem 399:152–161
154. Stefan RI, Van Staden JF, Bala C, Aboul-Enein HY (2004) J Pharm Biomed Anal 36: 889–892
155. Hutton EA, Ogorevc B, Smyth MR (2004) Electroanalysis 16:1616–1621
156. Roy PR, Okajima T, Ohsaka T (2004) J Electroanal Chem 561:75–82
157. Chailapakul O, Amatatongchai M, Wilairat P, Grudpan K, Nacapricha D (2004) Talanta 64:1253–1258
158. Limbut W, Thavarungkul P, Kanatharana P, Asawatreratanakul P, Limsakul C, Wongkittisuksa B (2004) Biosens Bioelectron 19:813–821
159. Wang J, Musameh M (2004) Anal Chim Acta 511:33–36
160. Guan JG, Miao YQ, Chen JR (2004) Biosens Bioelectron 19:789–794
161. Lin X, Gong J (2004) Anal Chim Acta 507:255–261
162. Alvarez-Romeo GA, Rojas-Hernandez A, Morales-Pérez A, Ramırez-Silva TM (2004) Biosens Bioelectron 19:1057–1065

163. Carvalho RM, Freire RS, Rath S, Kubota LT (2004) J Pharm Biomed Anal 34:871–878
164. Deo RP, Wang J (2004) Electrochem Commun 6:284–287
165. Yang CC, Kumar AS, Mc K, Chien SH, Zen JM (2005) Anal Chim Acta 554:66–73
166. Barroso MF, Paiga P, Vaz MCVF, Delerue-Matos C (2005) Anal Bioanal Chem 383:880–885
167. Nouws HPA, Delerue-Matos C, Barros AA, Rodrigues JA (2005) J Pharm Biomed Anal 39: 290–293
168. Nouws HPA, Delerue-Matos C, Barros AA, Rodrigues JA, Santos-Silva A (2005) Anal Bioanal Chem 382:1662–1668
169. Nouws HPA, Delerue-Matos C, Barros AA, Rodrigues JA (2006) J Pharm Biomed Anal 42: 341–346
170. Boonsong K, Chuanuwatanakul S, Wangfuengkanagul N, Chailapakul O (2005) Sens Actuators B 108:627–632
171. Hirakawa K, Katayama M, Nakano NSK, Imato T (2006) Anal Sci 22:81–86
172. Pfaffen V, Ortiz PI (2006) Anal Sci 22:91–94
173. Shaidarova LG, Gedmina AV, Chelnokova IA, Budnikov GK (2006) J Anal Chem 61: 601–608
174. Possari R, Carvalhal RF, Mendes RK, Kubota LT (2006) Anal Chim Acta 575:172–179
175. Felix FS, Yamashita M, Angnes L (2006) Biosens Bioelectron 21:2283–2289
176. Salimi A, Roushani M, Haghighi B, Soltanian S (2006) Biosens Bioelectron 22:220–226
177. Quintino MSM, Angnes L (2006) J Pharm Biomed Anal 42:400–404
178. Shaidarova LG, Davletshina LN, Budnikov GK (2006) J Anal Chem 61:502–509
179. Fanjul-Bolado P, Gonzalez-Garcia MB, Costa-Garcia A (2006) Anal Bioanal Chem 385: 1202–1208
180. Felix FS, Brett CMA, Angnes L (2007) J Pharm Biomed Anal 43:1622–1627
181. Rodriguez JA, Barrado E, Castrillejo Y, Santos JR, Lima JLFC (2007) J Pharm Biomed Anal 45:47–53
182. Guerrieri A, Lattanzio V, Palmisano F, Zambonina PG (2006) Biosens Bioelectron 21: 1710–1718
183. Viswanathan S, Liao WC, Huang CC, Hsu WL, Ho JA (2007) Talanta 74:229–234
184. Lowinsohn D, Bertotti M (2007) Anal Biochem 365:260–265
185. Bergamini MF, Santos AL, Stradiotto NR, Zanoni MVB (2007) J Pharm Biomed Anal 43: 315–319
186. Wu J, Yan F, Tang J, Zhai C, Ju H (2007) Clin Chem 53:1495–1502
187. Chuanuwatanakul S, Chailapakul O, Motomızu S (2008) Anal Sci 24:493–498
188. Murata M, Ivandini TA, Shibata M, Nomura S, Fujishima A, Einaga Y (2008) J Electroanal Chem 612:29–36
189. Hsieh BC, Hsiao HY, Cheng TJ, Chen RLC (2008) Anal Chim Acta 623:157–162
190. Wang Y, Luo J, Chena H, He Q, Gan N, Li T (2008) Anal Chim Acta 625:180–187
191. Messina GA, Panini NV, Martinez NA, Raba J (2008) Anal Biochem 380:262–267
192. Chiku M, Nakamura J, Fujishima A, Einaga Y (2008) Anal Chem 80:5783–5787
193. Chen PY, Vittal R, Nien PC, Ho KC (2009) Biosens Bioelectron 24:3504–3509
194. Li Y, Umasankar Y, Chen SM (2009) Anal Biochem 388:288–295
195. Chiu JY, Yu CM, Yen MJ, Chen LC (2009) Biosens Bioelectron 24:2015–2020
196. Su Y, Huang W, Hu R, Ding H, Hu K (2009) Biosens Bioelectron 24:2665–2670
197. Karuwan C, Wisitsoraat A, Maturos T, Phokharatkul D, Sappat A, Jaruwongrungsee K, Lomas T, Tuantranont A (2009) Talanta 79:995–1000
198. Daneshgar P, Norouzi P, Ganjali MR, Zamani HA (2009) Talanta 77:1075–1080
199. Jin J, Muroga M, Takahashi F, Nakamura T (2010) Bioelectrochemistry 79:147–151
200. Piano M, Serban S, Pittson R, Drago GA, Hart JP (2010) Talanta 82:34–37
201. Noda T, Ukai T, Yao T (2010) Anal Sci 26:675–679
202. Silva IS, Araújo MFA, Ferreira HA, Varela JJG, Tanaka SMCN, Tanakac AA, Angnes L (2010) Talanta 83:1701–1706
203. Ming ZJC, Caixia Y, Yifeng T (2011) Analyst 136:4070–4074

204. Chan-Eama S, Teerasonga S, Damwand K, Nacaprichaa D, Chaisuksant R (2011) Talanta 84: 1350–1354
205. Gaoa Q, Guoa Y, Zhanga W, Qia H, Zhang C (2011) Sens Actuators B 153:219–225
206. Hasebe Y, Wang Y, Fukuoka K (2011) J Environ Sci 23:1050–1056
207. Song W, Ding J, Liang R, Qin W (2011) Anal Chim Acta 704:68–72
208. Baskar S, Chang JL, Zen JM (2012) Biosens Bioelectron 33:95–99
209. Hayashi T, Sakurada I, Honda K, Motohashi S, Uchikura K (2012) Anal Sci 28:127–133
210. Akar IC, Özdokur KV, Demir B, Yavuz E, Demirkol DO, Koc S, Timur S, Ertas FN (2013) Sens Actuators B 185:331–336
211. Guerrier A, Ciriello R, Cataldi TRI (2013) Anal Chim Acta 795:52–59
212. Pereira PF, Marra MC, Cunha RR, Silva WP, Munoz RAA, Richter EM (2014) J Electroanal Chem 713:32–38

Chapter 8
Electroanalytical Method Validation in Pharmaceutical Analysis and Their Applications

Contents

8.1 Method Validation in Drug Assay Using Electroanalytical Techniques

Analytical chemistry, specifically electroanalytical chemistry, plays an important role in many branches of chemistry, biochemistry, pharmacy, food productions, life science, environment, etc. Many decisions are based on the results of quantitative analyses, and it is important to be aware of the quality of the results whenever analytical or electroanalytical methods are used. Analytical properties are quality indicators for a variety of systems, objects, tools, and outputs involved in chemical or biochemical measurements that allow one to compare and validate both analytical processes and the results that they provide. The objective of any analytical measurements is to obtain consistent, reliable, and accurate data. Validated analytical methods play a major role in achieving this goal. Analytical method validation is a major issue in the pharmaceutical industry for controlling drug quality, development, and registration. Simply, it is used to justify the analytical method used, in other words, to show that the method accomplishes what is claimed or intended. Validation is required for the development of new analytical methods, methods submitted as a part of a new drug application, bioequivalence and bioavailability studies, and for the analysis of drugs as raw material or in their dosage forms. Likewise, all laboratory tests must be validated before being introduced for patient testing to insure that the values reported will meet clinical expectations with a desired degree of reliability. Revalidation should be required, to a less or greater

S.A. Ozkan et al., *Electroanalysis in Biomedical and Pharmaceutical Sciences*,
Monographs in Electrochemistry, DOI 10.1007/978-3-662-47138-8_8

degree, following any change in reagents, supporting electrolytes, instruments, or experimental conditions. Evaluation and validation of an electroanalytical method performance is required to assess the degree of error expected due to inaccuracy and imprecision and to confirm that the degree of error meets the anticipated laboratory or clinical requirements. The procedures recommended for method validation differ with the type of test and the anticipated use. Experiments must be designed so that the correct data are obtained. The appropriate statistical methodologies should be used to correctly estimate errors with sufficient precision and to make the right decision about method's validity.

Electroanalytical method validation process can be performed basically in three steps as follows:

1. Identification of appropriate and necessary validation parameters
2. Design of experiments for parameter evaluation
3. Determination of acceptance criteria

Method validation was introduced as regulatory requirements in the USA in 1978 (for Federal Register), and it has become a part of registration applications for drug dosage forms in the other countries. The US Food and Drug Administration (FDA) issued a document entitled "Guideline for Submitting Samples and Analytical Data for Methods Validation" [1–4] dated on February 1987. This updated and accepted document reflects changes introduced by the International Conference on Harmonization (ICH) of Technical Requirements for the Registration of Drugs for Human Use guidelines, in Technology and current policy [5, 6]. The 3rd ICH has been published as a text for the use of applicants to bring a solution of some complications [7]. Additionally, some other useful protocols for method validation in the literature have derived, amongst others, from the Current Good Manufacturing Practice (cGMP), Food and Drug Administration (FDA), National Drug Administration (NDA), United States Pharmacopoeia (USP) Convention, and the American Public Health Association.

Method validation is an essential component of the measurements that a laboratory should implement to permit it to produce reliable analytical data. It should not be regarded as a singular activity, but should always be understood with respect to the life cycle of the analytical procedure. The ICH guidelines and USP requirements should be regarded as the basic background to analytical validation, not as a checklist. Validation process of the proposed method is an essential but time-consuming activity for most laboratories. It is therefore important to understand the requirements of method validation in more detail and the options that are available to allow for optimum utilization of analytical resources in the laboratories. Analytical validation includes the identification of the performance parameters relevant for the given procedure, the definition of appropriate acceptance criteria, and the appropriate design of the validation studies. In order to achieve this, the analyst must be aware of the fundamental meaning of these performance and validation parameters, as well as calculations, tests, and their relationship to the specific application.

An analytical method is a set of procedures involved in the collection, processing, storage, and analysis of a drug active compound and its dosage form medium. The availability of selective and sensitive analytical methods is a prerequisite for the generation of reliable data on the assay of pharmaceutical active compounds, bioavailability, and bioequivalence studies of drugs.

In the broad field of analytical techniques, electrochemical methods offer interesting alternatives with respect to sensitivity and selectivity criteria and limited sample treatment requirements for the analysis of drug compounds in their dosage forms and in biological samples such as plasma, urine, serum, and cerebrospinal fluid. These methods must, however, be validated with respect to their reliability for the intended use. The obtained results can be used to judge the quality, applicability, accuracy, and consistency of the method; it is an integral part of any analytical procedure.

8.2 Types of Method Validation

Electroanalytical methods need to be validated or revalidated according to the analyst or experimental requirements and in the following instances:

- Before initial use in routine testing of the developed method
- When transferring the developed method to another laboratory
- Whenever the method parameters or conditions for which the method has been validated change (such as an instrument with different characteristics or samples with a different matrix) and the changes are out of the first scope of the method.

The performance characteristics of an electroanalytical method should be based on the intended use of the method. It is not always necessary to check for all the analytical parameters that are available for a specific technique or required validation group. In this way, the experiments can be limited to what is really necessary [1–20]. Hence, types of method validation can be divided as three main groups:

1. Full validation
2. Partial validation
3. Cross validation

Full Validation This is necessary:

- When a new drug compound is analyzed.
- When the electroanalytical method is developed and implemented for the first time.
- When impurities or metabolites are added to an existing assay for quantification.

Partial Validation It can be performed if validated electroanalytical methods have been modified. It can be ranged from the within day precision and accuracy to a nearly full validation. Partial validation is required when:

- Instrument and/or software changes
- A method is transferred between laboratories and analysts
- Changes in species occur within the same matrix
- Changes in matrix occur within the same species
- Changes occur in the electroanalytical methodology
- Changes occur in sample preparing procedures

Cross Validation It is necessary when two or more electroanalytical or analytical methods are used to generate data within the same study. Using cross validation, two electroanalytical methods or one electroanalytical plus another independent analytical method results (preferably a reference method if available) for the same analyte and in the same matrix media can be compared. A cross validation should be considered:

- When sample analyses within a single study are conducted in more than one laboratory.
- When data obtained using different analytical or electroanalytical techniques in different studies are included in a regulatory submission.

The steps of electroanalytical method development and validation depend upon the type of method being developed. However, the following steps are commonly used:

- Method development plan
- Background information such as literature data and investigated compounds specifications
- Method development procedures
- Method validation steps
- Experimental test procedures for method validation
- Statistical analysis of obtained results
- Validation reports

Electroanalytical method validation criteria should be defined early in the process according to the following subjects [1–20]:

- The analyte
- The expected concentration level
- The matrix of the sample
- The required determination information
- Possible interferences (selectivity)
- Necessary and/or possible detection and determination limits
- The required concentration range
- The expected accuracy and precision
- The robustness and/or ruggedness of the proposed method, etc.

A well-developed electroanalytical method should be readily validated. As the method development and validation processes advance, the information gathered is captured in the design and used for subsequent improvement of the methods.

Fig. 8.1 Life cycle of an electroanalytical method

The electroanalytical method validation activity is not a one-time study. It has a life cycle procedure (Fig. 8.1).

Any chemicals used in electroanalytical methods (e.g., reagent, supporting electrolyte, standards, raw material) should be accurately identified, sufficiently stable, available in sufficient quantities, and as pure as possible. Any other materials and consumables such as working electrodes should be new or in good working order and be qualified to meet the electroanalytical performance criteria. After one set of experimental conditions can be ensured, the consumables may be used for most experiments.

During the method validation procedure, the detailed description of the electroanalytical method, the validation experiments, the parameters, acceptance criteria and limits, the validation results, and quality control checks should be written and defined in a form of a report by the laboratory responsible for drug analysis. During the optimization of validation parameters, the experiments should be defined with a minimum number of control analyses. Experiments must be designed so that the correct data are obtained. The appropriate statistical methodologies must be used to estimate errors with sufficient precision. After the development and validation of the electroanalytical method on drug analysis, the report should contain the following items:

1. The objective and scope of the method
2. The applicability and type of the method
3. Methodology and experiments
4. The compound and matrix details
5. The preparation of the working solutions, chemicals and reagents, etc.
6. Quality control procedures of chemicals and standard solutions
7. The performance requirements of the equipments and experiments such as applied potential, scan rate, working electrode, ionic strength, pH, and temperature
8. Critical and limiting parameters

9. Experimental conditions such as buffer, supporting electrolyte, pharmaceutical dosage form content, sample preparation, extraction, and evaporation
10. The calculation of the statistical parameters
11. Necessary experimental curves, voltammograms, and plots such as calibration curves, peak current–pH plot, peak potential–pH plot, and current–scan rate plots
12. The acceptance limits
13. Revalidation criteria
14. Conclusion and summary of the obtained results for the proposed method

With respect to method suitability, the performance characteristics should be checked by the detection limit, linearity and range, approximate precision, and accuracy. If this preliminary validation data is not appropriate, the used electroanalytical technique itself, the equipment, or acceptance limits can be changed.

Validation is a formal and systematic proof of the suitability of the analytical procedure for its intended purpose. Analytical validation also necessitates documentation prepared and approved for records keeping. The required validation parameters, also called "analytical performance characteristics," depend on the type of analytical method. This is based on electroanalytical techniques performed according to the validation protocols that comply with the ICH or other official pharmacopoeia guidelines on method validation on drug analysis. The validation of the overall system includes not only all the qualification and calibration activities but also extends to the software and its usage. That's why validation should be applied using preferably computerized controlled systems.

The validation processes should be required and realized according to the ICH and/or pharmacopoeia requirements. The desired quality level of the electroanalytical techniques should be stated at the beginning steps in order to be able to select the degree of validation requirements. While planning to test procedures, one has to take into account the statistical tests that are to be applied.

Typical analytical performance characteristics that should be considered in the validation of the types of procedures in this chapter are as follows:

1. Selectivity/Specificity
   (a) For raw material
   (b) For the investigated samples
2. Trueness
3. Precision
   (a) Repeatability
   (b) Intermediate precision
   (c) Reproducibility
4. Accuracy
5. Linearity
6. Range
7. Limit of Detection (LOD)

8. Limit of Quantitation (LOQ)
9. Robustness,
10. Stability
11. Applicability

Methods can be classified in different ways. However, in the present instance an important distinction should always be made between qualitative and quantitative techniques.

For qualitative methods on drug testing, the following validation parameters are required:

(a) Specificity/Selectivity
(b) LOD
(c) Precision
(d) Stability

For quantitative methods on drug analysis tests, the following set of validation parameters are required:

(a) Specificity/Selectivity
(b) Linearity and range
(c) LOD and LOQ
(d) Precision
(e) Accuracy
(f) Applicability

Additional parameters to be determined which are desirable include robustness and stability.

All analytical procedures are of equal importance from a validation perspective. In general, validated analytical procedures should be used, irrespective of whether they are for in-process, release, acceptance, or stability testing. Analytical methods used for the assay of a drug substance as a raw material or in its dosage form for stability studies should be stability indicating. Validation results should refer to an analytical system rather than an analytical method. The analytical system comprises a defined method protocol, a defined linearity and range, and other related parameters for the analyte and validation processes. During the past 20 years, it is quite often observed and accepted that validation represents different aspects of method performance in reference to separate items for different analytical techniques, and these guidelines reflect the above mentioned pattern to a considerable extent. Validation parameters are the formal and systematic proofs that an electroanalytical method complies with the requirements for testing a product when observing a defined procedure. However, these parameters considered necessary for the method validation of different types of analytical processes have some small differences [1–20]. Generally, validation is the process of collecting documented evidence that the method performs according to its intended purpose. Each validation parameters and each data may need to be addressed in a different manner, depending on the intended application of the method. The desired quality level of an electroanalytical method should be stated at the beginning steps in order to be able to select the degree of validation needed. Validation is a continuous process, and it should comprise at least five important steps for an electroanalytical method which are described below:

1. Planning of the electroanalytical tests
2. Performing the electroanalytical tests
3. Statistical evaluation of the results
4. Report of the validation parameters
5. Application of all information gained during full validation processes and their explanations

While planning the test procedures, one has to take into account the statistical tests that are to be applied. For these statistical calculations, some computer programs can be very useful such as Microsoft Excel®, SPSS®, Statview®, and Sigma Plot®.

8.3 Electroanalytical Method Validation Characteristics and Their Calculations

Validation parameters are determined based on the final aim of the analytical procedure. The method validation process becomes much easier to understand if the focus is directed towards the sources of potential analytical errors and how these errors can be investigated. The important point is "What type of experiments or experimental designs will be the best to provide precise and unbiased estimations of the analytical errors?" For experimental design "Which statistical method should be used to describe the extent of the error observed?" The acceptance limits of the proposed methods should be notified without affecting interpretation.

8.3.1 Validation Characteristics

The required laboratory tests and definitions for method validation have been described by different working groups of national and international committees and are largely reported in the literature.

Validation requirements depend upon the aim and type of electroanalytical method and working sample. Typical electroanalytical performance characteristics that should be considered in the validation procedure are described and discussed in this chapter. These parameters are summarized in Table 8.1.

Table 8.1 Parameters for electroanalytical method validation with reference to ICH guidelines, USP, and ISO 17025

Parameters	Comments		
	ICH guidelines [1, 2]	USP 23 [15]	ISO 17025
Accuracy	√	√	√
Precision – Repeatability – Intermediate precision (within-day precision) – Reproducibility (between-day precision)	√ √ √ √	√ – – (defined as ruggedness)	– √ – √
Specificity (Selectivity)	√	√	√
Linearity	√	√	√
Range	√	√	–
LOD	√	√	√
LOQ	√	√	√
Robustness	√	√	–
Ruggedness	(defined as reproducibility)	√	–
Stability	Recommended	Recommended	–

8.3.1.1 Specificity (Selectivity)

Specificity or selectivity is defined as the ability to assess unequivocally an analyte in the presence of components that may be expected to be present in the analyzed sample. Actually, the term "specific" generally refers to a method that produces a response for a single analyte only, while the term selective refers to a method that provides a response for a number of chemical entities that may or may not be distinguished from each other [21]. Specificity and selectivity can often be used interchangeably. However, the concept of selectivity is something that can be graded while specificity is an absolute characteristic of an analytical procedure [15–19]. According to this description, it can be said that specificity/selectivity results are proved to the power of differentiation between the analytes and closely related substances such as matrix components, impurities, degradants, and isomers. It can be said that it is a measure of the ability of the method to identity/quantify the analytes in the presence of other substances, either endogenous or exogenous, in a sample matrix under the stated conditions of a proposed electroanalytical method. The degree of discrimination between the working compound and other substances present in or extracted from the matrix must be carefully considered. It is necessary to be sure that the identity of the analyte is unequivocal. Actually, attention needs to be paid to the cleanup, extraction, precipitation, or evaporation procedures used and the discrimination power of the selectivity/specificity of the proposed method. If the obtained response is distinguished from all other responses, the method can be called as fully selective [1–7, 16–39].

In practice, a test mixture is prepared which contains the investigated compound and all potentially interferent of the sample. Effects of the interferences and chances of false identifications can be examined for blank samples and those spiked with proper amounts of studied compound which may interfere with the identification and/or determination of the analyte. The obtained results can also be compared with the response of the pure analyte. In the pharmaceutical industry, the text mixtures of components can come from synthesis intermediates, excipients, and degradation products. Hence, an electroanalytical method can be described based on its ability to measure accurately and specifically the investigated analyte in the presence of components that may be expected to be present in the sample matrix. For describing and showing the specificity of a proposed electroanalytical method, the analyte, the supporting electrolyte, the matrix with and/or without analyte, the impurities of the studied compound, the degradation products, and metabolites should all be analyzed. Interestingly, ingredients and excipients often encountered in dosage forms generally do not interfere in electroanalytical methods.

For identity test, compounds of closely related structures such as metabolites or degradation compounds which are likely to be present should be discriminated from each other. This process could be confirmed by obtaining positive results from samples containing the analyte, coupled with negative results from samples which do not contain the analyte and also by comparison with the pure analyte results. As a general rule, selectivity (or specificity) should be sufficiently good for any interference to be ignored [16–39].

The specificity (or selectivity) for an assay method can be approached from two ways depending on the availability of impurities:

(a) If impurities are available: Using this way, the specificity of an assay method is determined by comparing test results from an analysis of sample containing the impurities, degradation products, excipients, or ingredients with those obtained from an analysis of samples without impurities, degradation products, excipients, or ingredients. Especially for a stability-indicating method, degradation peaks need to be resolved from the response of the working drug compound. However, the impurity responses do not need to be resolved from each other. The selectivity studies can also be realized by comparing the results obtained from freshly prepared samples with and without impurity, excipients, and degradation compounds in the presence of the studied drug using adequate statistical methodologies.
(b) If impurities are not available: Specificity may be demonstrated by the comparison of the obtained test results of sample containing impurities or degradation products with an other well-characterized and validated analytical procedure. This should include samples stored under the same conditions and applied to the same samples. The sample matrix may contain some unknown compounds which interfere at the selected applied potential. These interfering compounds can affect the peak potential and peak current magnitude. If the matrix composition is unknown, the systematic matrix effect studies can be tested using the standard addition method. For this, two sample

sets are prepared containing the same amount of analyte with one set without the matrix, degradants, or impurities and the other set comprising the matrix, degradants, or impurities. After standard addition, the results should be plotted as a function of analyte concentration versus obtained peak response. The obtained slopes are compared for the identification of the matrix composition effect. If the slopes of the two sets are found similar or close to each other, it means that the matrix does not have effect on the results [22–40].

In particular, the specificity tests are necessary for analytical procedures if used for assessing stability. The forced (stress) degradation studies can be realized for obtaining the stability parameters. For the forced degradation studies, the electroanalytical methods can be realized after stress conditions such as acid, base, heat, oxidation, and UV light treatment, in order to find possible degradation products and their interferences. Forced degradation or stress testing is undertaken to demonstrate specificity when developing stability-indicating methods, particularly when little information is available about potential degradation products. Stability-indicating studies also provide some information about the degradation pathways and degradation products that could form during storage. In particular, the issue of how much stress is adequate in stress testing is not addressed specifically. Overstressing a molecule can lead to degradation profiles that are not representative of real storage conditions and perhaps not relevant to method development. Therefore, stress-testing conditions should be realistic and not excessive. The ICH guideline Q1A suggests some conditions for realizing forced degradation studies [41].

A stability-indicating method is defined as validated analytical procedure that accurately and precisely measures the drug active compounds free from potential interferences like degradation products, process impurities, inactive ingredients, and other potential impurities. The possible degradation product of the drug active compounds can be produced by certain stress conditions like hydrolytic (acidic, neutral, basic), oxidative, photolytic, thermal, etc. During validation, the samples are exposed to the forced degradation processes to produce approximately 10–30 % degradation products of the active compound. The nature of the forced degradation testing will depend on the individual drug substance and the type of drug product involved (Table 8.2).

1. *Hydrolytic Degradation*: This kind of stress tests can be studied by refluxing the compound in 0.1 M HCl and/or 0.1 M NaOH and/or phosphate buffer at pH 7.0, separately for about 6 or 8 h. If some degradation product can be obtained, the experiments can be stopped at this level. However, in case no degradation can be obtained under the selected conditions, hydrolytic conditions such as different concentrations (e.g., 0.1 or 1 M HCl or NaOH) and the longer duration period can be used.
2. *Oxidative Degradation*: For the oxidative stress studies, H_2O_2 can be used as between 3 % and 30 % concentration range. This study can also be realized under atmospheric O_2.

Table 8.2 Example conditions of the stress degradation studies

Stress-degradation study	Conditions
Acid hydrolysis	0.1 M HCl
Base hydrolysis	0.1 M NaOH
Neutral hydrolysis	Phosphate buffer at pH 7.0
Oxidative conditions	Between 3 % and 30 % H_2O_2 or Atmospheric O_2
Photolytic degradation	UV, Fluorescent, or white lamb
Temperature	Under selected temperatures such as 75 °C
Humidity	Under > 75 % humidity

3. *Photolytic Degradation*: These studies should be realized by exposure to light using UV, fluorescent, and/or white lamp or their combination. For this type of degradation studies, Xenon and metal halide lamps can also be used.
4. *Temperatures*: For the temperature degradation studies, the temperature should increment as 10 °C above the accelerated temperatures such as from 65 °C to 75 °C.
5. *Humidity*: It can be applied at or greater than 75 % humidity values where appropriate.

Each result which is obtained from stress conditions tests should be compared with the blank solutions stored under normal conditions. Stress-degradation studies can be time consuming and difficult because it is often difficult to generate the proper level of degradation. However, these validation parameters give the information about the selectivity of the proposed method. Actually, forced degradation studies are mostly recommended for new drug active compounds by ICH.

8.3.1.2 Trueness

Trueness is a concept related to systematic errors of the analytical procedures: The closeness of agreement between the average values obtained from a large series of test results and an accepted reference value. The measure of trueness is usually expressed in terms of bias. Trueness has been referred to as 'accuracy of the mean.' This usage is not recommended [21, 42]. Trueness expresses the distance from the average value of a series of measurements ($\overline{x}_i$) and a reference value μ_T. This concept is measured by a bias, relative bias, or recovery:

$$\text{Bias} = \overline{x}_i - \mu_T$$

$$\text{Relative Bias } (\%) = 100 \times \left(\frac{\overline{x}_i - \mu_T}{\mu_T}\right)$$

$$\text{Recovery } (\%) = 100 \times \left(\frac{\overline{x}_i}{\mu_T}\right) = 100 \; - \text{Relative Bias } (\%)$$

Some possible error sources causing biased measurements are:

- Sample impurities: Because of the contamination or low degree of purity or due to the presence of degradation products.
- Sample preparation steps: Inefficient extraction procedures from matrix of the sample, unsuitable or impure solvents, reagents or material, incomplete removal of ingredients, or endogenous substances.
- Inadequate calibration equation.
- Error in drug active compound analysis step: Due to the unstable measurement or unexpected matrix effects and interferences.

The true value for trueness assessment can be obtained in different ways [1–8, 18–20, 43–48]:

1. Using appropriate certified reference material
2. Using the results from an established reference method (such as pharmacopoeia or published literature method)
3. Spiking a blank matrix with known concentrations of compound
4. Standard addition of the compound to the studied matrix media

1. Appropriate Certified Reference Material

The use of a certified reference material is most appropriate. However, for newly synthesized compounds in commercial development, reference standards are usually unavailable. The important point is the necessity to know the purity of used reference material which should be as high as possible. The purification and characterization of the selected reference compound should be defined in detail by the different techniques such as mass spectrometry, spectrophotometry, IR, NMR, and DSC.

2. Using an Established Reference Method

The comparison of the obtained results from the electroanalytical method with the results obtained by an established validated or pharmacopoeia reference method can be used to assess trueness. Hence, trueness of the proposed electroanalytical method can be compared by the independent and validated reference method. The selected reference method should be known to be accurate, precise, well defined, and correct for the sample matrix of interest. Ideally, if there is a pharmacopoeia method available, it should be used as a reference method for comparison. If not, the reference method can be selected in the already published scientific literature or obtained from the pharmaceutical industry. The selected reference method should be applied to the determination of the studied compound in its dosage form or in a biological sample [18–20, 46–53].

Statistical theory allows to set limits around an experimentally determined mean $\bar{x}$ within which the population mean μ lies with a given degree of probability. These set limits are called confidence limits, and they define the confidence interval.

The limits of the confidence interval can be estimated with the sample standard deviation s and the important statistical parameter t (from Student's t-distribution; t as function of probability (P) and degrees of freedom ($n-1$), with n being the number of subsequent measurements):

$$\operatorname{cnf}(\mu) = \overline{x} \pm \frac{t(P,\ n-1)s}{\sqrt{n}}$$

Trueness should be reported as the mean percent recovery, *bias %*, with the confidence intervals. These values should be obtained at least from number of six experimental results.

3. Spiking a Blank Matrix with Known Concentrations of Compound
Spiking with known concentrations of the investigated compound can be used, if the electroanalytical method is used to measure a drug active compound in a complex and/or unknown matrix media such as biological samples or pharmaceutical dosage forms. For realizing this type of recovery studies, well-characterized, fully analyzed, and pure known standard drug active compound is required. If certified reference materials or control samples or the matrix composition are not available or not known, a blank sample matrix of interest can be spiked with a known concentration.

Using this method, pure compounds should be added to a blank matrix media which is called as placebo, at different concentration levels. Also, the spiked concentrations should be within the range of concern and should begin with the concentration close to the limit of quantification (LOQ; see Sect. 8.3.1.8). At each level, a minimum of three repeated experiments should be realized.

4. Standard Addition of the Compound to the Studied Matrix Media
This technique can be used if no matrix components or blank are available. In this technique, a known amount of the pure analyte is added to the sample media at different levels. The range of spiked samples amount can be between 50 % and 150 % of the nominal settings of compound level. This sample matrix already contains a previously determined amount of the studied compound. The proposed electroanalytical method is first applied to record the signal corresponding to studied compound in the matrix. Then, the same method is used for analyzing the spiked amount of the compound. The difference (or the ratio) between the spiked amount value and the measured amount due to the spiking is a measure of the bias (or recovery) of the analytical procedure.

Number of Experiments
The ICH guideline Q2 recommends assessing a minimum of nine determinations using at three different concentration levels covering the specified range, e.g., each concentration level at least three replicates. The trueness results should be reported as bias, relative bias, or recovery together with their confidence intervals. The concentration levels selected should hence cover the range of concern and should include concentrations close to the required LOQ value, one in the middle of the range and one at the high end of the working range. Hence, three different concentration levels can be used in this type of recovery studies. When relatively large concentration ranges are studied, more than three concentration levels are advised. This is the case for, e.g., impurity assays, dissolution assays, or in bioanalytical applications [18–20, 46–58].

8.3.1.3 Precision

ICH Q2 defines the precision of an analytical method as the closeness of an agreement between each measurements obtained from multiple sampling of the same homogeneous sample under described conditions. Precision is a criterion that is linked to the random errors of an analytical method. Precision can be considered at three levels: repeatability, *i*, ntermediate precision, and reproducibility. Precision studies indicate the dispersion or spread around the mean value. It is usually measured in terms of imprecision expressed as a variance, standard deviation, or relative standard deviation (RSD).

$$\mathrm{RSD}\% = \frac{s}{\bar{x}}100$$

Precision can be affected by the measuring technique (e.g., random noise due to reference electrode potential fluctuation, temperature change . . .), sample properties (e.g., inhomogeneities), solution impurities, and chemical effects or even by the operator.

The short-term variability is called as repeatability. It is also known as intra-assay precision. It includes, in addition to the system precision, the contributions from the sample preparation, such as weighing and dilution homogenization. The repeatability is obtained under same operating conditions using the same sample and solutions over a short time period. It is the lowest level degree of precision. Repeatability is an indication of how easy it is for an operator in a laboratory to obtain the same results for the same sample of material using the same measurement technique, the same equipment, and same reagents. Repeatability should be determined from a minimum of six measurements at 100 % of the target concentration in the same sample or nine measurements covering the specified range of the procedure for three different concentration levels and three replications of each. Hence, the same data used for the evaluation of trueness can be used to assess repeatability [1–7, 18–21].

Intermediate precision is defined as the variation within the same laboratory arising from the use of different stock solutions, different analysts, different days, different equipments, etc. In determining intermediate precision, experimental design can be employed so that the effect of the individual variables may be obtained. Experimental design can minimize the number of experiments that needs to be performed [18–21, 54, 55, 57, 59–62]. In the intermediate precision, more parameters can be changed and studied during the analysis period. Intermediate precision can also be called as between-day, between-run, or inter-assay precision. It shows the total random error of the analytical method under study under varied conditions that may be encountered during the routine application of this analytical method. Specific statistical methodologies should be used in order to obtain the repeatability variance and intermediate precision variance, such as random one way analysis of variance (ANOVA) [21, 63, 64]. For intermediate precision, ICH recommends to vary the relevant and important additional sources of random errors that will likely arise during the daily application of the method, typically: days, operators, and equipments. Intermediate precision should be reported as standard deviation or RSD together with their confidence interval. As done for repeatability, it should be determined preferably at three concentration levels at least covering the studied concentration range.

The third precision parameter is reproducibility which expresses the precision between laboratories. It is used especially for collaborative studies usually applied to standardization of analytical procedures. It is considered in case of the standardization of an electroanalytical method, for instance, for inclusion of procedures in pharmacopoeias. Reproducibility conditions can be described as: Conditions for obtaining analytical results independent of each other consisting of the use of a set of analytical processes or methods on an identical object (same sample, same reagents, same materials, etc.) by different analysts with different apparatus, electrodes, supplies, and materials in different laboratories. This is a critical parameter when using electroanalytical techniques due to the inherent heterogeneous process occurring at the electrode–solution interface. The working electrode and its surface pretreatment (renewing) step need to be perfectly controlled. The former needs to be well characterized and the latter is particularly critical when manual surface cleaning and smoothing have to be performed. Electrochemical "cleaning" methods and the use of disposable one shot miniaturized electrodes (e.g., commercially available screen printed electrodes) are less prone to interlaboratory fluctuations.

8.3.1.4 Accuracy

In document ICH Q2, accuracy is defined as: "...the closeness of agreement between the value which is accepted either as a conventional true value or an accepted reference value and the value found." In the ISO definition [42], a note is added specifying that accuracy is the combination of random error and systematic error or bias. From this, it has to be understood that accuracy strictly applies to results and not to analytical methods. The closeness of agreement observed is based on the sum of the systematic and random errors, namely the total error linked to the result. Consequently, the accuracy should be measured by a parameter related to total error. A first way to provide a native measure of total error and hence of the results accuracy is to sum the intermediate precision RSD value with the relative bias value obtained at each concentration level of the validation standards [65]. Other more informative and rigorous way to measure total error and accuracy is to use the methodology of statistical tolerance intervals [21, 56–58, 63–68]. The interested readers are referred to the literature for an extended coverage on this topic [63–68].

$$\text{Total Error} = \text{Bias}\% + 1.96 \times \text{CV}\%$$

where CV is the abbreviation of coefficient variation and it represents the ratio of the standard deviation to the mean. Accordingly, using a multiple of the coefficient of variation (CV), the above equation and the value of 1.96 (for some working media 2.0) is given for a 95 % confidence interval or limit of the possible analytic error.

Unlike the confidence interval, the tolerance interval estimates the range which contains a certain percentage of each individual measurement in the population. The tolerance interval can be determined by using the following equation:

$$\text{Tolerence interval} = \overline{x} \pm ks$$

where s is the standard deviation and k is the tolerance factor to adjust the width of the interval.

The range for the accuracy limit should be within the linear range. Typical accuracy of the recovery of the drug substance in the mixture is expected to be about 98–102 %. Values of accuracy of the recovery data beyond this range need to be investigated.

8.3.1.5 Linearity

The linearity of an analytical method is its ability to elicit test results that are directly proportional to the concentration of an analyte in a sample within a given range [1–7, 22–39]. ICH defines an analytical procedure as its ability, within a given range, to obtain test results that are directly proportional to the concentration of the analyte in the sample. The linearity validation criteria should not be confused with response functions or calibration curves. Linearity should be evaluated by visual inspection of a plot of signals as a function of analyte concentration or content. If there is a linear relationship, test results should be evaluated by appropriate statistical methods, for example, by calculation of a regression line by the method of least squares. This relationship can be any straight line or even not at all a straight line (quadratic, sigmoidal, exponential,. . .). In some cases, to obtain linearity between assays and sample concentrations, the test data may need to be subjected to a mathematical transformation prior to the regression analysis. Data from the regression line itself may be helpful to provide mathematical estimates of the degree of linearity. It is evident to understand that, if the analytical procedure does not commit any random and systematic error, the equation obtained between the concentration of the validation standards and the analytical results is the identity line $y = x$. While this is never observed in practice, linearity should be assessed, for example, by verifying that the slope of the equation is not too far away from "1" and the intercept sufficiently close to "0." The determination coefficient "r^2" can also be given. However, this parameter does not give any information about the suitability of the linearity criterion [21]. For the establishment of linearity, a minimum of five concentrations is recommended. Other approaches should be justified.

8.3.1.6 Range

The range is defined by ICH as the interval from the upper to the lower concentration of analyte in the sample for which it has been demonstrated that the electroanalytical procedure has a suitable level of precision, accuracy, and linearity. The range of the method is estimated from the linearity and is verified by confirming that the method provides acceptable precision and linearity when applied to a sample containing an analyte at the extremes of the range as well as within the range. It must include at least the expected or required range of analytical results, the latter being directly linked to the acceptance limits of the specification, or the target test concentration of the analytical procedure style. The working range is predefined by the purpose of the method and may reflect only a part of full linear range. The valid electroanalytical range of the method is that range of concentrations, which pass the linearity, precision, and trueness and hence accuracy criteria. The concentration range should cover the target amount in the sample to be measured. For the assay of a drug substance in a dosage form, it is normally recommended to operate in the range of 80–120 % of the nominal concentration. For content uniformity, a normal range would cover between 70 % and 130 % of the nominal concentration, unless a wider and more appropriate range is justified. For the dissolution testing, a normal range is ±20 % over the specified range.

8.3.1.7 Limit of Detection

The Limit of detection (LOD) is the lowest quantity of a substance that can be distinguished from the absence of that substance (a *blank value*) within a stated confidence limit. The LOD is usually expressed as a concentration of the analyte in the sample, such as M, μg mL^{-1}, ppm, and ppb. The LOD is an important criterion for the assay of a drug compound in a biological sample, for impurity, and degradation compound analysis in pharmaceutical dosage forms. It is of limited interest for the assay of the active drug compound in a drug formulation.

The LOD is frequently confused with the sensitivity of the method. The sensitivity is the slope of the calibration line. The LOD is not a robust or rugged parameter and can be affected by minor changes in the electroanalytical system such as temperature, supporting electrolyte, pH, purity of reagents, matrix effects, and instrumental conditions. It is therefore important that this parameter is always verified by laboratories adopting previously validated methods. Although various methods of LOD determination can be found, the most widely applied and accepted are given below [1–5, 69–79]. These approaches are depending on whether the procedure is realized using instrumental or noninstrumental methods. The mostly used type of LOD determinations are recommended by the official guidelines and pharmacopoeias.

A. Based on Visual LOD Inspection

The LOD is determined by the assay of compounds with known concentrations of analyte and by establishing the minimum level at which the analyte can be reliably detected [1–5, 69–79]. The analytical responses of blank and low concentration samples are compared to determine conclusively what concentration of analyte is necessary to distinguish its presence from its absence. This approach may be used for noninstrumental methods but may also be used with instrumental methods.

B. Based on Signal-to-Noise Ratio

When the final measurement is based on instrumental reading, the background response will have to be taken into account. For procedures which exhibit baseline noise, the LOD can be based on the signal-to-noise ratio (S/N). This method is mostly used in chromatographic, electrophoretic, and spectrometric techniques. In most of the electroanalytical studies, noise cannot be obtained clearly because of the nature of the method. Determination of the S/N (Fig. 8.2a) is performed by comparing measured signals from the samples with known low concentration of analyte with those of blank samples and by establishing the minimum concentration at which the analyte can be reliably detected. Often an S/N ratio of 3:1 (Fig 8.2b) or 2:1 is considered acceptable for estimating the LOD [1–5, 69–79]. Presentation of relevant experimental curves is sufficient for justification of the LOD. For using S/N technique, the apparatus sensitivity should be constant. However, this type of evaluation gives a far too optimistic value for LOD. Also it leads to a situation where the presence of an analyte remains undetected at higher concentrations than claimed.

This method is commonly applied to analytical methods that exhibit baseline noise. In electroanalytical studies, the noise corresponds to fluctuation of electrode interfacial pseudocapacitance, electrical perturbations, etc.

C. Based on the Standard Deviation of the Response Another and most widely used approach in electroanalytical chemistry estimates the LOD from the standard deviation of the response and the slope of the related calibration curve. A calibration curve is realized using samples containing the analyte in the range of the LOD. For this method, LOD can be expressed by the following equation:

$$\mathrm{LOD} = \frac{3s}{m}$$

where s is the standard deviation of blank value and m is the slope of the related calibration curve. By using the signal-to-noise method, the peak-to-peak noise around the analyte retention time (in chromatography) is measured, and subsequently, the concentration of the analyte that would yield a signal equal to certain value of noise-to-signal ratio is estimated. The noise magnitude can be measured either manually on the chromatogram printout or by auto-integrator of the instrument. A signal-to-noise ratio (S/N) of three is generally accepted for estimating LOD, and signal-to-noise ratio of ten is used for estimating LOQ. This method is commonly applied to analytical methods that exhibit baseline noise [5]. Hence, using the above equation the response can be converted to the concentration units.

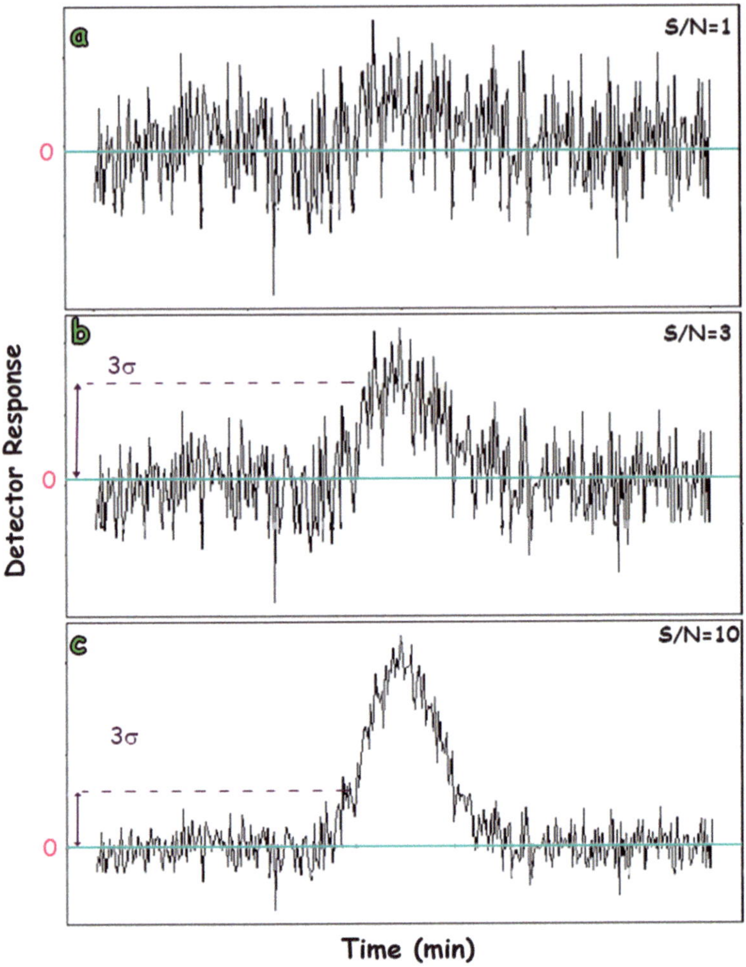

Fig. 8.2 (**a**) Limit of detection using S/N method; (**b**) For LOD: S/N: 3; (**c**) For LOQ: S/N:10

The relationship between LOD and sensitivity is also shown using this equation depending on the slope function. The estimate of s can be obtained by several approaches, which are given as follows:

1. The standard deviation s of the blank is determined by measuring the blank response at least ten times.
2. The standard deviation s of the regression line (slope) can also be used; however, in that case different formulae have to be used.
3. The standard deviation s of the intercept of the regression line can also be used.

Another alternative way for practically assessing the LOD is:

$$\text{LOD} = \bar{x} + 3s$$

where $\bar{x}$ is the mean blank value and s is the standard deviation, respectively. s is obtained from a set of blank measurements containing no analyte [1–5, 69–79]. In this way, at least ten independent blank measurements should be realized. For calculation of $\bar{x}$, ten independent sample blanks fortified at lowest acceptable concentration are measured.

In addition to the above methods, LOD can be calculated using another approach. In this way, the blank working medium is spiked with different analyte concentrations close to the LOD or low limit of the linearity range. At least six different concentrations and totally ten independent measurements should be realized for each concentration levels. The standard deviation of the obtained measurements should be calculated. The LOD value corresponds to those concentrations of analyte with RSD values of 33 % [1–10, 76–79].

The mode of LOD estimation should always be reported. If LOD is reported using visual evaluation or signal-to-noise ratio method, the presentation of the related working curve (chromatogram, voltammogram) is considered acceptable for justification. If the LOD value is calculated using standard deviation of blank, response, or intercept and slope method with related equations, this estimate may be validated by the independent assay of a suitable number of diluted samples which are close to the low limit of linearity range.

8.3.1.8 Limit of Quantification (LOQ)

The limit of quantification (LOQ) is a characteristic of a quantitative assay for a low level of analyte in a sample matrix. It can be described as the smallest concentration of the studied compound which gives a result that can be accurately quantified. The LOQ is also described as the lowest concentration at which the analyte cannot only be reliably detected but at which some predefined goals for bias and imprecision are met. The LOQ should be established by using an appropriate reference material and sample. This value should not be determined by extrapolation. The LOQ is always higher than the LOD and is often taken as a fixed multiple (generally 2 or 3) of the LOD and corresponds to the low limit of the range [1–10, 76–79]. In practice, the LOQ can be estimated analogously to the LOD using one of the procedures described in the above section. It has to be noted, however, that the verification of the trueness and precision of the method at this estimated LOQ should be fully assessed in order to completely validate the LOQ. Indeed the accuracy of results obtained at the LOQ has to be demonstrated as acceptable for the intended purpose of the analytical method [21, 57, 58, 66–68].

A. Based on Visual LOQ Inspection

This approach may be used for both noninstrumental and instrumental methods. Using this method, LOQ is determined by the electroanalysis of samples with known concentrations of analyte and by establishing the minimum level at which the analyte can be quantitated with acceptable accuracy and precision [1–10, 69–80]. As in the IUPAC method, testing of the accuracy and precision at the estimated LOQ will provide the final evidence for determination of the LOQ. If the CV % for the repeatability measurement at defined concentration is less than or equal to the accuracy and precision acceptance criteria, then the estimated LOQ is acceptable.

B. Based on Signal-to-Noise Ratio
This approach determines the signal-to-noise ratio (S/N) by comparing measured signals from samples with known low concentrations of analyte with those of blank samples. In this technique, the LOQ is the minimum measurement concentration at which the analyte can be reliably quantified for an S/N of 10:1 (Fig. 8.2c) [1–10, 69–80].

C. Based on the Standard Deviation of the Response
In electroanalytical studies, the mostly used way for the calculation of LOQ can be used by the following equation:

$$\mathrm{LOQ} = \frac{10s}{m}$$

In the above equation, s is the standard deviation of blank value (see Sect. 8.3.1.7C) and m is the slope of the related calibration line. Using the above equation, the response can easily be converted to the concentration units. Also, the relationship between LOQ and sensitivity is also confirmed using the above equation depending on the slope function. The estimation of s can be obtained using the same steps and with different approaches which are given in LOD section (8.3.1.7). This equation is also showing the relationship between LOQ and sensitivity, depending on the slope function.

For practically assessing the LOQ, another alternative way is:

$$\mathrm{LOQ} = \bar{x} + 10s$$

where $\bar{x}$ is the mean concentration of the blank and s is the standard deviation, respectively, and the same value used in LOD calculation. The only changing factor is "10" instead of "3" [1–5, 69–80].

LOQ can also be evaluated using some different methods. In one of this way, the blank working medium is spiked with different analyte concentrations, e.g., six different concentrations in the linearity range and close to the low limit of the linearity range or LOQ values. At least ten independent replicates should be realized for each concentration level, and the standard deviation should be calculated. The LOQ value is the concentration of analyte corresponding to a RSD value of 10 % [1–10, 76–79].

The obtained LOQ value and the calculation or evaluation method should be described with their related presentation or curves (such as voltammogram and chromatogram) for the justification. If the LOQ value is calculated using the standard deviation of blank response or intercept and slope method using with related equations, this estimate may be validated by the independent assay of a suitable number of diluted samples which are close to the low limit of the range. The LOQ cannot be lower than LOD.

8.3.1.9 Robustness

Robustness is a term introduced by USP/ICH [1–7, 81]. According to ICH, the robustness of a method is "a measure of its capacity to remain unaffected by small, but deliberate variations in method parameters and provides an indication of its reliability during normal usage." As a short description, the robustness is related to intra-laboratory influences. It is an attribute of electroanalytical methods that is gaining increasing importance through its impact on quality systems for analytical laboratories. The robustness of a method can be shown as the ability to reproduce an electroanalytical method in different circumstances without the occurrence of unexpected differences in the obtained results. The evaluation of robustness is normally considered during the development phase and depends on the type of procedure under study. Robustness tests examine the effect that operational parameters have on the analysis results. It is not given explicitly in the list of required validation parameters, but it is recommended as a part of method development to establish the critical measurement parameters. These tests should show the reliability of the analysis with respect to deliberate variations in method parameters [1–7].

The aim of the robustness studies is to find the method parameters that might lead to variations in the results when measurements are carried out under small different working conditions. Hence, the experimental weakness of the method is to detect and quantify, so that any critical factors can be anticipated and controlled in order to ensure that the operating conditions will fall within an undisturbed range. For the robustness tests of an electroanalytical method, the careful selection of the experimental variables that are crucial to the electroanalytical method such as supporting electrolyte, percentage of organic solvent, nature of the buffer, medium pH, ionic strength, scan rate, pulse period, pulse width, and temperature should be carefully studied. These studies should be realized under slightly different experimental conditions. Since the recorded peak current and/or peak potential can be significantly affected by a slight pH change, electrolytes with good buffering capacity should be selected in electroanalysis.

A test for robustness is an individual test, and it can be varied depending on the analytical method and equipment applied. The common way is first to define the parameters with reasonable maximum variation. Then each parameter is successively varied, whereas the others are held at nominal settings: this is a one factor at a time strategy. Nowadays, the statistical designs of experimentals (DoEs) methodology can often be preferred for obtaining as much as possible relevant information within a short time period and with limited experiments. Changes should be realized separately and the effect evaluated for each set of experimental condition on the precision and trueness [1–10, 82–89]. By combining changes in conditions

Table 8.3 Design to calculate the robustness of a proposed electroanalytical method

Parameters				
Experiments	Supporting electrolyte (P_1)	pH (P_2)	Applied potential (P_3)	Results
1	+	+	+	R_1
2	+	−	−	R_2
3	−	+	−	R_3
4	−	−	+	R_4

and performing a set of experiments, one can determine which factors have a significant or even critical influence on the electroanalytical results. For example, if three parameters can be evaluated, a design called fractional factorial design with only four experiments would be sufficient (Table 8.3).

In this table "+" and "−" signs denote the coded levels for the minimum and maximum values of the study. For each parameter, there should be two experiments at each level. After obtaining the results, the effect of parameters can be calculated as:

$$\text{Effect } P_1 = \left[\frac{(R_1 + R_2)}{2} - \frac{(R_3 + R_4)}{2}\right]$$

Using the above equation for each parameter results, a statistical and/or graphical analysis of the effects must be carried out in order to draw chemically relevant conclusions. Hence, the researchers can take an action to improve the performance of their developed method, if necessary.

The selected robustness parameters are specific for each method, and the scope of the investigations can vary depending on the method. Robustness is an essential subject in the validation parameters [1–10, 82–88]. The robustness study shows the estimation of the main total experimental effects and characterizes the behavior of the electroanalytical process. Allowable variations of each method parameter should be specified in the proposed technique.

8.3.1.10 Stability

The stability of drug active compounds in their dosage forms, biological fluids, or a raw material is a function of the storage conditions, the chemical properties of the drug, the matrix, and the container systems. A very important aspect to be considered in validation is the aging of the sample at the various stages of the process, in order to define the conditions appropriately. The stability of an analyte in a particular matrix and container system is relevant only to that matrix and container system and should not be extrapolated to other matrices and container systems. It can be defined as the, *"chemical stability of an analyte in a given matrix under specific conditions for given time intervals."* Many solutes readily decompose prior to the analytical investigations, for example, during the preparation of the sample solutions, extraction, cleanup, phase transfer, or storage of prepared tubes (in refrigerator or during the wait time). Under these circumstances, method development should investigate the stability of the analytes and standards. Hence, stability procedures should be evaluated as the stability of the analytes during sample collection and handling, after short-term (room temperature) and long-term (frozen at the intended storage temperature) storage and after going through freeze and thaw cycles and the analytical process. Conditions used in stability experiments should reflect situations likely to be encountered during actual sample handling and analysis. Under these circumstances, method development should investigate the stability of the analytes, standards, and stock solutions. All stability determinations should use a set of samples prepared from a freshly made stock solution of the analyte in the appropriate analyte-free, interference-free pharmaceutical and biological matrix.

The term system stability has been defined as the stability of the sample being analyzed in a sample solution. It is a measure of the bias in assay results generated during a preselected time interval, such as some selected hours up to 48 h or a week period, using single solution (Fig. 8.3).

System stability is appropriate if the results do not exceed 20 % of system precision. System stability should be determined by replicate analysis of the sample solutions, and it is considered appropriate when the calculated RSD values, at different time intervals, do not exceed more than 20 % of the corresponding value of the system precision. Stock solutions of the analyte for stability evaluation should be prepared in a suitable solvent at known concentration. To reduce the occurrence of changes in the analyte during the method development process, the following activities should be considered for the stock solution stability:

1. Stock solution stability
2. Determine the post-preparative stability (after 15–20 min and waiting measurements) of the prepared solutions
3. Determine the short-term stability (up to 6 h) of the prepared solutions
4. Determine the long-term stability (up to a week) of the prepared solutions
5. Freeze and thaw stability of the prepared solutions

The stability of the stock, standard, and working solutions should be evaluated to cover the time interval between preparation and use. The stability of all working solutions should be determined under the same conditions of light and/or dark, using the same temperatures, in the same matrix media and solvents. Solution stability is determined by comparing the response from aged standard, sample, or any other solutions to the response of freshly prepared standard, sample, or solutions. Studies to determine the stability of the analytes of interest should be conducted to replicate as closely as possible the actual in use conditions and the storage period should exceed the expected period of use. Hence, the stability studies that are conducted should provide evidence of how the quality of the drug substance and drug product changes over time when subjected to various environmental conditions, such as humidity, light, temperature, and oxidative irritant.

Fig. 8.3 Replicate analysis of the same sample solutions: (a) Consecutive replicates in 16 h; (b) Consecutive replicates in 1 month

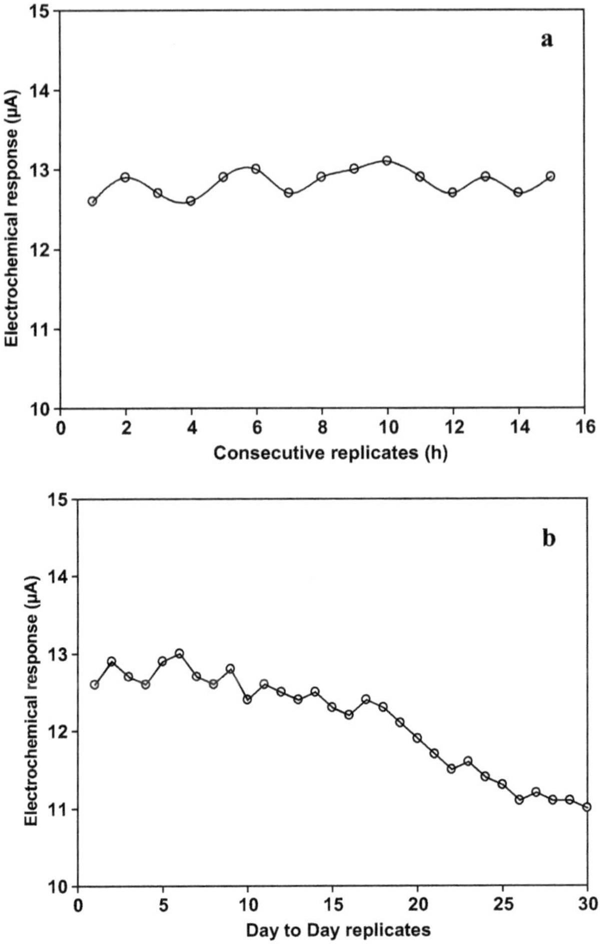

The aim of the stability test is to detect and if possible determine any degradation compounds which may appear during the experimental procedure, such as sample collection, processing, storing, preparing, and analyzing. It is also important to evaluate the proposed method's ability to measure drug products in the presence of its degradation products. For the precise, accurate, and stable experimental conditions of electroanalytical procedure, the samples, stock solutions, standards, reagents, the types, and pH of the supporting electrolytes and working electrodes must be stable for a reasonable time or at least within the experimental range. The necessary time period is depending on the experimental processes and should be indicated in the report.

A stability-indicating method is defined as a validated electrochemical analysis that accurately and precisely measure drug active compounds free from potential interferences like impurities, degradation products, excipients, endogenous substances, and other impurities. Also, the sample solutions should be checked for the decomposition and degradation products within a reasonable time period. This parameter is particularly important for biological samples and some of drug dosage forms, which may be stored frozen for several times prior to assay. The stability of this type of samples should be constant for at least two freeze–thaw cycles [1–10, 90–93].

Stock solutions of the analyte for stability evaluation should be prepared in an appropriate solvent (compatible with the electroanalytical method) at known concentrations. Stability of the stock solution studies should be evaluated during sample collection, preparation, handling experimental process, after short- and long-term storage, and after freeze–thaw cycles. A minimum of two cycles at two different concentrations should be studied in duplicate. After the repeated determinations at each storage level, both repeatability and reproducibility should be calculated. The criteria for the definition of stability should be clearly defined. Stability is a measure of the bias in accuracy results generated during a preselected time interval, using a single solution. All studies should be realized using the same solutions which prepared for each step of stability studies. Hence, it is considered that the appropriate results are obtained when the RSD values calculated on the assays obtained at different time intervals. System stability results of the sample and reference solutions are proved to the stability at least over 2 days, this will be sufficient for the drug assay [90–101]. Hence, at least a few hours stability is generally required for the standard and sample solutions stability. The calculated RSD value is a good indicator of the analyte stability. If the stock and standard solutions are not stable at room temperature, the solutions should be protected in the refrigerator and the experiments can be realized at lower temperature than the room temperature (such as between 2 and 8 °C) using cold water bath. If the stability of the stock solution does not require the system stability conditions, the standard stock solution may need to be prepared freshly every day or every time the voltammetric measurement is performed. Hence, the measurement system stability can be tested using freshly prepared solutions.

1. Stock Solution Stability

The stability of stock solutions of drug active compounds and/or the internal standard solution should be evaluated at room temperature for at least 6 h. This stability is evaluated to confirm that analyte degradation does not occur during the preparation, extraction or evaporation steps of working samples prior to their assay. If the stock solutions are put in a refrigerator or frozen for this period, the stability should also be documented. After completion of the desired storage time, the stability should be tested by comparing the freshly prepared solution results. The stability results of the stock solutions are determined by replicate analysis of the sample solution and calculation of the amount found and the RSD % of the responses [96–104].

2. Short-Term Stability

This type of stability studies should be designed and conducted to cover the type of storage conditions that are to be expected for studying the samples. Three aliquots of each, between low and high concentration ranges, should be thawed at room temperature from 6 to 24 h and analyzed. Stability results should be performed against freshly prepared standard samples. The short-term stability should be calculated and reported after calculation of the amount found and the RSD % of the responses. System stability is considered appropriate when the RSD does not exceed more than 20 % of the corresponding value of the short-term system precision [96–104].

3. Long-Term Stability
The storage time in a long-term stability evaluation should exceed the time between the date of first sample collection and the date of last sample analysis. In electroanalytical studies, these types of measurements are initiated during method validation, possibly evaluating analyte stability for a few weeks period, for example, 4 weeks or 2 months. This type of stability studies should be determined by storing at least three aliquots of each of the low, medium, and high concentration levels under the same conditions as the study samples. The volume of samples should be sufficient for the assay on three separate occasions. Long-term stability tests should cover at least the minimum period of sample storage, especially working with biological sample and injectable dosage forms.

The obtained results should be compared with freshly prepared samples which prepare at the beginning of the storage. Nominal concentrations can be used for the comparison purposes with the related RSD %.

4. Freeze–Thaw Stability
This type of stability studies are mostly used in biological samples. Freezing process may cause loss of analyte by several causes such as adsorption of the drug active compound to precipitates or onto endogenous substances of plasma, proteins or crystallization from urine samples, degradation by enzymes, by dissolved oxygen, etc. Analyte stability should be determined after three freeze and thaw cycles. Reasons necessitating analysis after multiple freeze–thaw cycles include failed analytical runs due to the use of incorrect dilutions or sample preparation steps. At least three aliquots at each of low, medium, and high concentration of samples should be stored at the intended storage temperature and conditions for 24 h and thawed unassisted at room temperature. After completely thawed, the samples should be frozen for 12–24 h under the same conditions. This step should be repeated two more times and then analyzed on the third cycle. The results are compared to the initial amount of the fresh unfrozen or freshly prepared control samples [96–104].

5. Post-preparative Stability
This type of stability of the processed samples, including the resident time in the container, should also be determined. The previously prepared and waited samples can be shown same stability problems or produce unwanted additional responses. That is why the stability of the compound should be assessed over the anticipated working time for the batch size in validation samples by determining concentrations on the basis of original linearity standards.

For establishing the stability of a proposed electroanalytical study, the mean analyzed results for all investigated samples should be within 15 % of the nominal sample concentrations. The recommended value is that the coefficient of variation (CV) for the replicate assay does not exceed 15 %. Stability studies are, however, not mandatory parameters for the validation of pharmaceutical analyses; they are only recommended by the ICH guidelines

8.3.1.11 Applicability

Determination of drug active compounds using electroanalytical techniques includes also identification of the analyte. Polarographic and voltammetric methods using solid electrodes and also potentiometric method by ion selective electrodes seem to be the most popular during these days [18, 105, 106]. These methods are suitable for the qualitative and quantitative analysis of electrochemically reducible (such as nitro, nitroso, and azo) and oxidizable (aromatic amines, phenols, indoles, etc.) compounds [105, 106]. In these measurements, the voltammetric peak potential is used for the qualitative identification, and the peak response (height or area) is used for the quantitative assay. Bare and modified electrodes are suitable for voltammetric or amperometric (coupled to LC, CZE) assays [105–108]. The assay should be completed within the time period for which stability data are available. After full validation of the proposed electroanalytical method, the analysis of the samples of interest can be performed. When a new method is proposed or when there is a choice of methods, it may be useful to provide an indication or description of the ease or tediousness of the selected method. The problems are in most cases related to the availability and maintenance of certain equipments and the required staff or skills. Also, the supply of required parts, solvents, electrodes, reagents, and impurities is not always assured nor the uninterrupted supply of a stable electrical power supply.

References

1. Ermer J, Miller JHMB (eds) (2005) Method validation in pharmaceutical analysis. Wiley-VCH, Weinheim
2. Riley CM, Rosanske TW (eds) (1996) Development and validation of analytical methods. Pergamon, Oxford
3. The United States Pharmacopoeia, USP 23 (1994) Validation of compendial methods. General information <1225>. The United States Pharmacopoeial Convention Inc., Taunton, MA
4. Guidelines for Submitting Samples and Validation Data for Methods Validation US Food and Drug Administration (1987) Center for Drug and Biologics. Department of Health and Human Services
5. ICH Guideline Q2B (1996) Validation of analytical procedures: methodology. November
6. ICH Guideline Q2 (R1) (2005) Validation of analytical procedures: text and methodology. November

7. The Third International Conference on Harmonization of Technical Requirements for Registration on Pharmaceuticals for Human Use Validation of Analytical Procedures: Methodology (1995) Announcements, Yokohama, Japan
8. Chan CC, Lam H, Lee YC, Zhang XM (eds) (2004) Analytical method validation and instrument performance verification. Wiley-Interscience, Hoboken, NJ
9. Ahuja S, Scypinski S (eds) (2001) Handbook of modern pharmaceutical analysis. Academic, London
10. De Bievre P, Günzler H (eds) (2005) Validation in chemical measurement. Springer, Heidelberg
11. Lunte SM, Radzik DM (1996) Pharmaceutical and biomedical applications of capillary electrophoresis. Pergamon, Oxford
12. Haider SI (ed) (2002) Pharmaceutical master validation plan. CRC Press, New York, NY
13. Thompson M, Ellison SLR, Wood R (2002) Pure Appl Chem 74:835–855
14. Berry IR, Harpaz D (eds) (2001) Validation of active pharmaceutical ingredients. CRC Press, Boca Raton, FL
15. Bliesner DM (ed) (2006) Validating chromatographic methods: a practical guide. Wiley-Interscience, Hoboken, NJ
16. Taverniers I, De Loose M, Van Bockstaele E (2004) Trac Trends Anal Chem 23:535–552
17. Konieczka P (2007) Crit Rev Anal Chem 37:173–190
18. Ozkan SA (2012) Electroanalytical methods in pharmaceutical analysis and their validation, 1st edn. HNB Publishing, New York, NY
19. Prichard E, Barwick V (2007) Quality assurance in analytical chemistry. Wiley, Chichester
20. Kellner R, Mermet JM, Otto M, Valcarcel M, Widmer HM (eds) (2004) Analytical chemistry, a modern approach to analytical science, 2nd edn. Wiley-VCH, New York, NY
21. Rozet E, Ceccato A, Hubert C, Ziemons E, Oprean R, Rudaz S, Boulanger B, Hubert P (2007) J Chromatogr A 1158:111–125
22. Branch SK (2005) J Pharm Biomed Anal 38:798–805
23. Bakshi M, Singh S (2002) J Pharm Biomed Anal 28:1011–1040
24. Ermer J, Ploss HJ (2005) J Pharm Biomed Anal 37:859–870
25. Furlanetto S, Orlandini S, Mura P, Sergent M, Pinzauti S (2003) Anal Bioanal Chem 377: 937–944
26. Bruce P, Minkkinen P, Riekkola ML (1998) Microchim Acta 128:93–106
27. Grdinic V, Vukovic J (2004) J Pharm Biomed Anal 35:489–512
28. Taylor JK (1983) Anal Chem 55:600A–608A
29. Gorog S (2007) Trac Trends Anal Chem 26:12–17
30. Jeya Chandra M (ed) (2001) Statistical quality control. CRC Press, Boca Raton, FL
31. Nilsen CL (ed) (1996) Managing the analytical laboratory: plain and simple. Interpharm/CRC Press, London
32. Stefan RI, Aboul-Enein HY (1998) Accred Qual Assur 3:194–196
33. Hibbert B (1999) Accred Qual Assur 4:352–356
34. Garfield FM (1992) Quality assurance principles for analytical laboratories, 2nd edn. AOAC International, Gaithersburg, MD
35. Ermer J (2001) J Pharm Biomed Anal 24:755–767
36. Hajicostas E (2003) Validation of analytical methods. Springer, Berlin
37. Huber L (ed) (1998) Validation and qualification in analytical laboratories. Interpharm Press, East Englewood, CO
38. Vessman J (1996) J Pharm Biomed Anal 14:867–869
39. Chanbers D, Kelly G, Limentani G, Lister A, Lung KR, Warner E (2005) Pharm Tech 9: 64–80
40. Peters FT, Maurer HM (2002) Accred Qual Assur 7:441–449
41. ICH Topic Q1A (R2) (2003) Stability testing of new drug substances and products. August
42. JCGM 200:2008 International vocabulary of metrology – basic and general concepts and associated terms (VIM)

43. Milman BL (ed) (2011) Chemical identification and its quality assurance. Springer, Berlin
44. Funk W, Dammann V, Donnevert G (eds) (2007) Quality assurance in analytical chemistry, 2nd completely revised and enlarged edn. Wiley-VCH, Weinheim
45. Reichenbacher M, Einax JW (2011) Challenges in analytical quality assurance. Springer, Berlin
46. Hartmann C, Massart DL, McDowall RD (1994) J Pharm Biomed Anal 12:1337–1343
47. Hartmann C, Smeyers-Verbeke J, Massart DL, McDowall RD (1998) J Pharm Biomed Anal 17:193–218
48. Castro M, Bolfarine H, Galea-Rojas M, Castilho MV (2005) Anal Chim Acta 538:375–381
49. Westgard JO (1995) Clin Lab Sci 9:277–283
50. Westgard JO (1999) Basic method validation. Westgard Quality Corporation, Ogunquit, ME
51. Stöckl D, Reinauer H (1993) Scand J Clin Lab Invest 53:16–18
52. Stöckl D (1995) Clin Chem 41:120–121
53. Pollock MA (1992) Ann Clin Biochem 29:556–560
54. Green JM (1996) Anal Chem News Features 305A–309A
55. Winslow PA, Meyer RF (1997) J Validation Technol 14:361–367
56. Bouabidi A, Talbi M, Bourichi H, Bouklouze A, El Karbane M, Boulanger B, Hubert P, Rozet E (2012) Drug Test Anal 4:1014–1422
57. Rozet E, Marini Djang'Eing'A R, Ziemons E, Boulanger B, Hubert P (2011) J Pharm Biomed Anal 55:848–858
58. Rozet E, Ziemons E, Marini RD, Boulanger B, Hubert P (2012) Anal Chem 84:106–112
59. Agalloco JP, Carleton FJ (eds) (2007) Validation of pharmaceutical process, 3rd edn. CRC Press, New York, NY
60. Christian GD (ed) (2004) Analytical chemistry, 6th edn. Wiley, Hoboken, NJ
61. Meier PC, Zünd RE (eds) (2000) Statistical methods in analytical chemistry, 2nd edn. In: Winefordner JD (Series ed) Chemical analysis: a series of monographs on analytical chemistry and its applications, vol 153. Wiley, New York, NY
62. Rosslein M, Rezzonico S, Hedinger R, Wolf M (2007) Accred Qual Assur 12:45–47
63. Rozet E, Wascotte V, Lecouturier N, Préat V, Dewé W, Boulanger B, Hubert P (2007) Anal Chim Acta 591:239–247
64. Rozet E, Bouabidi A, Fillet M, Ziemons E, Chapuzet E, Mertens B, Klinkenberg R, Ceccato A, Talbi M, Streel B, Bouklouze A, Boulanger B, Hubert P (2010) J Chromatogr A 1217:3180–3192
65. Rozet E, Dewe W, Rudaz S, Marini Djang'Eing'A R, Ziemons E, Boulanger B, Hubert P (2011) Trends Anal Chem 30:797–806
66. Hubert P, Nguyen-Huu JJ, Boulanger B, Chapuzet E, Chiap P, Cohen N, Compagnon PA, Dewé W, Feinberg M, Lallier M, Laurentie M, Mercier N, Muzard G, Nivet C, Valat L (2004) J Pharm Biomed Anal 36:579–586
67. Hubert P, Nguyen-Huu JJ, Boulanger B, Chapuzet E, Chiap P, Cohen N, Compagnon PA, Dewé W, Feinberg M, Lallier M, Laurentie M, Mercier N, Muzard G, Nivet C, Valat L, Rozet E (2007) J Pharm Biomed Anal 45:70–81
68. Hubert PH, Nguyen-Huu JJ, Boulanger B, Chapuzet E, Cohen N, Compagnon PA, Dewé W, Feinberg M, Laurentie M, Mercier N, Muzard G, Valat L, Rozet E (2007) J Pharm Biomed Anal 45:82–96
69. Danzer K, Currie LA (1998) Pure Appl Chem 70:993–1014
70. Asuero AG, Sayago A, Gonzalez AG (2006) Crit Rev Anal Chem 36:41–59
71. Van Loco J, Elskens M, Croux C, Beernaert H (2002) Accred Qual Asur 7:281–294
72. Hibbert DB (2005) Accred Qual Assur 10:300–301
73. Huber W (2004) Accred Qual Assur 9:726–726
74. Ellison SLR (2006) Accred Qual Assur 11:146–152
75. Gonzalez AG, Herrador MA, Asuero AG, Sayago A (2006) Accred Qual Assur 11:256–258
76. Kaus R (1998) Accred Qual Assur 3:150–154
77. Armbruster DA, Tillman MD, Hubbs LM (1994) Clin Chem 40:1233–1238

78. Currie LA (1995) Pure Appl Chem 67:1699–1723
79. Switaj-Zawadka A, Konieczka P, Przyk E, Namiesnik J (2005) Anal Lett 38:353–376
80. Kuselman I, Sherman F (1999) Accred Qual Assur 4:124–128
81. Krull IS, Swartz M (1999) Anal Lett 32:1067–1080
82. Cuadros-Rodriguez L, Romero R, Bosque-Sendra JM (2005) Crit Rev Anal Chem 35:57–69
83. Vander Heyden Y, Hartmann C, Massart DL, Michel L, Kiechle P, Erni F (1995) Anal Chim Acta 316:15–26
84. Vander Heyden Y, Luypaert K, Hartmann C, Massart DL, Hoogmartens J, De Beer J (1995) Anal Chim Acta 312:245–262
85. Van Leeuwen JA, Vandeginste BGM, Kateman G, Mulholland M, Cleland A (1990) Anal Chim Acta 228:145–153
86. Mulholland M (1988) TrAc, Trends Anal Chem 7:383–389
87. Nijhuis A, Vander Knaap HCM, de Jong S, Vandeginste BGM (1999) Anal Chim Acta 391: 187–202
88. Vander Heyden Y, Massart DL, Zhu Y, Hoogmartens J, De Beer J (1996) J Pharm Biomed Anal 14:1313–1326
89. VonHolst C, Müller A, Björklund E, Anklam E (2001) Eur Food Res Technol 213:154–160
90. Shah VP, Midha KK, Dighe S, Mc Gilveray IJ, Skelly FP, Yacobi A, Layloff T, Viswanathan CT, Cooke CE, Mc Dowall RD, Pittman KA, Spector S (1992) Pharm Res 9:588–592
91. Grimm W (1993) Drug Dev Ind Pharm 19:2975–2982
92. Cloud PA (1997) BioPharm 30:39–42
93. Toennes SW, Kauert GF (2001) J Anal Toxicol 25:339–343
94. Rosing H, Man WY, Doyle E, Bult A, Beijnen JH (2000) J Liq Chromatogr Relat Technol 23: 329–354
95. Dadgar D, Burnett PE, Choc MG, Gallicano K, Hooper JW (1995) J Pharm Biomed Anal 13: 89–97
96. Dadgar D, Burnett PE (1996) J Pharm Biomed Anal 14:23–31
97. Walz R, Knauf M (2008) Pharm Ind 70:1135–1139
98. Dantus MM, Wells ML (2005) J Liq Chromatogr Relat Technol 27:1413–1442
99. Braggio S, Barnaby RJ, Grossi P, Cugola M (1996) J Pharm Biomed Anal 14:375–388
100. Timm U, Wall M, Dell D (1985) J Pharm Sci 74:972–977
101. Buick AR, Doig MV, Jeal SC, Land GS, Mc Dowall RD (1990) J Pharm Biomed Anal 8: 629–637
102. Taylor RB, Shivji ASH (1987) Pharm Res 4:177–180
103. ICH Harmonized Tripartite Guideline (2003) Q1E Evaluation for stability Data
104. Ruberg SJ, Stegeman JW (1991) Biometrics 47:1059–1069
105. Filipiak M (2001) Anal Sci Suppl 17:i1667–i1670
106. Barek J, Moreira JC, Zima J (2005) Sensors 5:148–158
107. Flanagon RJ, Perrett D, Whelpton R (eds) (2005) Electrochemical detection in HPLC. Analysis of drugs and poisons. Royal Society of Chemistry, Cambridge
108. LaCourse WR (ed) (1997) Pulsed electrochemical detection in high performance liquid chromatography. Wiley, New York

Chapter 9
Applications for Drug Assays

Contents

Abbreviations

$[Cu_4(apyhist)_4]^{4+}$	Cyclic-tetrameric copper (II) species containing the ligand (4 imidazolyl) ethylene-2-amino-1-ethylpyridine
5-HT	5-Hydroxytriptamine (serotonin)
AC	Alternating current
AChE	Acetylcholinesterase
ACP	Alternating current polarography
ACS	Alumina-coated silica
ACV	Alternating current voltammetry
AdSV	Adsorptive stripping voltammetry
AgLAF-AgSAE	Silver liquid amalgam film-silver solid amalgam annular band electrode
AgNPs	Silver nanoparticles
AHNSA	4-amino-3-hydroxynaphthalene sulfonic acid
ARS	Alizarine Red S
ASV	Adsorptive stripping voltammetry
BDDE	Boron-doped diamond electrode
BPPGE	Basal plane pyrolytic graphite electrode

S.A. Ozkan et al., *Electroanalysis in Biomedical and Pharmaceutical Sciences*, Monographs in Electrochemistry, DOI 10.1007/978-3-662-47138-8_9

CA	Chronoamperometry
CAdSV	Cathodic adsorptive stripping voltammetry
CC	Chronocoulometry
CCE	Carbon-ceramic electrode
CD	Cyclodextrin
CDP	Cyclodextrin prepolymer
CE	Carbon electrode
CFME	Carbon fiber microelectrode
CILE	Carbon ionic liquid electrode
Cl	Cardiolipin
CNF	Carbon nanofiber
CNT	Carbon nanotube
CNT-mer	Multiwalled carbon nanotubes bearing terminal monomeric unit
CNTPE	Carbon nanotube paste electrode
$Co(DMG)_2ClPy$	Chloro(pyridine)bis(dimethylglyoximato)cobalt(III)
CoHCNFe	Cobalt hexacyanoferrate nanoparticles
CPC	Controlled potential coulometry
CPE	Carbon paste electrode
CP-TNMCPE	Carbon paste electrode modified by meso-tetrakis (3-methylphenyl) cobalt porphyrin and TiO_2 nanoparticles
CSV	Cathodic stripping voltammetry
CTS CHIT	Chitosan
CV	Cyclic voltammetry
Cys	Cysteine
D50wx2	Dowex50wx2
DCP	Direct current polarography
DDP	3,4-dihydroxybenzaldehyde-2,4-dinitrophenylhydrazone
DME	Dropping mercury electrode
DMF	Dimethyl formamide
DNA	Deoxyribonucleic acid
DPAdASV	Differential pulse adsorptive anodic stripping voltammetry
DPAdCSV	Differential-pulse adsorptive cathodic stripping voltammetry
DPAdSV	Differential pulse adsorptive stripping voltammetry
DPASV	Differential pulse anodic stripping voltammetry
DPAV	Differential pulse anodic voltammetry
DPE	Diamond paste electrode
DPP	Differential pulse polarography
DPSV	Differential pulse stripping voltammetry
DPV	Differential pulse voltammetry
dsDNA	Double stranded deoxyribonucleic acid
DTPS SAM-ME	Au DTPS self-assembled monolayer modified electrode

EBNBHCNPE	2 2′-[1,2-ethanediylbis (nitriloethylidyne)]-bis-hydroquinone carbon nanotube paste electrode
ECD	Electrochemical detection
EIS	Electrochemical impedance spectroscopy
$EMIMBF_4$	1-ethyl-3-methylimidazolium tetrafluoroborate
EPPGE	Edge plane pyrolytic graphite electrode
ERGO	Electrochemically reduced graphene oxide
FCDCNTPE	Carbon nanotube paste electrode modified with ferrocene dicarboxylic acid
FCMCNTPE	Ferrocene modified carbon nanotubes paste electrode
Fe (III)-SBMCPE	Fe (III) schiff base modified carbon paste electrode
FeNAZ-CH	Iron ion-doped natrolite zeolite
FFT CV	Fast Fourier transform cyclic voltammetry
FIA	Flow injection analysis
FSCV	Fast scan cyclic voltammetry
GAMCS NPs	Gold-aminomercaptothiadiazole core-shell nanoparticles
GCE	Glassy carbon electrode
GCPE	Glassy carbon paste electrode
GE AuE	Gold electrode
GluOx	Glutamate oxidase
GNMCPE	Gold nanoparticles modified carbon paste electrode
GNP AuNP	Gold nanoparticle
GNSs	Graphene nanosheets
GO	Graphene oxide
GPE	Graphene paste electrode
GPGE	Gold modified pencil graphite electrode
Gr GrO	Graphene
GS	Graphene sheets
GS-PTCA	Graphene sheets and 3,4,9,10-perylenetetracarboxylic acid
GSR	Graphite silicone rubber
HMDE	Hanging mercury drop electrode
HNCMS	Hollow nitrogen-doped carbon microspheres
HRP HP	Horseradish peroxidase
IL	Ionic liquid
ITO	Indium tin oxide
LAdCV	Linear adsorptive cyclic voltammogram
LC	Liquid chromatography
lcys	L-cysteine
LOD	Limit of detection
LOQ	Limit of quantification
LSAdCSV	Linear sweep adsorptive cathodic stripping voltammetry
LSP	Linear sweep polarography
LSV	Linear sweep voltammetry
m-AgSAE	Mercury meniscus modified silver solid amalgam electrode

MCCE	Modified carbon composite electrode
MCPE	Modified carbon paste electrode
ME	Mercury electrode
Me-CoSal	Cobalt methyl salophen
MIP	Molecularly imprinted polymer
MMT-Ca	Montmorillonite-Ca
MRT	Main residence time
MWCNT	Multiwalled carbon nanotube
MWCNT-COOH cMWCNT	Carboxylated multiwalled carbon nanotubes
MWCNTILEE	Multiwall carbon nanotubes ionic liquid electrode
MWCNTPE	Multiwalled carbon nanotube paste electrode
NCPE	Nafion-incorporated carbon paste electrode
NdHCF	Neodymium hexacyanoferrate film
N-DHPBMCNTPE	*N*-(3,4-dihydroxyphenethyl)-3,5-dinitrobenzamide modified carbon nanotube paste electrode
Nf	Nafion
NGCE	Nafion-coated glassy carbon electrode
NiHCF-NP	Nickel hexacyanoferrate nanoparticle
NiHCF-NP	Nickel hexacyanoferrate nanoparticles
Ni-NP	Nickel phosphate nanoparticles
NiONPs	Nickel oxide nanoparticles
NIP	Non-imprinted polymer
nPt-MWCNTPE	Pt-nanoparticles modified multiwalled carbon nanotube paste electrode
NSAIDs	Non-steroidal antiinflammatory drugs
o-MWCNT	Oxidized multiwall carbon nanotube
OPG-CoPc	Ordinary pyrolytic graphite-cobalt phthalocyanine
p-(AHNSA)	Poly(4-amino-3-hydroxynaphthalene sulfonic acid)
PAA	Poly(acrylic) acid
p-ABA	*p*-aminobenzoic acid
PA-MNPs	PEGylated arginine functionalized magnetic nanoparticles
p-AMTa	3-Amino-5-mercapto-1,2,4-triazole
PANI PA	Polyaniline
p-APMCNTPE	Multiwall carbon nanotubes paste electrode modified with *p*-aminophenol
p-CACNTPE	*p*-chloranil-carbon nanotubes paste electrode
p-CAMCNTPE	Carbon nanotubes paste electrode modified with *p*-chloranil
PCFE	Nanocrystalline graphite-like pyrolytic carbon film electrode
PCPE	Pseudo-carbon paste electrode
PdNPs	Palladium nanoparticles
PDR-PGE	Dendrimer modified disposable graphite electrode
PEDOT	Poly(3,4-ethylene dioxythiophene)

PFMCPE	Polymer film modified carbon paste electrode
PGA	Poly-glutamic acid
PGE	Pencil graphite electrode
PIGE	Paraffin wax impregnated graphite electrode
PMB	Poly(methylene blue)
PMWCNTPE	Pretreated multiwalled carbon nanotube paste electrode
PNT	L-diphenylalanine nanotubes
PoAP	poly(o-aminophenol)
poly(DA)	Diglycolic acid polymer
poly(PVC)	Poly(pyrocatechol violet)
Polyhis	Polyhistidine
polyTTBA/PS	Poly [2,2′:5′,2″-terthiophene-3′-(*p*-benzoic acid)] conducting polymer—phosphatidylserine
polyTTBA	Carboxylic acid group-functionalized conducting polymer
PPGE	Pretreated pencil graphite electrode
PSS	poly(sodium 4-styrenesulfonate-co-acrylic acid)
PtE	Platinum electrode
PVF	Poly(vinylferrocenium)
PyCE	Pyrolytic carbon electrodes
rGO	Reduced graphene oxide
RTIL	Room temperature ionic liquids
SBAP	Poly(2-hydroxy-5-[(4-sulfophenyl)azo]benzoic acid) film
SDS	Sodium dodecyl sulfate
SiAlNb	Sol–gel silica
SiC-NPs	Silicon carbide nanoparticles
SMDE	Static mercury drop electrode
SPCE	Screen-printed carbon electrodes
SPE	Screen-printed electrode
SPGCE	Screen-printed glassy carbon electrode
SWAdASV	Square wave adsorptive anodic stripping voltammetry
SWAdCSV	Square-wave adsorptive cathodic stripping voltammetry
SWAdSV	Square wave adsorptive stripping voltammetry
SWASV	Square wave anodic stripping voltammetry
SWCNT	Single walled carbon nanotube
SWP	Square wave polarography
SWV	Square wave voltammetry
TNGCE	Tellurium-nanowire-coated glassy carbon electrodes
TP	Tast polarography
Tyr	Tyrosinase
ZCP	Zero current potentiometry
ZnSNP	ZnS nanoparticle
ZnTNP	Zn-tetranaphthaloporphyrin
ZnTPP	Zn-tetraphenyl porphyrin
ZnTSPP	Zn-5,10,15,20-tetra(4-sulfophenyl)porphyrin
β-CD	Beta-cyclodextrin

9.1 Introduction

The choice of an appropriate technique for the determination of a drug or a mixture of drug compounds is mainly governed by the sample matrix complexity, the analyte concentration, and the sample treatment time. Compared to optical, mass, and thermal methods, electrochemical techniques are especially attractive because they offer unique detection capabilities with remarkable selectivity and sensitivity for a moderate cost [1–17]. The final selection of an analytical technique depends on the intrinsic structure and physicochemical properties of the studied analyte. Potentiometric electrodes have been described for the selective determination of pharmaceuticals, but voltammetric methods are most generally developed due to their inherent high sensitivity and versatility provided that the studied drug compound is electroactive.

9.2 Assay of Drugs

As with other instrumentations, electroanalytical methods have largely benefited from progress in material sciences, miniaturization, and computerized processing of analytical data. Electroanalytical methods are widely developed at the academic level for the assay of pharmaceutically active compounds in their dosage forms and in biological samples. Their routine use in the pharmaceutical industry and clinical laboratories is, however, not as popular. This is unfortunate since many pharmaceutical analytical problems could readily be solved with a high degree of accuracy and precision using electrochemical methods. In the analysis of a drug compound in its dosage form, electroanalytical methods have been shown to be highly suitable due to the simplicity of sample preparation and lack of interferences from excipients [1–8]. The nowadays most commonly used electrochemical techniques are cyclic, differential pulse (DPV) and square-wave voltammetry (SWV), stripping voltammetry, amperometry, and coulometry. Cyclic voltammetry (CV) is usually the first experiment to be performed in order to check for the electroactivity of a molecule of interest. It enables a wide potential range to be scanned rapidly for the screening of the electroactivity of a drug compound (Tables 9.1 and 9.2). Different electrodes are available, yet most frequently used are glassy carbon, carbon paste, gold, platinum, and boron-doped diamond electrodes. The experiments can be performed in a few mL or in μL volumes of solution when using screen-printed electrodes. The latter have reached substantial commercial success, they are based on carbon composite material, and can be modified by different reagents, metal, or carbon particles. They are mass produced with good reproducibility (RSD of geometric working electrode area below 5 %) and may serve for multiple or single use. It is expected that their use will stimulate routine analyses in various pharmaceutical areas in the near future. In addition to the possible quantification of the studied compound, CV provides mechanistic information on the redox pattern

Table 9.1 Redox potentials and chemical structures of some oxidizable groups

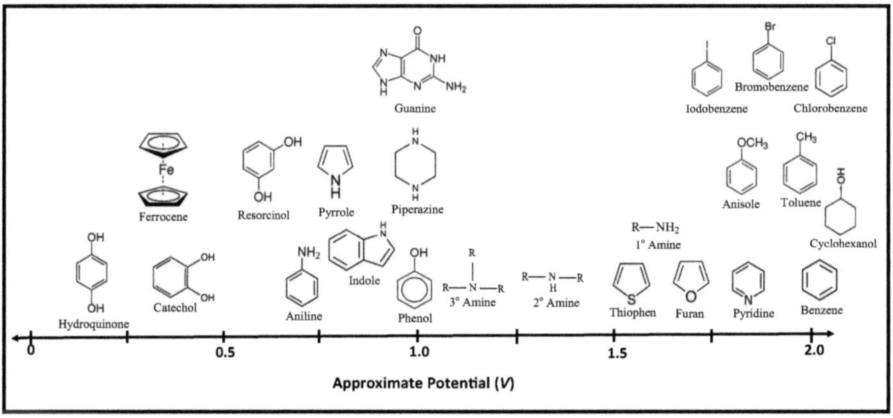

Table 9.2 Redox potentials and chemical structures of some reducible groups

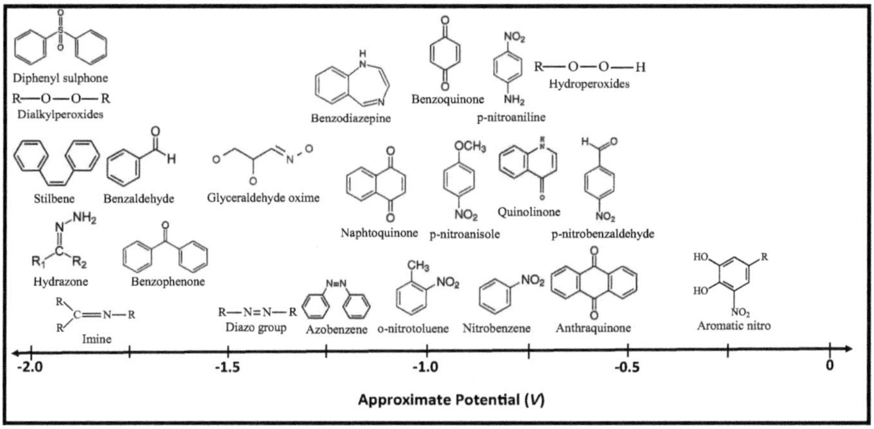

occurring at the electrode–solution interface. Multiple voltammetric cyclings on the same surface provide also information on the extent of drug adsorption and surface fouling phenomena. This information is useful for orienting subsequent quantitative electroanalyses especially regarding surface cleaning or renewing. DPV or SWV can be directly applied to a sample solution provided that selectivity criteria are achieved. In biological samples, however, since the drug of interest is usually present at low concentration and due to the complexity of the matrix, a sample pretreatment and a separation step are mandatory. Liquid–liquid or solid phase extraction with subsequent dissolution in a suitable buffer can be considered but usually an additional separation step is needed such as liquid chromatography or capillary electrophoresis with electrochemical detection (conductivity or amperometry) [1–17].

A dosage form can be dissolved in an aqueous medium, filtered, and analyzed directly by voltammetry or, if interfering species are encountered, by LC with amperometric detection. Biological samples are too complex for allowing the direct analysis of a drug compound; an extraction step can be realized by preconcentrating the analyte of interest onto the electrode surface followed by medium exchange and subsequent electroanalysis. Generally, however, liquid–liquid or solid phase extraction is required for sample cleanup and analyte preconcentration as mentioned above. The most demanding of the pharmaceutical analytical assays are those that require the measurement of a drug and eventually its metabolites in biological fluids such as plasma and urine following the administration of low doses. These assays must be of high selectivity with sensitivity in the "ng mL^{-1}" range [1–17]. Biological fluids such as plasma, serum, and whole blood are most commonly analyzed for pharmacokinetic parameter evaluation. Whole blood presents a unique challenge in method development and validation because of the viscous nature of blood and complexity of its composition. A major step in the challenging analysis of whole blood samples has recently been achieved by the development of a highly innovative flow system with a microfluidic chip for in vivo continuous monitoring of a drug compound. The biosensor chip contained a series of gold working electrodes with an immobilized electroactive aptamer selective to the drug to be monitored (e.g., doxorubicin or kanamycin) [18].

When dealing with biological samples, liquid chromatography with mass spectrometric detection (LC-MS/MS) is first applied in method development due to high sensitivity and selectivity achieved. Once the LC-MS/MS is validated, the LC separation methodology can subsequently be applied in routine analysis using other detectors such as UV–Vis or amperometry [17–22]. The innovative combination of flow-through electrochemical cells on-line with mass spectrometry is an additional attractive feature of electrochemistry. This hyphenated and purely instrumental configuration has allowed pushing electrochemistry several steps forward in the arsenal of analytical tools for drug biotransformation studies. Coupling EC with LC/TOF-MS permits structural identifications with a high degree of certainty, and better mechanistic information is obtained about the oxidation or reduction products generated at solid electrodes. The EC-MS or LC-EC-MS methodology applied to drug compounds allows microsynthesis and on-line identification of low amounts of products, and these processes can mimic to some extent in vivo biotransformation [19].

As with any analytical procedure, the reproducibility and accuracy of an electroanalytical assay must be verified for the parent compound and/or a chemical derivative or a metabolite. Methods developed should be uncomplicated and analytically robust for use in a routine basis. This chapter centers on coulometric and voltammetric procedures to demonstrate their utility in solving analytically oriented pharmaceutical problems in drug development [17–22].

Electrochemical biosensors are electroanalytical tools which combine, in an intimate contact, a biological compound of recognition with a physical transducer [1–8]. Their unique dual structure is particularly attractive in some specific niches in drug research and development [1–5]. An interesting concept is to integrate metabolically active enzymes as bio-recognition element. Two major difficulties, however, are encountered:

(a) The complexity to measure the enzymatic activity
(b) The enzyme stability

These problems can be solved to some extent using different strategies such as electrode surface modification for fast electron transfer to occur and immobilization techniques [1–5]. The high degree of miniaturization offered by screen-printed electrodes with an all in one three electrodes configuration is substantially stimulating the biosensors research field [17]. Cell-based biosensors and DNA-based biosensors can be considered as unique research tools in drug research. A cell-based biosensor represents an advantageous compromise between a purely in vitro model and a whole organism. The use of cells takes into account a more faithful perspective of the complexity of the in vivo environment. DNA-based immobilized electrodes have emerged as useful tools for drug–DNA interactions studies and for detecting damages caused by metabolically generated species which can affect the integrity of the genetic material. In recent years, there has been a growing interest in the electrochemical investigation of interaction between anticancer or antiviral drugs and DNA. Observing the electrooxidation of surface immobilized DNA (through its guanine and adenine residues) in the absence and in the presence of a drug compound provides good evidence for the interaction mechanism to be elucidated. Also this interaction could be used for the quantification of these drugs and for the determination of new drugs targeting DNA [23].

The aim of this chapter is to give some examples and information to aid the use of recent electroanalytical and LC-EC methods and electrochemical DNA biosensors in the analysis of drugs in their dosage forms and in biological samples [1–4, 8–15, 17–26]. The selected pharmaceutical active compounds that can be determined using different electroanalytical, LC-EC, LC-EC-MS techniques, and DNA–drug interaction studies are reported in tables together with some details. The available information such as supporting electrolyte, pH, measuring or detection potential, column, mobile phase, extraction procedure, internal standard, sensitivity, selectivity, precision, accuracy, etc. is presented as follows and in the tables. Chemical names or structural formulae are given to aid identification of electroactive moieties.

9.3 Examples of Electrochemical Application

9.3.1 Cyclic, Pulsed, and Stripping Voltammetric Methods Using Solid Electrodes in Pharmaceutical Analysis

CV, LSV, DPV, SWV, and stripping voltammetric methods are appropriate for the determination of numerous classes of pharmaceutically active compounds as reported in tables.

9.3.1.1 Antibacterial, Antifungal, Antiviral, and Antiparasitic

Ashrafi et al. developed an electrooxidation method that allows the determination of the antiviral drug abacavir in pharmaceutical dosage forms and human serum samples using a MWCNT-EPPG electrode [27]. Cyclic voltammograms showed two oxidation peaks located at +0.91 V and +1.10 V. Dependency of peak potential and peak current on pH was investigated in detail. The oxidation potentials were affected by the pH indicating that protons are involved in the electrochemical oxidation of abacavir. Mass transport to the electrode surface was found due to a combination of diffusion and adsorption which was concluded by studying the potential scan rate dependency of both peak currents. A linear relation was observed upon the abacavir concentration within a range of 0.1–20 μM with ($r = 0.999$) for both peak responses. The method was validated with respect to selectivity, sensitivity, precision, and accuracy. Abacavir was also determined with the proposed method in its pharmaceutical dosage forms and human serum samples. Another antiviral drug fosamprenavir was studied in aqueous-alcohol medium in the presence of TritonX-100 at a carbon paste electrode [28]. Linearity was in the range of 1–50 μM fosamprenavir. Thanks to the low detection limit (0.48 μM) and good repeatability and reproducibility of the results, the method was used for the determination of the drug in a pharmaceutical dosage form. Dogan-Topal et al. developed an electroanalytical method using MWCNT-GCE for the determination of valganciclovir (antiviral drug against cytomegalovirius infections) in pharmaceutical dosage forms [29]. The calibration curve was linear in the concentration range of 7.5−1,000 nM with the detection limit of 1.52 nM. A sensitive electrochemical sensor was developed for the determination of trimethoprim (a drug used for the treatment and prevention of urinary tract infection) using a MIP-graphene modified GC electrode [30]. The results were accurate, precise, and the detection limit was 0.13 μM. The new sensor was applied to quantify trimethoprim in urine samples. Norfloxacin is a synthetic fluoroquinolone antibiotic. A MWCNT modified pyrolytic graphite electrode was prepared and applied to detect norfloxacin based on its electrochemical reduction [31]. The dynamic range for the norfloxacin analysis ranged between 1.2 and 1,000 μM with a detection limit of 40.6 nM. The determination of norfloxacin and caffeine was selective and the method was successfully applied in biological fluids and pharmaceutical dosage forms.

Other selected examples of assays of dosage forms and biological samples using various solid electrodes are listed in Table 9.3.

9.3.1.2 Antineoplastic Drugs

Although chemotherapeutic drugs are often thought of as drugs that kill or suppress cancer cells, some also target organism that invade the body and are used to treat microbial infection. Antineoplastic agents are a group of specialized drugs used primarily to treat cancer. Antineoplastic agents are classified by origin and by how they work to destroy cancer cells. They can be administered to patients alone or in combination with other antineoplastic drugs. These agents diffuse in the body and destroy cancer cells. Many of the side effects associated with antineoplastic agents occur because treatment destroys the body's normal cells in addition to cancerous cells.

The interest in developing electrochemical sensing devices for use in clinical and pharmaceuticals group is growing rapidly. Hence, sensitive, selective, and rapid methods are required for the assay of antineoplastic agents in their dosage forms and biological samples. The selected studies in this area are given as follows and in Table 9.4.

Zhang et al. presented a sensitive voltammetric method for determination of baicalein by using a TRGO modified GCE in 100 mM KCl-10 mM sodium phosphate buffer solution. Under physiological conditions, the modified electrode showed a linear voltammetric response from 10 nM to 10 μM, with a detection limit of 6.0 nM [76].

Bozal-Palabiyik et al. developed a sensitive electroanalytical method for the determination of the anticancer drug etoposide (ETP) using AdSDPV at a MWCNT-GCE. The calibration curve was linear in the concentration range of 20 nM−2 μM with a detection limit of 5.4 nM [77]. The repeatability of the peak current was found at RSD = 1.55 % ($n = 5$) in pH 6.0 Britton–Robinson buffer. The method was successfully utilized for the determination of ETP in a pharmaceutical dosage form, and a recovery of 99.55 % was obtained. This method using a MWCNT modified GCE appears to be the most sensitive electrochemical method reported for the determination of ETP. The possible oxidation mechanism of etoposide was also discussed.

The effect of surface modification on the electrooxidation behavior of the anticancer drug idarubicin was studied at MWCNT-GC and MWCNT-EPPG electrodes [78]. The method was validated and successfully applied for the determination of idarubicin in pharmaceutical dosage forms. Under the optimized conditions, idarubicin gave a linear response in the range 93.6 nM–1.87 μM for the modified glassy carbon and 93.6 nM–0.936 μM for the modified edge plane pyrolytic graphite electrodes. The detection limits were found as 18.7 nM and 37.5 nM based on modified glassy carbon and edge plane pyrolytic graphite electrodes, respectively.

Table 9.3 Electrochemical studies on antibacterial, antifungal, antiviral, and antiparasitic drugs using solid electrodes

Drug group	Drug name	Electrochemical behavior	Electrode type	Technique	LOD/LOQ	Applications	References
Tetracycline antibiotic	Doxycycline	Oxidation	MIP and NIP modified electrodes	DPV	43.5 μM	Pharmaceutical dosage forms	[32]
Antiprotozoal	Secnidazole Tinidazole Ornidazole	Reduction	Activated GCE	DPV	0.2 nM 0.3 nM 0.25 nM	Pharmaceutical dosage forms	[33]
Antihelmintic	Mebendazole	Oxidation	Graphene nanosheets and carbon nanospheres/ Chitosan modified GCE	DPAdSV	10.5 nM	Tablets, human serum	[34]
Fluoroquinolone antibiotics	Ciprofloxacin Ofloxacin Norfloxacin Gatifloxacin	Oxidation	β-cyclodextrin and L-arginine modified CPE	DPV	0.05 μM 0.1 μM 0.1 μM 0.06 μM	Pharmaceutical dosage forms, serum	[35]
Penicillin antibiotic	Amoxicillin	Oxidation	Polyaniline film modified CPE	CV, DPAdSV, SWAdSV	0.73 nM (DPAdSV) 0.35 nM (SWAdSV)	Pharmaceutical dosage forms, biological samples	[36]
Macrolide antibiotic	Azithromycin	Oxidation	MWCNT decorated with $MgCr_2O_4$	DPV	0.07 μM	Pharmaceutical dosage forms	[37]
Fluoroquinolone antibiotic	Norfloxacin	Oxidation	CuO nanoleaves MWCNTs/GC composite film modified electrode	CV, DPV	0.321 μM	Tablets	[38]
Cephalosporins antibiotic	Cefixime	Oxidation	MWCNT decorated $NiFe_2O_3$ magnetic nanoparticles	LSV	0.02 μM	Pharmaceutical dosage forms	[39]
Fluoroquinolone antibiotics	Ofloxacin Gatifloxacin	Oxidation	Cysteic acid modified CPE	CV, DPV	0.02 μM 0.02 μM	Pharmaceutical dosage forms, human serum	[40]

Antiviral drug	Nevirapine	Oxidation	Uracil/CPE	CV, DPV, EIS	0.05 μM	Human serum	[41]
Antiviral drug	Lamivudine	Reduction	Thin-film ME	DPV, Linear scan and adsorptive linear cyclic voltammetry	0.57 nM	Active compound	[42]
Antiviral drug	Acyclovir	Electrocatalytic oxidation	Fullerene-C_{60} modified GCE	CV, DPV	14.8 nM	Tablets, human urine, serum	[43]
Penicillin antibiotic	Penicillin V	Oxidation	BDDE	CV, DPV	0.25 μM	Tablets, human urine	[44]
Tetracycline antibiotic	Minocycline	Oxidation	CPE	DPV	0.7 μM	Pharmaceutical dosage forms	[45]
Fluoroquinolone antibiotic	Levofloxacin	Electrocatalytic oxidation	PoAP/MWCNTs/GCE	DPV	1 μM	Capsules	[46]
Penicillin antibiotic	Amoxicillin	Electrocatalytic oxidation	$FeCr_2O_4$/MWCNTs/GCE	EIS, DPV	0.05 μM	Urine, blood plasma, capsules	[47]
Antimalarial	Quinine	Electrocatalytic oxidation	p-(AHNSA)/GCE	CV, SWV	14.2 nM	Pharmaceutical dosage forms, human urine	[48]
Antiprotozoal	Metronidazole	Electrocatalytic reduction	Gr/IL/GCE	CV, DPV	47 nM	Tablets	[49]
Antiviral	Acyclovir	Electrocatalytic oxidation	Microspheres modified CPE/Nf	CV, CA, Amperometry	0.04 mM	Tablets, topical cream	[50]
Fluoroquinolone antibiotic	Moxifloxacin HCl	Oxidation	Chloranil/CPE	DPV	67.4 nM	Tablets	[51]
Fluoroquinolone antibiotic	Gatifloxacin	Oxidation	MWCNTPE	CV, DPV	4.5 nM	Biological fluids	[52]
Antiprotozoal	Nitazoxanide Tizoxanide	Oxidation	CPE	CV, DPV, DPAdASV	75 nM 0.943 μM	Pharmaceutical dosage forms, human serum	[53]

(continued)

Table 9.3 (continued)

Drug group	Drug name	Electrochemical behavior	Electrode type	Technique	LOD/LOQ	Applications	References
Sulfonamide antibiotic	Sulfamethoxazole	Oxidation	MWCNTPE	CV CA	0.395 nM	Tablets	[54]
Macrolide antibiotic	Azithromycin	Electrocatalytic Oxidation	Gr/IL/GCE	CV	0.254 nM	Pharmaceutical dosage forms	[55]
Cephalosporin antibiotic	Cefpirome	Reduction	MWCNT/GCE	DPV, SWV	10.8 mM (DPV) 1.26 μM (SWV)	Pharmaceutical dosage forms	[56]
Antiviral drug	Efavirenz	Reduction	Mercury film electrode	LAdCV, DPAdSV	3 nM	Active compound	[57]
Macrolide antibiotic	Midecamycin	Oxidation	GNP-MWCNT/GCE	CV	0.5 μM	Tablets	[58]
Penicillin antibiotic	Benzylpenicillin	Oxidation Reduction	MIP/ Fe_3O_4@SiO_2_MWCNTs-CTS/CE	CV, DPV	1.5 nM	Human serum	[59]
Tetracycline antibiotic	Doxycycline Hyclate	Oxidation	CPE, GCE	DPV	65.6 nM (CPE) 95.5 nM (GCE)	Pharmaceutical dosage forms	[60]
Antiviral drug	Oseltamivir	Reduction	Thin-film mercury electrode	AdSV	5.3 nM	Pharmaceutical dosage forms	[61]
Antibiotic	Clindamycin hydrochloride	Oxidation	CPE	DPAdSV, SWAdSV	68.3 nM (DPAdSV) 190 nM (SWAdSV)	Pharmaceutical dosage forms, human urine	[62]
Antibacterial, antifungal	Clioquinol	Oxidation	GCE	CV, DPV SWV	–	Mechanistic study	[63]

Antiprotozoal	Furazolidone	Reduction	MWCNT/GCE	CV, CA	2.30 μM	Pharmaceutical dosage forms	[64]
Macrolide antibiotic	Roxithromycin	Oxidation	SWCNT/GCE	CV, CC	0.5 μM	Pharmaceutical dosage forms	[65]
Antiviral	Nevirapine	Reduction	Thin-film mercury electrode	Linear cyclic scan stripping voltammetry	3 nM	Active compound	[66]

Table 9.4 Electrochemical assay on antineoplastic drugs using solid electrodes

Drug name	Electrochemical behavior	Electrode type	Technique	Linearity range	LOD/LOQ	Applications	References
6-mercaptopurine	Electrocatalytic oxidation	p-APMCNTPE-TiO_2	CV, DPV	0.09–350 μM	0.065 μM	Tablets, human urine	[67]
Gemcitabine HCl	Oxidation	CPE	DPV	0.05 μM–0.3 mM	8.2 nM	Pharmaceutical dosage forms, human serum	[68]
Doxorubicin	Oxidation	MWCNT-COOH/SPE	CV, DPV, Amperometry	36.8 nM–2.35 μM	18.4 nM	Rabbit and rat blood	[69]
Berberine	Reduction	GCE	CV	1–80 μM	0.8 μM	Tablets	[70]
Phenoxazine Diphenoxazine derivatives	Oxidation	GCE	CV, CPC	–	–	Oxidation mechanism studies	[71]
5-Fluorouracil	Oxidation	GCE	CV, DPV	20 nM–0.6 μM	20.13 nM	Pharmaceutical dosage forms, human urine Human serum	[72]
Daunomycin	Reduction	GCE/AuNPs/polyTTBA/PS-aptamer/AuNPs	CV DPV	0.1–60.0 nM	52.3 ± 2.1 pM	Human urine	[73]
Daunorubicin	Oxidation	PGE PDR-PGE	DPV, EIS	– –	317 nM (PGE) 128 nM (PDR-PGE)	Active compound	[74]
Methotrexate	Reduction	m-AgSAE	CV, DPV	2 nM–1 μM	1.8 nM	Pharmaceutical dosage forms	[75]

Interfering species such as ascorbic acid, dopamine, and aspirin showed no interference allowing the selective determination of idarubicin.

The voltammetric behavior of the anticancer drug irinotecan was investigated at a PMB/MWCNT modified GCE [79]. The calibration curve was linear in the concentration range between 8 and 80 μM with a detection limit of 0.2 μM by DPV. Another anticancer drug picoplatin was studied using a graphene-MWCNT modified GCE using DPV [80]. It can be used to detect as low as 0.21 μM of drug. Other selected examples, which include dosage forms and biological samples using various solid electrodes, are listed in Table 9.4.

9.3.1.3 Drugs Acting on the Central Nervous System

Central nervous system agents can be used as analgesics, anesthetics, anti-emetics, anti-convulsants, sedatives, and have many more therapeutic uses such as for Alzheimer's disease, depression, Parkinson's disease, etc. These drugs act on the brain and spinal cord and are used widely to treat psychiatric disorders, induce anesthesia, or reduce pain [81]. This group is widely worked using electrochemical methods because of their electroactivity. In this section some of the selected works are discussed as follows and in Table 9.5.

Shahrokhian et al. studied the electrooxidation of the neuroleptic drug clozapine on the surface of a GCE modified with a thin film of MWCNTs/new coccine-doped polypyrrole modified electrode by using LSV [117]. The pH of the supporting electrolyte (D), drop size of the deposited MWCNTs suspension (E), and accumulation time of CLZ on the surface of modified electrode (F) were considered as effective experimental factors. By using factorial-based response-surface methodology, the optimum values of factors were obtained as 5.44, 10 μL, and 300 s for D, E, and F, respectively. Under optimized condition, a significant increase (~14 times) was observed in the anodic peak current of CLZ on the surface of the modified electrode relative to the bare GCE. Oxidation peak currents increased linearly with CLZ concentration in the range of 0.01–5.00 μM with a detection limit of 3.00 nM. The RSD value for the peak current of CLZ was obtained as 4.5 %. The modified electrode was used for the determination of clozapine concentration in pharmaceutical and clinical preparations with satisfactory results. Mashhdizadeh and Afshar developed a new carbon paste electrode chemically modified with TiO_2 nanoparticles for the same drug. Using adsorptive DPV at optimum parameters, a linear dynamic range of 0.5–45 μM clozapine was observed with detection limit of 61.0 nM [118].

Agin et al. developed a selective and sensitive electroanalytical method for the determination of the antidepressant drug mirtazapine at a GCE and a BDDE [119]. In cyclic voltammetry, mirtazapine showed one sharp oxidation peak and one additional wave in acidic media in the positive potentials; at pH 5.50, the mirtazapine peak was single and sharp. Under optimized conditions, the peak current showed a linear dependence with concentration in the range 0.8–100 μM

Table 9.5 Voltammetric assay on central nervous system drugs using solid electrodes

Drug group	Drug name	Electrochemical behavior	Electrode type	Technique	LOD/LOQ	Applications	References
Opioid analgesic	Methadone	Electrocatalytic oxidation	MWCNT/GCE	CV, DPV	0.25 μM	Pharmaceutical dosage forms, saliva, human urine	[82]
Antidepressant	Escitalopram	Electrocatalytic oxidation	Bi_2O_3 modified GCE	CV	0.165 μM	Pharmaceutical dosage forms	[83]
Antipsychotic	Olanzapine	Electrocatalytic oxidation	Amine-functionalized TiO_2-MWCNT modified GCE	CV, SWV, CA	0.09 μM	Pharmaceutical dosage forms, blood serum samples	[84]
Antiepileptic	Carbamazepine	Oxidation	Ag/TiO_2 nanocomposite modified CPE	CV, DPV	0.86 μM	Pharmaceutical dosage forms	[85]
Antipsychotic	Trifluoperazine	Oxidation	MWCNT/GCE	DPAdSV	0.749 nM	Pharmaceutical dosage forms	[86]
Psychostimulant	Caffeine	Oxidation	MWCNT/GCE	SWV	3.52 nM 10 nM	Pharmaceutical dosage forms, urine, tea leaves, coffee, cold drink	[87]
Anti Parkinson's	L-dopa Carbidopa	Oxidation	Graphene/GCE	DPV	0.8 μM 1.8 μM	Pharmaceutical dosage forms	[88]
Opioid analgesic	Codeine	Oxidation	BDD film electrode	CV, DPV	0.08 μM	Tablets, urine	[89]
Antipsychotic	Thioridazine	Electrocatalytic oxidation	ZnSNP-MCPE	CV, DPV	65 nM	Tablets	[90]
Opioid analgesic	Morphine	Electrocatalytic oxidation	PPGE	DPV, CV, hydrodynamic amperometry	0.26 μM	Human urine, street drug samples	[91]
Anti Parkinson's	L-Dopa	Electrocatalytic oxidation	$Co(DMG)_2ClPy$/MWCNT/BPPGE	CV, SWV	0.86 μM	Pharmaceutical dosage forms	[92]
Myorelaxant	Chlorzoxazone	Oxidation	AuE	CV, LSV, SWV	45 nM	Tablets, urine	[93]

Ani Parkinson's	Carbidopa	Electrocatalytic oxidation	FCDCNTPE	CV, LSV, SWV	65 nM	Urine	[94]
Anti Parkinson's	Levodopa	Electrocatalytic oxidation	CP-TNMCPE	CV, DPV	69 nM	Tablets, urine, human blood serum, water	[95]
Antiepileptic	Lamotrigine	Reduction	PGE	CV, EIS, AdSV	80 nM	Tablets, human blood serum	[96]
Antiepileptic	Carbamazepine	Reduction	GCE	DPV	0.593 μM	Rabbit serum	[97]
Myorelaxant	Chlorzoxazone	Oxidation	GCE	CV, SWV	44.1 nM	Tablets, human plasma, urine	[98]
Antiepileptic	Gabapentin	Oxidation	MCPE	CV, Amperometry	0.3 μM	Capsules, human serum, urine	[99]
Opioid analgesic	Morphine Noscapine Heroin	Electrocatalytic oxidation	GNSs/GCE	DPV	0.4 μM 0.2 μM 0.5 μM	Pharmaceutical dosage forms	[100]
Psychostimulant	Caffeine	Oxidation	PCPE	CV, SWV	0.348 μM	Pharmaceutical dosage forms, tea, coffee, cola	[101]
Antiepileptic	Carbamazepine	Oxidation	Gr-AuNPs-Au	CV, LSV, EIS	3.03 μM	Active compound	[102]
Psychostimulant Antioxidant	Caffeine (−) Epigallo-catechin gallate	Oxidation	MIP modified electrode	CV	–	Active compound	[103]
Psychostimulant	Xanthine Hypoxanthine	Oxidation	poly(PCV)/MWCNTs-COOH/GCE	DPV	0.05 μM 0.20 μM 0.16 μM	Human serum	[104]
Anti Parkinson's	L-dopa Benserazide	Oxidation	GCE	CV, DPV, SWV	2.52 μM (DPV) 2.89 μM (SWV) 2.99 μM (DPV) 2.62 μM (SWV)	Tablets	[105]

(continued)

Table 9.5 (continued)

Drug group	Drug name	Electrochemical behavior	Electrode type	Technique	LOD/LOQ	Applications	References
Opioid analgesic	Morphine	Oxidation	GNPs/NCPE	DPV	1.33 nM	Human urine	[106]
Anti-Parkinson's	Levodopa Carbidopa Tryptophan	Oxidation	EBNBHCNPE	CV, DPV	0.094 μM 7.2 μM 12.3 μM	Human urine, water	[107]
Antipsychotic	Quetiapine fumarate	Oxidation	SBAP/GCE	CV, SWV	19 nM	Pharmaceutical dosage forms, human urine	[108]
General anesthetics	2,6-Diisopro-pylphenol (Propofol)	Oxidation Reduction	GCE	CV, ASV	3.2 ± 0.1 μM (CV) 5.5 ± 0.4 μM (ASV)	Active compound	[109]
Myorelaxant	Methocarbamol	Oxidation	MWCNT/GCE	DPV	0.104 nM	Human serum	[110]
Opioid analgesics	Morphine	Oxidation	Gold nanotube arrays/GCE	CV, DPV	40.6 nM	Human serum	[111]
Anti Parkinson's	Benserazide	Oxidation	GCE	DPV	7.12 μM	Tablets	[112]
Antipsychotic	Clozapine	Oxidation Reduction	MWCNT-SDS/CPE	CV	1 mM	Active compound	[113]
Antidepressant	Venlafaxine Desvenlafaxine	Oxidation	Nf-CNT-GCE	CV, CC, EIS, DPAdSV	12.4 nM 21.1 nM	Pharmaceutical dosage forms, human urine, human serum	[114]
Antipsychotic	Chlorpromazine	Oxidation	GPE	CV, DPV	6.0 nM	Tablets	[115]
Antipsychotic	Chlorpromazine	Oxidation	ARS/GCE	LSV	5.16 μM	Pharmaceutical dosage forms	[116]

for the GCE and in the range 4–100 μM for the BDDE using DPV and SWV. The possible oxidation mechanism of mirtazapine was also discussed.

Raj and John applied a new electrochemically reduced graphene oxide modified electrode for the simultaneous determination of uric acid (UA), xanthine (XN), hypoxanthine (HXN), and caffeine (CAF) in human blood serum and urine samples [120]. Selective determination of one purine derivative in the presence of low concentrations of the other three purine derivatives was also realized at this modified electrode. Using a pulsed voltammetric method, detection limits of 88 nM, 0.11 μM, 0.32 μM, and 0.43 μM were obtained for UA, XN, HXN, and CAF, respectively. The practical application of the modified electrode was demonstrated by simultaneously determining the concentrations of UA, XN, HXN, and CAF in human blood plasma and urine samples.

Other selected examples, which include dosage forms and biological samples using various solid electrodes, are listed in Table 9.5.

9.3.1.4 Drugs Acting on the Peripheral Nervous System

Drug affecting the peripheral nervous system alter voluntary (through the somatic motor system) and involuntary (though the autonomic nervous system) processes in the body [81]. In this section, some sensitive and selective voltammetric methods are reported and tabulated for the assay of peripheral nervous system drugs in their dosage forms and biological samples (Table 9.6).

A lichen-like nickel oxide nanostructure was synthesized and characterized [129]. The nanostructure was then applied to modify a CPE and for the fabrication of a sensor for the "enzyme less" electrocatalytic oxidation of acetylcholine (ACh). The mechanism and kinetics of the electrooxidation process were investigated by cyclic voltammetry, steady-state polarization curve, and chronoamperometry. The catalytic rate constant and the charge transfer coefficient of ACh electrooxidation by the active nickel species and the diffusion coefficient of ACh were reported. A sensitive and time-saving hydrodynamic amperometric method was developed for the determination of ACh. ACh was determined with a sensitivity of 392.4 mA M^{-1} cm^{-2} and a limit of detection of 26.7 μM.

Prasad et al. developed a CNT-mer (multiwalled carbon nanotubes bearing terminal monomeric unit) dispersed sensor for determination of epinephrine [130]. A linear dynamic range of 0.49–32.12 nM epinephrine was observed with detection limit of 0.108 nM. The proposed sensor was advantageous in obtaining enhanced differential pulse anodic stripping voltammetric current vis-a-vis the corresponding imprinted sensor modified with randomly dispersed flocculated multiwalled carbon nanotube bundles.

Other selected examples, which include dosage forms and biological samples using various solid electrodes, are listed in Table 9.6.

Table 9.6 Electrochemical studies on peripheral nervous system active drugs using solid electrodes

Drug group	Drug name	Electrochemical behavior	Electrode type	Technique	LOD/ LOQ	Applications	References
Adrenergic agonist (Cardiac stimulant)	Norepinefrine	Oxidation	Ionic-liquid-carbon nanotubes paste electrode	CV, SWV	0.08 μM	Pharmaceutical dosage forms, urine	[121]
Muscarinic antagonist (Antispasmodic)	Tolterodine	Oxidation	GCE	CV, DPV	–	Mechanistic study	[122]
Adrenergic antagonist (α blocker)	Terazosin	Electrocatalytic oxidation	GNMCPE	CV, EIS, DPV	0.12 nM	Tablets, urine	[123]
Adrenergic agonist (Nasal decongestant)	Ephedrine	Electrocatalytic oxidation	PCPE-PAA	CV, SWV	0.35 μM	Pharmaceutical dosage forms, urine	[124]
Adrenergic agonists (Nasal decongestant)	Ephedrine Pseudoephedrine	Electrocatalytic oxidation	MWCNT/GCE	DPAdSV, CAdSV	0.75 μM 0.82 μM	Urine, pharmaceutical dosage forms	[125]
Adrenergic agonist (Bronchodilator)	Isoproterenol	Oxidation	MWCNTILEE	CV, EIS, DPV	0.85 μM	Ampoules, human urine	[126]
Adrenergic agonist (Bronchodilator)	Isoproterenol	Electrocatalytic oxidation	p-CACNTPE	CV, CA, EIS	0.009 μM 2.3 μM	Ampoules, human urine	[127]
Adrenergic agonist (Cardiac stimulant)	Norepinephrine	Electrocatalytic oxidation	DDP/CPE	CV, DPV	77 ± 2 nM	Pharmaceutical dosage forms	[128]

9.3.1.5 Cardiovascular and Renal Active Drugs

These drugs are used to treat conditions related to the heart, blood vessels, and kidneys, such as hypertension, heart failure, ischemia, cardiac arrhythmias, and abnormalities in fluid volume and blood composition [81]. Cardiovascular drugs encompass a large number of prescription medications that are used to control heart disease. It is a complicated group of drugs with many being used for different heart problems. Renal drugs increase urine output by the kidney. Most diuretics produce diuresis by inhibiting the reabsorption of sodium at different segments of the renal tubular system. Sometimes a combination of diuretics with cardiovascular drugs is given because this can be significantly more effective than either compound alone (synergistic effect).

In this section, selected voltammetric applies and their validation parameters on the cardiovascular, renal drugs, or their combinations are reported and tabulated for the assay of peripheral nervous system drugs in their pharmaceutical dosage forms and biological samples as follows and in Table 9.7.

Kaur et al. fabricated electrochemical AgNPs/rGO sensor for the simultaneous determination of ascorbic acid, dopamine, uric acid, and tryptophan [175]. Electrochemical studies were carried out by using cyclic voltammetry, linear sweep voltammetry, and chronoamperometry. AgNPs/rGO modified electrode exhibited excellent electrocatalytic activity, stability, sensitivity, and selectivity with well-separated oxidation peaks for ascorbic acid, dopamine, uric acid, and tryptophan allowing the simultaneous determination of their quaternary mixture. The analytical performance of this sensor was demonstrated for the determination of ascorbic acid and dopamine in commercial pharmaceutical samples such as vitamin C tablets and dopamine injections, respectively. The applicability of this sensor was also extended in the determination of uric acid in human urine samples.

Khoobi et al. developed MWCNT/CPE for simultaneous determination of betaxolol and atenolol using DPV coupled with multivariate curve resolution-alternating least squares [176]. Characterization of the modified electrode was carried out by electrochemical impedance spectroscopy and cyclic voltammetry. Operating conditions were improved with central composite rotatable design and response surface methodology, involving several chemical and instrumental parameters. Then second order data were built from variable pulse heights of DPV and after correction in potential shift analyzed by MCR-ALS. Analytical parameters such as linearity, repeatability, and stability were also investigated and a detection limit of 0.19 and 0.29 μM for betaxolol and atenolol was achieved, respectively. The proposed method was successfully applied for the simultaneous determination of the two analytes in human plasma.

The determination of bezafibrate (BZF) in pharmaceutical dosage forms by SWV at a cathodically pretreated BDDE was proposed [177]. CV results showed one irreversible oxidation peak for BZF at 1.20 V at pH 2.0. Under optimized SWV conditions, a linear analytical curve was obtained for BZF concentration range 0.10–9.1 mM in the BR buffer solution (pH 2.0), with detection limit of 0.098 mM.

Table 9.7 Electrochemical studies on cardiovascular and renal drugs using solid electrodes

Drug group	Drug name	Electrochemical behavior	Electrode type	Technique	LOD/LOQ	Applications	References
Adrenergic agonist (Cardiac stimulant)	Dopamine	Oxidation	Graphene anchored with PdPt nanoparticles	DPV	0.61 μM 0.04 μM 0.10 μM	Human urine, blood serum samples	[131]
Adrenergic agonist (Cardiac stimulant)	Dopamine	Oxidation	MWCNT/polyglycine modified CPE	DPV	0.012 μM	Pharmaceutical dosage forms	[132]
Adrenergic agonist (Cardiac stimulant)	Dopamine	Oxidation	SiO_2-coated graphene oxide and molecularly imprinted polymers electrode	DPV	0.03 μM	Pharmaceutical dosage forms, urine	[133]
Hyperlipidemic	Atorvastatin	Oxidation	CPE	CV, DPV	4.08 nM	Pharmaceutical dosage forms, urine	[134]
Antianginal	Nifedipine	Reduction	MWCNT-β-CD modified CPE	CV, DPV	0.025 μM	Pharmaceutical dosage forms, blood serum	[135]
Antiarrhythmic	Quinidine	Oxidation	Nanographene-based sensor	CV	0.573 nM	Pharmaceutical dosage forms	[136]
Vasodilators	Dipyridamole	Oxidation	MIP modified CPE	DPAdSV	0.05 nM	Pharmaceutical dosage forms, serum	[137]
Antihypertensive	Captopril	Oxidation	Supported manganese-based voltammetric sensor	CV, LSV	0.09 μM	Pharmaceutical dosage forms, urine	[138]
Hyperlipidemic	Gemfibrozil	Oxidation	GCE modified with MWCNT within dihexadecyl hydrogen phosphate film	SWAdSV	0.053 μM	Pharmaceutical dosage forms, urine samples	[139]
Antihypertensive	Methyldopa	Electrocatalytic oxidation	p-CAMCNTPE	CV, EIS, SWV	0.2 μM	Pharmaceutical dosage forms, human serum, urine	[140]

Antihypertensive	Sotalol Alprenolol Atenolol	Oxidation	Carbon nanofiber paste electrode	CE-amperometric detector	≤10 nM	Urine	[141]
Peripheric vasodilator	Phentolamine	Oxidation	BDDE	CV, SWV	2.52 μM	Pharmaceutical dosage forms	[142]
Adrenergic agonist (Cardiac stimulant)	Dopamine	Electrocatalytic oxidation	PMB-Gr/CILE	CV, DPV	5.6 nM	Injection sample, human urine	[143]
Vasoprotective	Rutin	Electrocatalytic oxidation	MCCE- $Cu_3(PO_4)_2$	CV, SWV	0.012 μM	Pharmaceutical dosage forms	[144]
Antihypertensive /Diuretic	Acetazolamide	Reduction	GPE	CV, DPV	2.2 nM	Human serum, urine	[145]
Hyperlipidemic	Simvastatin	Electrocatalytic oxidation	SDS/GCE	CV	0.25 μM	Pharmaceutical dosage forms	[146]
Antianginal	Benidipine	Oxidation	GCE, BDDE	CV, DPV, SWV	55.3 nM (DPV, GCE) 75.6 nM (SWV, GCE) 0.594 μM (DPV, BDDE) 0.605 μM (SWV, BDDE)	Tablets	[147]
Adrenergic agonist (Cardiac stimulant)	Dopamine	Oxidation	PGA-SWCNT film electrode	CV, DPV	0.38 μM	In the presence of ascorbic acid and uric acid	[148]
Adrenergic agonist (Cardiac stimulant)	Dopamine	Electrocatalytic oxidation	TNGCE	CV, DPV	1 nM	Biological samples	[149]
Adrenergic agonist (Cardiac stimulant)	Dopamine	Electrocatalytic oxidation	AuE-Au_{nano}-Cys-SDS	CV, EIS, SWV	16 nM	In the presence of ascorbic acid, dopamine, and acetaminophen	[150]
Antihypertensive	Captopril	Electrocatalytic oxidation	GCE	CV, Single step CA, EIS	4.8 μM	Tablets, urine	[151]

(continued)

Table 9.7 (continued)

Drug group	Drug name	Electrochemical behavior	Electrode type	Technique	LOD/LOQ	Applications	References
Antitrombotic	Cilostazol	Electrocatalytic oxidation	MWCNTs-Bi_2O_3/GCE	CV, SWV	2.06 μM	Tablets, human plasma	[152]
Antianginal	Amlodipine besylate	Oxidation	o-MWCNT/AuE	CV, SWASV	2.65 μM	Tablets	[153]
Peripheric vasodilator	Pentoxifylline	Oxidation	MWCNTPE	CV, DPV	0.169 μM	Tablets, human urine	[154]
Antihypertensive /Diuretic	Furosemide	Oxidation	MWCNTPE	CV, DPV	0.29 μM	Tablets, human urine	[155]
Adrenergic agonist (Cardiac stimulant)	Dopamine	Electrocatalytic oxidation	GS-PTCA/GCE	CV, DPV	5.60 μM 0.92 μM 0.13 μM 0.06 μM	In the presence of interferents	[156]
Adrenergic agonist (Cardiac stimulant)	Dopamine	Electrocatalytic oxidation	PCFE	DPV	2.9 μM 0.04 μM 0.03 μM	Dopamine HCl injection, human urine	[157]
Antianginal	Atenolol Carvedilol Propranolol	Oxidation	CuO/CPE	CV, DPV	1.12 μM 3.01 μM 2.91 μM	Human serum	[158]
Antianginal	Bisoprolol fumarate Propranolol hydrochloride	Oxidation	NGCE	SWV, SWAdSV	0.71 μM (SWV) 0.010 μM (SWAdSV) 0.43 μM (SWV) 0.013 μM (SWAdSV)	Pharmaceutical dosage forms, human urine	[159]

Adrenergic agonist (Cardiac stimulant)	Dopamine Norepinephrine	Oxidation	CFME	FSCV	1.35 μM	Rat brain	[160]
Adrenergic agonist (Cardiac stimulant)	Dopamine	Oxidation	PNT-[Cu_4 (apyhist)$_4$]$^{4+}$/Nafion/GCE	CV, SWV	2.80 μM	Biological sample	[161]
Adrenergic agonists (Cardiac stimulant)	Dopamine	Electrocatalytic Oxidation	MWCNTs/CILE	CV, SWV	0.010 μM	Human serum	[162]
Adrenergic agonist (Cardiac stimulant)	Dopamine	Oxidation	Gr/p-ABA/GCE	CV, DPV, CA	20 nM	Human serum	[163]
Antihypertensive	Captopril	Oxidation	N-DHPBMCNTPE	CV, CA, EIS	0.034 μM	Human urine, tablets	[164]
Adrenergic agonist (Cardiac stimulant)	Dopamine	Oxidation Reduction	GCE/MWCNT-FeNAZ-CH	CV, LSV	1.11 μM 1.05 μM 0.033 μM 0.011 μM	Human urine, human serum	[165]
Adrenergic agonist (Cardiac stimulant)	Dopamine	Oxidation	SWNTs-DMF/GCE	DPV	0.137 μM	Pharmaceutical dosage forms	[166]
Adrenergic agonist (Cardiac stimulant)	Dopamine Nitrite	Oxidation	CDP-GS-MWCNTs/GCE	CV, DPV	1.65 μM 0.05 μM 1.65 μM	Human urine	[167]
Adrenergic agonist (Cardiac stimulant)	Dopamine	Oxidation	CNF	FSCV	–	Active compound	[168]
Adrenergic agonist (Cardiac stimulant)	Dopamine	Electrocatalytic Oxidation	DTPS SAM-ME	CV, CC, EIS	0.030 μM	Pharmaceutical dosage forms, human urine	[169]

(continued)

Table 9.7 (continued)

Drug group	Drug name	Electrochemical behavior	Electrode type	Technique	LOD/LOQ	Applications	References
Vasodilators	Nicorandil	Reduction	GCE	CV, SWV	31.2 μM	Pharmaceutical dosage forms	[170]
Adrenergic agonist (Cardiac stimulant)	Dopamine	Electrocatalytic Oxidation	HNCMS/GCE	CV, DPV	0.02 μM 0.04 μM 0.91 μM	Human urine	[171]
Adrenergic agonist (Cardiac stimulant)	Dopamine	Oxidation	PEDOT/Pt	LSV	61 nM	Human urine	[172]
Antianginal	Propranolol	Oxidation	GSR	CV, DPV	1.1 μM	Pharmaceutical dosage forms	[173]
Hyperlipidemic	Ezetimibe	Oxidation	GCE, BDDE	DPV, SWV	0.247 μM (DPV, GCE) 0.299 μM (SWV, GCE) 0.227 μM (DPV, BDDE) 0.245 μM (SWV, BDDE)	Pharmaceutical dosage forms, human serum	[174]

The obtained recoveries ranged from 93.4 to 108 %. The proposed method was successfully applied to the determination of the BZF content in tablets, and the results were in close agreement with those obtained using a comparative spectrophotometric method.

Prasad et al. developed a dual-template imprinted polymer modified carbon-ceramic electrode for ultra trace simultaneous analysis of ascorbic acid and dopamine [178]. The detection limits realized by the proposed sensor, under optimized conditions, were found to be as low as 12.72 nM for ascorbic acid and 1.37 nM for dopamine.

9.3.1.6 Endocrine Drugs

Endocrine drugs are hormones or hormone derivatives, or drugs that may modify the synthesis or action of normally secreted hormones. Endocrine drugs mimic or block the effect of endogenous hormones. Endocrine dysfunctions may be related to the hypothalamus or pituitary gland or to the target tissues that function under the control of hypothalamus and/or pituitary hormone [81]. The electroanalytical performance and determination parameters of some endocrine drugs using solid electrodes were demonstrated and discussed as follows and in Table 9.8.

Rafiee and Fakhari developed a nickel oxide nanoparticles modified nafion-MWCNT screen-printed electrode for the determination of insulin by CV [188]. Amperometry was also used to evaluate the analytical performance of the modified electrode in the quantitation of insulin. Excellent analytical features, including high sensitivity (1.83 μA μM^{-1}), low detection limit (6.1 nM), and satisfactory dynamic range (20.0–260.0 nM) were achieved under optimized conditions.

In the search for new and mercury-free electrode material, different bismuth-based electrodes have been developed by Lima and Spinelli for the determination of progesterone in four pharmaceutical samples [189]. Two peaks were observed by CV, at −1.68 V and −1.47 V in 0.1 M BR solutions at pH 12.0. Square-wave adsorptive stripping voltammetry was used after accumulation of progesterone for 60 s at −1.0 V. The peak current at −1.63 V, related to the reduction of progesterone, increased linearly with the hormone concentration in the range of 0.40–7.90 μM.

Typical endocrine drugs determined voltammetric methods are listed in Table 9.8.

9.3.1.7 Anti-inflammatory, Anti-allergy, and Immunoregulatory Drugs

Anti-inflammatory, anti-allergy, and immunoregulatory drugs constitute a large group that regulate autacoid (prostaglandins, leukotrienes, cytokines, histamine, 5-HT, and endogenous peptides)-related dysfunctions in the body.

Table 9.8 Electrochemical assay on endocrine drugs using solid electrodes

Drug group	Drug name	Electrochemical behavior	Electrode type	Technique	LOD/LOQ	Applications	References
Antidiabetic	Metformin	Oxidation	Copper-loaded activated charcoal modified electrode	DPV	9 nM	Pharmaceutical dosage forms	[179]
Antithyroid	Methimazole	Oxidation	Fluorine-doped tin oxide electrode	CV, SWV	1.98 μM	Pharmaceutical dosage forms	[180]
Estrogen antagonist	Raloxifene	Oxidation	GCE	CV, DPV	0.075 μM	Tablets, plasma	[181]
Gonadal hormone	Tamoxifen	Oxidation	GCE	CV, DPAdASV, SWAdASV	1.8 μM (DPAdASV) 2.4 μM (SWAdASV)	Tablets, human urine, serum	[182]
Gonadal hormone	Flutamide	Electrocatalytic reduction	PFMCPE	CV, DPV	0.181 μM	Tablets	[183]
Antithyroid	Methimazole	Electrocatalytic oxidation	MWCNTs-Salen-Co(III)-MIP/GCE	CV, DPV	0.420 μM	Tablets, human urine	[184]
Gonadal hormone	Fulvestrant	Oxidation	BDDE	DPAdSV	0.236 μM	Pharmaceutical dosage forms, human serum	[185]
Gonadal hormone	Tamoxifen	Oxidation	AuE	CV, DPV, SWV	6.97 nM	Pharmaceutical dosage forms	[186]
Antidiabetic	Glimepiride	Oxidation	GCE, CPE	CV, DPV	2.0 μM (GCE) 0.75 μM (CPE)	Tablet	[187]

Some voltammetric methods are discussed as follows and tabulated in Table 9.9 for the assay of these drugs in their dosage forms and in biological samples.

Electrochemical reduction and determination of montelukast, a drug prescribed for the treatment of hay fever (chronic allergic rhinitis) and asthma, was studied by Yildiz et al. using CV and DPV at chitosan modified CPE [218]. The linear range was 0.17−18.3 μM by DPV. Limit of detection and limit of quantification were 0.0532 μM and 0.161 μM, respectively. The proposed method was applied to the determination of the drug in pharmaceutical dosage forms and spiked human plasma.

Ghadimi et al. fabricated and characterized a poly(4-vinylpyridine)/MWCNT modified GCE (P4VP/MWCNT GCE) [219]. The electrode demonstrated excellent electrochemical activity toward paracetamol oxidation compared to the bare GCE and MWCNT/GCE. The anodic peak currents of paracetamol at the P4VP/MWCNT GCE were about 300 fold higher than at the non-modified electrode. By applying DPV under optimized experimental conditions, a good linear relation between oxidation peak current and concentration of paracetamol over the range of 0.02−450 μM with a limit of detection of 1.69 nM were achieved. This novel electrode was stable for more than 60 days, and reproducible responses were obtained at 99 % of the initial current of paracetamol without any influence of physiologically common interferences such as ascorbic acid and uric acid. This electrode was useful for determination of paracetamol in pharmaceuticals and urine samples.

A simple, sensitive, and reproducible silver nanoparticles/MWCNT modified GCE was developed by Karadas et al. for the voltammetric determination of zolmitriptan prescribed for the treatment of migraine headaches [220]. The voltammetric determination of zolmitriptan was performed between the linear range of 10 nM and 0.8 μM with a detection limit of 1.47 nM.

Typical anti-inflammatory, antiallergy, and immunoregulatory drugs determined voltammetric methods. Their application details are listed in Table 9.9.

9.3.1.8 Miscellaneous

In this section, some selected voltammetric examples of electroactive drugs with different pharmacologically effects are collected and reported. Some of them are discussed as follows and the others are reported in Table 9.10.

Silva et al. developed a MWCNT/gold nanorod modified GCE for the electrochemical detection of L-cysteine [253]. The sensor presented a linear response range from 5.0 up to 200.00 μM, detection limit of 8.25 nM, and a sensitivity of 120 nA μM^{-1}.

Levodropropizine is a cough suppressant and its electrochemical determination was investigated on GCE and BDDE by Agin et al. using different voltammetric techniques [254]. The electrooxidative behavior was investigated by cyclic, linear sweep, differential pulse, and square wave voltammetry. The dependence of the peak current and peak potentials on pH, concentration, nature of the buffer, and

Table 9.9 Electroanalytical assay of some selected anti-inflammatory, antiallergy, and immunoregulatory drugs using solid electrodes

Drug group	Drug name	Electrochemical behavior	Electrode type	Technique	LOD/LOQ	Applications	References
Analgesic and antipyretic	Paracetamol Tyrosine	Oxidation	CNT-graphene nanosheet nanocomposite modified GCE	DPV	0.10 μM 0.19 μM	Pharmaceutical dosage forms, blood serum	[190]
Analgesic and antipyretic	Paracetamol	Oxidation	Microcrystalline natural graphite-polystyrene composite film electrode	DPV	0.034 μM	Pharmaceutical dosage forms, urine	[191]
Analgesic and antipyretic	Paracetamol	Oxidation	Poly(Patton and Reeder's reagent)-CPE	DPV	0.53 μM	Pharmaceutical dosage forms	[192]
Analgesic and antipyretic	Ascorbic acid Paracetamol	Electrocatalytic oxidation	GCE/MWCNT-Polyhis	CV, DPV	0.76 μM 32 nM	Tablets	[193]
NSAID	4-aminophenazone	Oxidation	PGE	CV, DPV	0.046 μM	Human plasma, urine	[194]
Antihistaminic	Chlorpheniramine	Electrocatalytic oxidation	MCPE	CV, DPV	0.08 μM	Tablets, synthetic serum	[195]
Analgesic and antipyretic	Paracetamol	Electrocatalytic oxidation	$EMIMBF_4$-NiHCF-NP-Gel modified electrode	CV, DPV, Amperometry	0.51 μM	Tablets	[196]
Analgesic and antipyretic	Paracetamol	Electrocatalytic oxidation	poly(DA)/GCE	CV, EIS	6.7 nM	Active compound	[197]
NSAID	Diclofenac	Electrocatalytic oxidation	MWCNTs/$Cu(OH)_2$/IL/GCE	CV, DPV	0.04 μM	Ampoule, tablet, human & fish blood serum	[198]
Analgesic and antipyretic	Paracetamol	Electrocatalytic oxidation	ERGO/NdHCF/GCE	CV, EIS, LSV	14.22 μM	Pharmaceutical dosage forms	[199]
Antirheumatic	Sulfasalazine	Reduction	MIP/CPE	CV, DPV	4.6 nM	Tablets, human serum	[200]

Antirheumatic	D-penicillamine	Oxidation	CPE	CV, DPV, double step potential CA	58 μM (CV) 50 μM (DPV)	Tablets	[201]
NSAID	Nimesulide	Catalytic reduction	SiC-NPs/GCE	CV, DPV, CA	30 nM	Human blood serum	[202]
NSAIDs	Mefenamic acid Indomethacin	Electrocatalytic oxidation	Fe (III)-SBMCPE	CV, DPV	0.02 μM 0.08 μM	Tablets	[203]
Analgesic and antipyretic	Acetaminophen Acetylsalicylic acid Dipyrone	Oxidation	PEDOT/GCE	CV, DPSV	0.066 μM 0.055 μM 0.060 μM	Pharmaceutical dosage forms, urine	[204]
Antiflu and anticold	Acetylsalicylic acid Caffeine	Oxidation	BDDE	CV, DPV	0.23 μM 0.16 μM	Tablets	[205]
Analgesic	Acetaminophen Tramadol	Oxidation	D50wx2-GNP-GCPE	CV, CC, EIS, SWAdSV	4.71 nM 0.011 μM	Pharmaceutical dosage forms, human urine, human serum	[206]
Analgesic and antipyretic	Paracetamol	Oxidation	MWCNT/GCE	CV	0.02 μM	Tablets	[207]
Analgesic and antipyretic	Acetaminophen	Oxidation	SPEs	CV	–	Active compound	[208]
Analgesic and antipyretic	Paracetamol	Oxidation	CPE-Au nano with Nafion	CV, DPV	7.7 nM	Pharmaceutical dosage forms, human urine	[209]
Analgesic and antipyretic	Paracetamol	Oxidation	TiO_2-Gr/CPE	CV	0.21 μM	Tablets	[210]
Antiulcer	Ranitidine	Reduction	GCE	CV, CC, SWAdSV	0.105 μM	Human urine, tablets	[211]

(continued)

Table 9.9 (continued)

Drug group	Drug name	Electrochemical behavior	Electrode type	Technique	LOD/LOQ	Applications	References
Antimigraine	Sumatriptan	Electrocatalytic oxidation	Me-CoSal/ MWCNT/ GCE	DPV	0.3 μM	Pharmaceutical dosage forms, human serum	[212]
Analgesic and antipyretic	Paracetamol	Electrocatalytic oxidation	MWCNT-ACS/GCE	CV, SWV	0.05 μM	Pharmaceutical dosage forms	[213]
Analgesic	Paracetamol Ascorbic acid	Electrocatalytic Oxidation	SWCNT/CCE	DPV	0.12 μM 3.0 μM	Tablets, human serum	[214]
Antiflu and anticold	Paracetamol Phenyl-ephrine hydrochloride Chlorpheniramine maleate	Electrocatalytic Oxidation	Ni-NP/CPE	DPV	0.24 mM 0.0064 mM 0.016 mM	Pharmaceutical dosage forms	[215]
Antihistaminic	Cetirizine dihydrochloride	Electrocatalytic oxidation	MWCNT/GCE	CV	0.071 μM	Pharmaceutical dosage forms, human urine	[216]
NSAID	Piroxicam	Oxidation	CD/CPE	FFT CV	5.0 nM	Human urine, human serum	[217]

Table 9.10 Electrochemical studies of other group of drugs using solid electrodes

Drug group	Drug name	Electrochemical behavior	Electrode type	Technique	LOD/LOQ	Applications	References
Respiratory system	Theophylline	Oxidation	Aligned carbon nanotubes thin-film electrode	DPV	16 nM	Pharmaceutical dosage forms	[221]
Peripheric vasodilator	Papaverine	Reduction	Mercury Film Electrode	DPAdSV	0.7 nM	Pharmaceutical dosage forms, urine	[222]
Bronchodilator	Doxofylline	Oxidation	BDDE	CV, SWV	13.07 μM	Pharmaceutical dosage forms	[223]
Vitamin	Vitamin B_{12}	Oxidation	BiFE	CV, SWAdSV	33.1 nM	Pharmaceutical dosage forms	[224]
Vitamin	Ascorbic acid	Electrocatalytic oxidation	PdNPs-GO/GCE	CV, Amperometry	0.02–2.28 mM	Tablets	[225]
Respiratory system	Theophylline	Electrocatalytic oxidation	MWCNTPE	CV, DPV	19.7 nM	Tablets, urine	[226]
Respiratory system	Theophylline	Electrocatalytic oxidation	GNP/lcys/Gr/Nf/GCE	CV, DPV	0.4 nM	Tablets, tea	[227]
Respiratory system	Theophylline	Electrocatalytic oxidation	Gr/Nf/GCE	CV, EIS, DPV	6 nM	Tablets	[228]
Vitamin	α-tocopherol	Oxidation	GCE	CV, LSV	1.02 μM	Pharmaceutical dosage forms, cosmetic creams	[229]
Respiratory system	Theophylline	Electrocatalytic oxidation	poly(AHNSA) / GCE	CV, DPV	47 nM	Pharmaceutical dosage forms	[230]
Treatment of liver	Oleanic acid	Oxidation	MIP/CoHCNFe/MWCNTs/PSS/CTS/CE	CV, EIS, SWV	2 nM	Tablets	[231]
Erectile dysfunction	Sildenafil citrate	Oxidation	SPGCE	CV, SWV	55 nM	Human urine, tablets	[232]

(continued)

Table 9.10 (continued)

Drug group	Drug name	Electrochemical behavior	Electrode type	Technique	LOD/LOQ	Applications	References
Erectile dysfunction	Sildenafil citrate	Reduction	BiFE	CV, SWAdCSV	18 nM	Active compound	[233]
Antiepileptic Analgesic	Phenobarbital Paracetamol	Electrocatalytic oxidation	nPt-MWCNTPE	CV, DPV	0.1 μM 0.17 μM	Tablets, human urine	[234]
Adrenergic agonist (Cardiac stimulant), Analgesic	Dopamine Paracetamol	Oxidation	PyCE	CV, DPV	2.3 μM 1.4 μM	Pharmaceutical dosage forms, human serum	[235]
Analgesic, Penicillin antibiotic	Paracetamol Penicillin V	Oxidation	BDDE	CV, SWV	0.21 μM 0.32 μM	Human urine	[236]
Adrenergic agonist (Cardiac stimulant), Hormone	Norepinephrine Serotonin	Electrocatalytic oxidation	p-AMTa/GCE	CV, DPV, Amperometry, single potential CA	16.5 pM 13.2 pM	Human blood plasma	[237]
Erectile dysfunction	Sildenafil citrate	Oxidation	ZnTPP/CPE ZnTNP/CPE ZnTSPP/CPE ZnTPP/DPE ZnTNP/DPE ZnTSPP/DPE	DPV	2.46 pM	Tablets	[238]
Amino acid	L-Cysteine	Oxidation	GAMCS NPs/GCE	CV, CA	240 nM	Human serum, human urine	[239]
Metabolic breakdown product	Uric acid	Oxidation	RTIL-NiHCF-NP-Gel modified PIGE	CV, DPV, EIS	0.33 μM	Human urine	[240]
Respiratory system	*N*-acetylcysteine	Oxidation	FCMCNTPE	CV, DPV, CA	0.6 μM	Pharmaceutical dosage forms, human urine, human serum	[241]

Hematologic	Deferasirox Deferiprone	Electrocatalytic oxidation	CoO/Pt	CV, CA	6.67 μM 2.52 μM	Tablets, human urine, human serum	[242]
Vitamin	Vitamin B_{12}	Reduction	SWCNT-Chitosan PGE	CV, EIS, SWV	0.89 nM	Pharmaceutical dosage forms	[243]
Amino acids	L-Tryptophan	Oxidation	PGA/CNTPE	CV	0.01 μM	Pharmaceutical dosage forms	[244]
Vitamin, Metabolic breakdown product	Ascorbic acid (AA) Uric acid (UA)	Oxidation	Organoclay film modified GCE	CV, CA	44 μM (AA)−10 μM (UA) with CV 6.4 μM (AA)–2.2 μM (UA) with CA	Pharmaceutical dosage forms, human urine	[245]
Amino acid	5-Hydroxy-L-tryptophan	Electrocatalytic oxidation	GPGE	DPV, CA	10 μM (DPV) 0.5 μM (CA)	Tablets	[246]
Vitamins	Vitamin C Vitamin B_1 Vitamin B_2	Reduction	AgLAF-AgSAE	DPAdSV	114 nM 9.97 nM 23.9 nM	Tablets, fruit juices	[247]
Analgesic, Adrenergic agonist (Respiratory system)	Paracetamol Epinephrine	Electrocatalytic oxidation	AuNP/MWCNT/ITO	SWV	4.6 nM 4.2 nM	Human urine	[248]
Amino acids	Cysteine Methionine	Oxidation	BDDE GCE	CV, DPV	–	Oxidation behavior study	[249]
Liver protector	Silymarin	Oxidation	CPE MMT-Ca/CPE	CV, SWAdSV	30 nM (CPE) 2.1 nM (MMT−Ca/CPE)	Pharmaceutical dosage forms, human serum	[250]

(continued)

Table 9.10 (continued)

Drug group	Drug name	Electrochemical behavior	Electrode type	Technique	LOD/LOQ	Applications	References
Respiratory system	*N*-acetylcysteine	Reduction	OPG-CoPc	CV, CA	0.9 μM	Pharmaceutical dosage forms	[251]
Vitamin	Ascorbic acid	Electrocatalytic oxidation	GCE, Gold, BDDE, Indium tin oxide electrode	Fourier Transform AC voltammetry	–	Mechanistic study	[252]

scan rate was examined. The proposed method was successfully applied for the determination of levodropropizine in pharmaceutical dosage forms.

Drugs not mentioned in the above-described pharmacologically section are collected in this part and listed in Table 9.10.

9.3.2 Applications of Polarography in Pharmaceutical Analysis

Polarography is dealing with a mercury-based electrode and is especially well suited for the electrochemical reduction of a drug compound. Theoretically, each drug active compound that is either electroreducible or electrooxidizable within the potential range of the dropping-mercury electrode (between +0.5 and −2.5 V) can be determined polarographically. If several electroreducible substances are present together in the same solution, they can all be estimated if their half-wave potentials are more than 0.2 V apart. Steps that are less than 0.2 V apart can sometimes be separated by either changing the pH value of the base solution or by converting the reactants into complexes from which they deposit at potentials far enough apart for each step to be measured separately. Some application examples of polarographic methods on drug analysis are discussed as follows and tabulated in Table 9.11 for all types of pharmaceutical active compounds.

DPP of atorvastatin, a drug used for high cholesterol and triglyceride levels, in pure and pharmaceutical dosage forms has been investigated in borax buffer using the SMDE by Ramadan et al. [342]. Regression analysis showed a good correlation coefficient between I_p and concentration over the studied range with LOD and LOQ of 19.8 and 58.58 nM, respectively.

Sun et al. investigated the electrochemical behavior of the anti-tumor drug capacitabine (CPT) using the second order derivative linear-sweep polarography [343]. A well-defined reduction peak of CPT was observed at the potential of −1.19 V in pH 5.0 Britton–Robinson (BR) buffer, which was attributed to the reduction of carbonyl group with the N–C bond cleavage on the mercury electrode. Under the selected optimized conditions, the reduction peak current was found to be linearly dependent on the CPT concentration in the range from 0.1 to 10 μM with a detection limit of 85.3 nM. The proposed method was further applied to determine the CPT contents in tablets.

Typical application and analysis of polarographic methods on drugs are listed in Table 9.11.

Table 9.11 Electrochemical studies on pharmaceutical compounds using polarographic techniques

Drug group	Drug name	Electrode type	Technique	Linearity range	LOD/LOQ	Applications	References
Macrolide antibiotic	Clarithromycin	DME	DPP	–	1.79 μM (C_{max} for plain) 2.65 μM (C_{max} for suspension)	Albino rat serum	[255]
Hyperlipidemic	Rosuvastatin Calcium	HMDE	CV, CA, SWV	0.199–9.985 μM	69.9 nM	Pharmaceutical dosage forms	[256]
Cephalosporin antibiotic	Cefoperazone	HMDE	CV, DPAdCSV SWAdCSV	15.5–108.5 nM	15.5 nM	Pharmaceutical dosage forms, human urine	[257]
Fluoroquinolone antibiotic	Ciprofloxacin	HMDE	CV, DPAdCSV, SWAdCSV	150.9–452.7 nM 150.9–452.7 nM	30.18 nM (DPAdCSV) 30.18 nM (SWAdCSV)	Pharmaceutical dosage forms, urine	[258]
Antihypertensive	Telmisartan	HMDE	SWAdSV	0.097–5.83 μM	0.084 μM	Tablets, human plasma	[259]
Antihypertensive	Olmesartan medoxomil	HMDE	SWV	1.79–26.06 μM	0.895 μM	Pharmaceutical dosage forms	[260]
Antidiabetic	Glyburide	DME	CV, DCP, SWP, DPP	1–112 μM	0.194 μM	Tablets	[261]
Antitrombotic	Heparin	DME	DPP	0.1–2.0 units mL^{-1}	2.04 units mL^{-1}	Human blood plasma samples, Pharmaceutical dosage forms	[262]
Gonadal hormone	Raloxifene HCl	HMDE	LSAdCSV SWAdCSV	2–100 nM (LSAdCSV) 0.05–70 nM (SWAdCSV)	0.6 nM (LSAdCSV) 0.015 nM (SWAdCSV)	Tablets, human serum	[263]
Fluoroquinolone antibiotic	Lomefloxacin	DME	DPP	1.0–100 μM	1.0 μM	Pharmaceutical dosage forms	[264]

Anti-obesity	Sibutramine HCl	HMDE	CV, DPV, SWV	5–118 mM	1.3 mM	Pharmaceutical dosage forms	[265]
Antidiabetic	Rosiglitazone	HMDE	SWAdSV, DPP, CV	0.05–0.8 μM	0.032 nM	Human urine, plasma	[266]
Adrenergic antagonist (β blocker)	Acebutolol	HMDE	CV, DPAdSV, SWAdSV	0.5–6 μM	0.17 μM	Pharmaceutical dosage forms, human plasma, urine	[267]
Antimicrobial	Ofloxacin Ornidazole	ME	DPP	39.5–276.67 μM 10.84–75.86 μM	3.32 μM 0.592 μM	Pharmaceutical dosage forms	[268]
Immunomodulator	Azathioprine	ME	DPP	0.5–50 μM	0.28 μM	Tablets	[269]
Antimicrobial	Nitazoxanide Ofloxacin	HMDE	DDP	–	0.270 M 0.575 M	Tablets	[270]
Sedative-hypnotic	Clonazepam	DME	DCP, CV, ACP, DPP	0.24–10 μM	0.24 μM	Tablets	[271]
Anticoagulant	Coumarin	ME	DCP, DPP	–	–	Albino mice, anticoagulant studies	[272]
Opioid analgesic	Naltrexone HCl	HMDE	DCP, CV, LSAdCSV, SWAdCSV	6 nM – 1 μM (LSAdCSV) 0.8 nM – 0.5 μM (SWAdCSV)	1.8 nM (LSAdCSV) 0.24 nM (SWAdCSV)	Tablets, human body fluids	[273]
Vitamins	Vitamin K_3	DME	DPP, DCP	0.5–30 μM	0.15 μM	Clinical injection solution, blood serum	[274]
Anticancer	Cisplatin	DME	DPP	8.3–57 nM	8.3 nM	Cancer cell culture	[275]
Antibacterial and antifungal	Nitroxoline	HMDE	CV, DCP, LSAdCSV, DPAdCSV, SWAdCSV	1 μM–9 mM (DCP) 9–100 nM (LSAdCSV) 0.03–0.5 μM (DPAdCSV) 0.9 nM–0.3 μM (SWAdCSV)	305 nM (DCP) 2.7 nM (LSAdCSV) 9.15 nM (DPAdCSV) 0.274 nM (SWAdCSV)	Tablets, human serum	[276]
Antihypertensive	Irbesartan	HMDE	CV, SWAdCSV	3 nM–0.5 μM	0.9 nM	Tablets, human serum	[277]

(continued)

Table 9.11 (continued)

Drug group	Drug name	Electrode type	Technique	Linearity range	LOD/LOQ	Applications	References
Anticancer	Nelarabine	HMDE	CV, AdSV	0.1 μM–10 mM	0.1 μM	Pharmaceutical dosage forms, urine	[278]
Anticancer	Dacarbazine	HMDE	CV, SWV	0.398–21.1 nM	0.277 nM	Vial, human urine, serum	[279]
Adrenocortico-steroids	Dexamethasone	HMDE	CV, SWAdSV	49.8 nM – 0.61 μM	2.54 nM	Pharmaceutical dosage forms	[280]
Antihypertensive	Minoxidil	HMDE	CV, DPAdCSV	–	–	Pharmaceutical topical solution, shampoo, cream	[281]
Hyperlipidemic	Ezetimibe	HMDE	CV, SWV, CA, CC	0.807–15.98 μM	0.242 μM	Tablets	[282]
Adrenocortico-steroids	Prednisolone	HMDE	CV, DPAdCSV	20.0–400.1 nM	10.96 nM	Tablets, human urine	[283]
NSAID	Dipyrone	DME SMDE HMDE	DPP	1.0–50.0 μM (DME) 3.8–50.0 μM (SMDE) 3.8–50.0 μM (HMDE)	1.0 μM (DME) 3.8 μM (SMDE) 3.8 μM (HMDE)	Pharmaceutical dosage forms	[284]
Digestive system and metabolism	Drotaverine HCl	HMDE	CV, SWAdCSV, DPAdCSV	1.84–16.56 μM (SWAdCSV) 1.84–16.56 μM (DPAdCSV)	4.148 nM (SWAdCSV) 18.66 nM (DPAdCSV)	Tablets	[285]
Fluoroquinolone antibiotic	Gatifloxacin	DME HMDE SMDE	DPP	10.65–532.5 μM 42.6–532.5 μM 42.6–532.5 μM	10.65 μM 42.6 μM 42.6 μM	Tablets	[286]
Macrolide antibiotic	Spiramycin	HMDE	DPP SWP	23.76–95.04 μM (DPP) 2.24–95.04 μM (SWP)	10.1 μM (DPP) 0.546 μM (SWP)	Tablets	[287]

Gonadal hormone	Cyproterone acetate	DME	DCP DPP	7.68–76.8 μM (DCP) 1.2–33.6 μM (DPP)	0.36 μM (DCP) 0.312 μM (DPP)	Tablets	[288]
Antihelminthic	Pyrantel pamoate	HMDE	CV, SWAdCSV, DPAdCSV	0.420–107.62 μM (SWAdCSV) 1.68–215.04 μM (DPAdCSV)	14.75 nM (SWAdCSV) 42.14 nM (DPAdCSV)	Tablets, urine	[289]
Cholinesterase inhibitor	Pyridostigmine bromide	HMDE	CV, SWAdCSV, DPAdCSV	0.383–275.76 μM (SWAdCSV) 3.83–306.4 μM (DPAdCSV)	0.079 μM (SWAdCSV) 0.123 μM (DPAdCSV)	Tablets, human urine	[290]
Antiprotozoal	Nitazoxanide	HMDE	CV, LSAdCSV, DPAdCSV, SWAdCSV	0.5 nM–0.2 μM (LSAdCSV) 0.8 nM–0.1 μM (DPAdCSV) 0.1 nM–0.2 μM (SWAdCSV)	0.15 nM (LSAdCSV) 0.24 nM (DPAdCSV) 0.03 nM (SWAdCSV)	Pharmaceutical dosage forms, human serum	[291]
Cephalosporin antibiotic	Cefixime	HMDE	CV, SWAdCSV, DPAdCSV	0.11–56.32 μM (SWAdCSV) 0.22–112.64 μM (DPAdCSV)	8.8 nM (SWAdCSV) 17.6 nM (DPAdCSV)	Tablets, human urine	[292]
Antiepileptic	Topiramate	DME SMDE HMDE	DPP	4–56 μM (DME) 12–80 μM (SMDE) 12–80 μM (HMDE)	4 μM (DME) 12 μM (SMDE) 12 μM (HMDE)	Tablets	[293]
Antiviral	Zidovudine	HMDE	CV, DPV	1–5 μM	0.01 μM	Pharmaceutical dosage forms	[294]

(continued)

Table 9.11 (continued)

Drug group	Drug name	Electrode type	Technique	Linearity range	LOD/LOQ	Applications	References
Peripheric vasodilator	Buflomedil HCl	HMDE	DPAdCSV, LSAdCSV, SWAdCSV	0.08–5 μM 0.05–1 μM 0.004–0.4 μM	24 nM 15 nM 1.2 nM	Tablets, human serum	[295]
Anti-Parkinson's	Entacapone	HMDE	CV, SWAdCSV DPAdCSV	16–160 μM (SWAdCSV) 18–500 μM (DPAdCSV)	0.425 nM (SWAdCSV) 40.21 nM (DPAdCSV)	Tablets	[296]
Antihypertensive	Candesartan silexetil	HMDE	SWAdCSV	0.82–2.90 μM	0.164 μM	Pharmaceutical dosage forms	[297]
Antiepileptic	Zonisamide	HMDE	CV, DPV, DCP	0.61–80.14 μM	0.146 μM	Capsules, human serum	[298]
Antihelminthic	Mebendazole	DME	DPP	3.38–27.1 μM	1.35 μM	Tablets	[299]
Cephalosporin antibiotic	Cefpodoxime proxetil	DME	DPP, Controlled potential coulometry	0.71–280 μM	0.57 μM	Tablets	[300]
Bipolar disorder	Quetiapine	DME	DCP DPP, ACP	20.8–114.4 μM (DCP) 10.4–114.4 μM (DPP)	0.156 μM (DCP) 0.104 μM (DPP)	Tablets	[301]
Antiulcer	Lansoprazole Rabeprazole	DME	DPP	1–70 μM 1–70 μM	1 μM 1 μM	Active compound	[302]
Sedative-hypnotic	Nitrazepam	DME	CV, DPP	0.13–0.5 mM	47 nM	Tablets	[303]
Antiprotozoal	Ornidazole	DME	CV, DPP	2 mM–4 μM	0.36 μM	Tablets	[304]
Antiviral	Acyclovir	ME	DCP DPP	4.44–35.52 μM 3.55–35.52 μM	0.84 μM 0.44 μM	Pharmaceutical dosage forms	[305]
Antispasmodic	Tegaserod maleate	HMDE	DPP	–	0.240 nM	Tablets	[306]
Aminocoumarin antibiotic	Novobiocin	HMDE	CV SWV SWAdSV	–	–	Mechanistic study	[307]

Antiviral drug	Ganciclovir	ME	SWV LSV	–	0.13 μM 0.14 μM	Active compound	[308]
Alzheimer's disease drug	Donepezil HCl	HMDE	CV, SWAdCSV	3–100 nM	0.93 nM	Tablets, human serum	[309]
Antihypertensive	Irbesartan	HMDE	CV, DPV, SWV	8–100 μM (DPV) 8–100 μM (SWV)	0.768 μM (DPV) 0.604 μM (SWV)	Tablets, human serum	[310]
Antihistaminic	Carbinoxamine maleate	DME	DCP DPP	29.49–584.98 μM 2.34–1170.45 μM	29.49 μM 2.34 μM	Pharmaceutical dosage forms	[311]
Cephalosporin antibiotic	Cefadroxil	HMDE	CV, SWAdSV	0.6–2 μM	0.002 μM	Capsulate	[312]
Antiprotozoal	Ornidazole	HMDE	CV, SWAdSV DPAdSV	0.8–10 μM (SWAdSV) 0.8–10 μM (DPAdSV)	0.034 μM (SWAdSV) 0.26 μM (DPAdSV)	Capsules	[313]
Fluoroquinolone antibiotic	Ofloxacine	HMDE	CV, SWAdSV	0.5–1.7 μM	0.011 μM	Tablets, human urine, serum	[314]
Antihypertensive	Candesartan cilexetil	HMDE	SWAdSV, CV, CA	0.409 – 2.19 μM	0.016 μM	Tablets	[315]
Fluoroquinolone antibiotic	Enrofloxacine	HMDE	SWAdCSV	10.0–80.0 nM	0.33 nM	Pharmaceutical dosage forms, blood serum	[316]
Postmenopausal osteoporosis	Ipriflavone	HMDE	SWAdCSV	5–400 nM	1.5 nM	Tablets	[317]
Digestive system and metabolism	Drotaverine HCl	HMDE	CV, DPAdCSV	50.01–593.1 nM	7.26 nM	Tablets, human urine	[318]
Adrenocorticosteroid	Dexamethasone	HMDE	CV, SWAdCSV	–	3.1 nM	Pharmaceutical dosage forms, biological fluids	[319]
Antibiotics	Pipemidic acid Ofloxacin	HMDE	CSV	32.96–329.68 μM 27.67–276.72 μM	0.164 μM 2.76 nM	Pharmaceutical dosage forms, human urine	[320]

(continued)

Table 9.11 (continued)

Drug group	Drug name	Electrode type	Technique	Linearity range	LOD/LOQ	Applications	References
Vitamin	Vitamin K_3	DME	DPP	–	0.7 μM	Clinical injection solution	[321]
Antihypertensive	Losartan Triamterene	ME	DCP, SWAdCSV	30–270 nM 0.5–200 nM	9.7 nM 0.3 nM	Pharmaceutical dosage forms, urine	[322]
Hyperlipidemic	Etofibrate (E) Fenofibrate (F) Atorvastatin (A)	DME	DPP, SWV, CV	5.49–27.45 μM (DPP, E) 1.37–6.85 μM (SWV, E) 1.38–6.9 μM (DPP, F) 0.552–2.76 μM (SWV,F) 5.37–26.85 μm (DPP,A) 5.37–26.85 μm (SWV,A)	0.522 μM (DPP, E) 0.109 μM (SWV, E) 0.138 μM (DPP, F) 0.102 μM (SWV, E) 0.358 μM (DPP, A) 0.376 μm (SWV, A)	Pharmaceutical dosage forms, human plasma	[323]
Gonadal hormone	Finasteride	DME	DPP, TP	0.05–0.5 mM	7.59 μM	Tablets	[324]
Antiulcer	Lansoprazole Omeprazole	DME	ACP	1.08–54 μM 0.58–29 μM	0.054 μM 0.029 μM	Capsules	[325]
Myorelaxant	Tetrazepam	ME	DPP, DPAdCSV, LSAdCSV, SWAdCSV	–	5 μM 0.13 μM 0.01 μM 0.003 μM	Tablets, human serum	[326]
Respiratory system Antihypertensive	Acetylcysteine Captopril	DME	DPP	2.45–8.58 μM 1.1–7.7 μM	0.33 μM 0.11 μM	Tablets	[327]
Cephalosporin antibiotic	Cefdinir	DME	CV, DPP	0.01–0.2 mM	0.3 μM	Pharmaceutical dosage forms	[328]
Antidiabetic	Rosiglitazone	DME	DCP DPP, ACP	11.2 – 67.2 μM (DCP) 0.28–44.8 μM (DPP)	0.42 μM (DCP) 0.196 μM (DPP)	Tablets, human plasma	[329]

Antiviral	Abacavir	HMDE	CV, DPV SWV, DCP	0.1–10 μM (CV) 0.1–10 μM (DPV) 0.1–10 μM (SWV)	0.0211 μM (CV) 0.0241 μM (DPV) 0.0269 μM (SWV)	Tablets	[330]
Macrolide antibiotic	Clarithromycin	HMDE	CV, LSAdCSV, SWAdCSV	0.1–1 μM (LSAdCSV) 0.05–0.4 μM (SWAdCSV)	0.03 μM (LSAdCSV) 0.015 μM (SWAdCSV)	Tablets, human urine	[331]
Hyperlipidemic	Simvastatin	HMDE	CV, SWV, SWAdSV	1–10 μM 0.01–0.75 μM	0.35 μM 4.5 nM	Tablets, human serum	[332]
Antihypertensive	Quinapril	HMDE	CV, SWV, CA	1.14–19.72 μM	0.50 μM	Tablets	[333]
Antihypertensive	Irbesartan	HMDE	First harmonic alternating current (AC) stripping sweep	0.372–3.724 μM	0.256 μM	Oral dosages	[334]
Antihypertensive	Losartan	HMDE	DCP(Ads), DPAdCSV, SWAdCSV, ACP(Ads)	0.378–2.835 μM	0.354 μM 0.307 μM 0.354 μM 0.284 μM	Pharmaceutical dosage forms	[335]
Carbapenem antibiotic	Imipenem	HMDE	DPP, DC tast polarography, CV, AdSV	0.018–1.2 μM	5.43 nM	Pharmaceutical dosage forms, human urine	[336]
Hyperlipidemic	Fluvastatin	HMDE	SWAdSV, Amperometry	0.01–2.7 μM	9.9 nM	Pharmaceutical dosage forms	[337]
Antihistaminic	Astemizole	HMDE	SWAdCSV	0.5–2.5 μM	0.014 μM	Tablets, biological fluids	[338]
Antihypertensive Anticancer	Captopril Thioguanine	HMDE	DPAdCSV	0.5–100 nM 0.15–180 nM	0.3 nM 0.08 nM	Pharmaceutical dosage forms, biological fluids	[339]
Antihypertensive	Valsartan	HMDE	Harmonic alternating current stripping sweep	0.184–1.472 μM	0.046 μM	Pharmaceutical dosage forms	[340]
Antihypertensive/ Diuretic	Spironolactone	HMDE	CV, SWAdSV	0.01–0.25 μM	0.172 nM	Pharmaceutical dosage forms, biological fluids	[341]

9.3.3 Applications of LC-EC Detection in Pharmaceutical Analysis

LC-EC with amperometry detection can advantageously be considered for highly sensitive assays provided that the drug compound is electroactive [20, 344]. Glassy carbon is the most used electrode, yet other working electrode material can be considered such as platinum or gold, e.g., for aminoglycoside antibiotics using pulsed amperometric detection (PAD). A general rule of thumb for proper use of an amperometric detector is that once the appropriate working electrode potential is applied, the detector should be maintained polarized under hydrodynamic conditions for weeks without interruption in order to keep low background currents. In this section, some selected applications are given below and in Table 9.12.

Jia et al. studied the influence of nanogold particles in the mobile phase on the LC-amperometric response of a glassy carbon electrode for the pharmacokinetic study of danshensu (DSS), protocatechuic acid (PA), protocatechuic aldehyde (PAH), hydroxy safflower yellow A (HSYA), and ferulic acid (FA) [377]. Some enhanced signal (electron transfer) was noticed thanks to the presence of the nanogold particles. Linear regression analysis for the four phenolic compounds (DSS, PA, HSYA, and FA) were performed in a range of spiking levels from 19.7 nM to 1.263 μM, 24–719.35 nM, 13.9−81.76 nM, and 37.03 nM−3.06 μM in rat plasma, respectively. The linear ranges of DSS, HSYA, and FA expanded from 19.7 nM to 1.263 μM, 13.9−204.41 nM, 36–578.57 nM in urine and feces, respectively. The sensitivity was evaluated by determining the LOQ, having a precision smaller than 10 % and signal/noise ratio higher than 10: DSS 19.7 nM, PA 24 nM, HSYA 13.9 nM, FA 36 nM (S/N = 10). The extraction recoveries of PA, HSYA, and FA from rat plasma were evaluated. The mean recoveries of the samples were more than 70 % which indicated that the extraction recoveries of DSS, PA, HSYA, and FA from the biological samples (plasma, urine, and feces) were concentration-independent in the concentration range evaluated.

Wall-jet/ thin layer amperometric electrochemical detection coupled with LC was used for the simultaneous analysis of dopamine (DA) and homovanillic acid (HVA) at a GCE by Lou et al. [378]. Under optimized conditions, the LC-ECD calibration curve showed good linearity from 0.01 to 100 μM for DA and HVA, and the limit of detection obtained were 1.1 nM for DA and 0.7 nM for HVA. Also the method was tested on human urine with satisfactory results.

A recently available BDD working electrode has been developed for use with LC to aid in the detection of molecules with high redox potentials by Birbeck and Mathews [379]. Hydrodynamic voltammograms were constructed for dopamine and adenosine, and the optimal potentials were determined to be +740 and +1,200 mV versus a palladium reference electrode, respectively. The detection range achieved with the BDD electrode was from the low nanomolar to high millimolar levels. To determine the practical utility of the BDD electrode, tissue content was analyzed for seven monoamine and two purine molecules, which were resolved in a single run in less than 28 min.

Table 9.12 Liquid chromatographic studies on pharmaceutical compounds using electrochemical detection

Drug group	Drug name	Electrode type	Detector Potential (mV)	Column	Mobile phase composition	Analysis time (min)	LOD/ LOQ	Applications	References
Flavonoids	Myricetin Quercetin Luteolin Kaempferol Apigenin	Nano-metal oxide modified carbon paste	1000	RP Luna C18 (250 mm × 4.6 mm, 5 μm)	Methanol:phosphate buffer (55:45, 60:40, 50:50, v/v) (pH:2.5)	20	47.1 pM 59.6 pM 87.3 pM 69.9 pM 148 pM	Human urine	[345]
Macrolide antibiotic	Clarithromycin	GCE	850	Waters Rheodyne symmetry C8 (100 mm × 4.6 mm, 3.5 μm)	Acetonitrile:0.045 M H_3PO_4 (37:63, v/v) (pH:6.7)	14	66.8 nM	Human plasma	[346]
Polycyclic aromatic hydrocarbon	1-Hydroxy pyrene	BDDFE	1000	LiChroCART 125–4 100 RP-18 (119 mm × 4 mm, 5 μm)	Methanol:0.05 M phosphate buffer (80:20, v/v) (pH:5.0)	5	0.013 μM	Human urine	[347]
Antioxidants	Glutathione Glutathione disulfide	Carbon porous	400, 700, 750, 800	Synergy Hydro-RP (150 mm × 4.6 mm, 5 μm)	1 % acetonitrile in 20 mM potassium phosphate buffer (pH:2.7)	15	0.011 μM 0.156 μM	Human blood	[348]
Traditional Chinese medicine	Tetrandrine Fangchinoline	GCE	900	WondaSil C18-WR (250 mm × 4.6 mm, 5 μm)	Acetonitrile:ammonium acetate buffer (32:68, v/v) (pH 6.5)	55	0.26 M 0.27 M	Herbal medicine	[349]
Isoquinoline alkaloids	Jatrorrhizine Coptisine Palmatine Berberine	GCE	1300	WondaSil C18-WR (250 mm × 4.6 mm, 5 μm)	40 mM phosphate buffer:acetonitrile (40:60, v/v) (pH:7.0)	12	0.02 μM 0.03 μM 0.03 μM 0.01 μM	Rat plasma	[350]
Phenolic acids	Gallic acid Protocatechuic acid Gentisic acid 4-Hydroxy benzoic acid Caffeic acid	Carbon fiber microelectrode	1200 1200 1200 1400 1200	Ascentis C18-A (100 mm × 2.1 mm, 3 μm)	50 mM disodium hydrogen phosphate +20 mM sodium perchlorate:acetonitrile (85/15, v/v) (pH:3.0)	6	0.266 nM 0.259 nM 0.306 nM 8.73 nM 0.320 nM	Salvia	[351]

(continued)

Table 9.12 (continued)

Drug group	Drug name	Electrode type	Detector Potential (mV)	Column	Mobile phase composition	Analysis time (min)	LOD/ LOQ	Applications	References
Amino acid and metabolites	D,L-tryptophane L-Kynurenine	MWCNT modified glassy carbon	1000	Reversed-phase ODS (250 mm × 4.6 mm, 5 μm)	Acetate buffer:methanol (4:1, v/v) (pH:5.0)	10	0.4 μM 0.5 μM	Human plasma	[352]
Flavonoids	Bavachin Iso-bavachalcone	GCE	800	Sepaxamethyst C18-P (250 mm × 4.6 mm, 5 μm)	0.03 mol/L acetate buffer solution:acetonitrile (2:3, v/v) (pH:5.17)	19	8.81 nM 0.117 nM	Fructus Psoraleae	[353]
Antioxidants	Cystine Cysteine Ascorbic acid Homocysteine Glutathione Methionine *N*-acetylcysteine Oxidized glutathione	GCE	900	Discovery HS C18 RP (250 mm × 4.6 mm, 5 μm)	0.05 % TFA:methanol (97:3, v/v)	17	49.93 nM 4.95 nM 4.54 nM 2.96 nM 29.3 nM 6.7 nM 4.29 nM 81.4 nM	Human plasma	[354]
Antioxidants	Lipoic acid Dihydrolipoic acid	GCE	1000	Discovery HS C18 RP (250 mm × 4.6 mm, 5 μm)	Acetonitrile: 0.05 M phosphate buffer (52:48, v/v) (pH:2.4)	9.5	2.42 nM 14.4 nM	Human plasma	[355]
Antihypertensive	Captopril	Porous graphite	950	Luna C18 (150 mm × 4.1 mm, 5 μm)	Phosphate buffer: acetonitrile (70:30, v/v) (pH:3.0)	9	2.76 μM	Dosage forms	[356]
Steroids	Lathosterol Cholesterol	GCE	2800	YMC-Pack ODS-AL (250 mm × 2.0 mm, 5 μm)	Acetonitrile containing 30 mmol/ L LiClO4	60	0.03 μM 0.03 μM	Human, rat, and mouse serum	[357]
Anti-Parkinson's	L-dopa 3-C-methyldopa Carbidopa	GCE	800	Phenomenex C18 (250 mm × 4.6 mm, 5 μm)	Methanol:aqueous solution (8:92, v/v) (pH:2.88)	25	20.3 nM 18.9 nM 17.7 nM	Human plasma	[358]

Erectile dysfunction	Sildenafil Vardenafil *N*-desmethyl sildenafil *N*-desethyl vardenafil	BDD	1520	Ascentis C18 (100 mm × 2.1 mm, 3 μm)	20 mM sodium dihydrogen phosphate with 40 mM sodium perchlorate: acetonitrile (70:30, v/v) (pH:3.5)	12	6.32 nM 6.14 nM 8.68 nM 4.25 nM	Human plasma	[359]
Oxidative damaged DNA	8-hydroxy-2′-deoxyguanosine	Nanocarbon film Glassy carbon	1000 800	Hydrosphere C18 (150 mm × 4.6 mm or 2.0 mm)	Phosphate buffer: MeOH (95/5, v/v) (pH:7.5)	20	3 nM 7.2 nM	Urinary sample	[360]
Antioxidant	Lipoic acid	GCE	1000	Discovery HS C18 RP (250 mm × 4.6 mm, 5 μm)	0.05 M phosphate buffer:acetonitrile (50:50, v/v) (pH:2.5)	6	0.969 nM	Human plasma	[361]
Neuro-transmitters and their metabolites	Dopamine Dihydroxyphenylacetic acid Homovanillic acid Serotonine 5-hydroxyindole acetic acid	–	−100 300	ESA MD (150 mm × 3.2 mm)	Acetonitrile: aqueous with 75 mM monobasic sodium phosphate buffer, 0.5 mM EDTA, 0.81 mM sodium octylsulfonate, and 5 % tetrahydrofuran (5 : 95, v/v) (pH:3.1)	15	6.53 nM 5.95 nM 5.49 nM 4.70 nM 5.23 nM	Human cerebrospinal fluid	[362]
Amino acids	Aspartate Glycine Glutamate Glutamine γ-aminobutyric acid	GCE	700	C18 Adsorbosphere OPA-HR (150 mm × 4.6 mm, 5 μm)	Gradient 0.1 M monosodium dihydrogen phosphate and 0.05 mM EDTA with 2 % MeOH:pure MeOH (pH:5.0)	30	5 μM 15 μM 5 μM 2.5 μM 5 μM	Suncus murinus brain tissue	[363]
Fluoro-quinolone antibiotics	Danofloxacin Difloxacin Sarafloxacin	Porous graphite	800	Novapak C18 (150 mm × 3.9 mm, 4 μm)	0.05 M sodium perchlorate-2 % acetic acid:acetonitrile (80:20, v/v)	16	0.8 ng g^{-1} 3.6 ng g^{-1} 4.0 ng g^{-1}	Chicken tissues	[364]
Antioxidants	Glutathione Glutathione disulfide	BDD	1475	Shiseido C18 (250 mm × 3.0 mm, 5 μm)	25 mM sodium dihydrogen phosphate, 1.4 mM of the ion-pairing reagent 1-octanesulfonic acid (pH:2.65), and 6 % acetonitrile	25	6.25 μM 0.16 μM	Mice and rat liver	[365]

(continued)

Table 9.12 (continued)

Drug group	Drug name	Electrode type	Detector Potential (mV)	Column	Mobile phase composition	Analysis time (min)	LOD/LOQ	Applications	References
Phytosterols	β-sitosterol Campesterol Stigmasterol Brassicasterol	GCE	2800	Develosil C30-UG-3 conventional (150 mm × 4.6 mm, 3 μm)	Acetonitrile:10 mM $LiClO_4$	45	3.40 μM 2.01 μM 1.06 μM 0.82 μM	Human and rat serum	[366]
Amino mono-saccharide	Glucosamine	Gold	100	CarbopackTM PA (3 mm × 150 mm, 6.5 μm)	Gradient water:200 mM sodium hydroxide	12	11.2 nM	Human plasma	[367]
Antipsychotic	Risperidone (−)-9-OH-Risperi-done (+)-9-OH-Risperidone	–	500 650 950	Chiral-AGP (100 mm × 4.0 mm, 5 μm)	Methanol:phosphate buffer (15:85, v/v) (pH:6.2)	11	1.22 nM 1.22 nM 1.22 nM	Human plasma	[368]
Vitamins Metabolic break-down product	Ascorbic acid Uric acid	–	0.0 450	YMC ODS-AQ (150 mm × 3.0 mm, 3.0 μm)	Gradient metha-nol:150 mM chloroacetic acid, 2 mM disodium EDTA (pH:3.0)	12	0.2 μM 0.2 μM	Human plasma	[369]
Short-chain fatty acids	Lactic acid Acetic acid Propionic acid Butyric acid Isovaleric acid Valeric acid	GCE	−900	Shim-pack SPR-H (250 mm × 7.8 mm)	0.1 mM $HClO_4$:6 mM VK_3 and 100 mM $LiClO_4$	35	2 μM	Rat and human feces	[370]
Antibiotic	Lincomycin	Screen-printed carbon	900	Symmetry C8 (100 mm × 2.1 mm, 5 μm)	Acetonitrile:metha-nol:Phosphate buffer (5:20:75, v/v/v) (pH:6)	6	0.08 μM	Feeds, honey, milk and urine	[371]
Sulfonamide antibiotics	Sulfa-methoxazole Trimethoprim	BDD	1250	Luna octyl silica (150 mm × 4.6 mm, 10 μm)	0.05 mol/L phosphate buffer solution:aceto-nitrile (82:18, v/v) (pH 5)	32	98.7 nM 51.7 nM	Bovine milk	[372]

Neuro-transmitters and their metabolites	5-hydroxyindole acetic acid Homovanillic acid Dopamine 5-hydroxy-tryptamine Norepinephrine	GCE	700	ODS C18 (500 mm × 3.9 mm, 5 μm)	Methanol:acetonitrile:water (19:3:78, v/v/v) (pH:3.3)	12	20.9 nM 22 nM 34.7 nM 22.7 nM 23.6 nM	Brain tissue	[373]
Vitamins	Pyridoxamine-5-′-phosphate Pyridoxamine Pyridoxal-5-′-phosphate Pyridoxal Pyridoxine 4-pyridoxic acid	Porous graphite electrode	350 800	Hypersil Gold C18 (250 mm × 4.6 mm, 5 μm)	35 mM sodium phosphate containing 2.5 mM heptanesulfonic acid (pH:3.2): methanol 78:12 (v/v)	18	0.8 nM 1.1 nM 1.5 nM 1.3 nM 1.1 nM 2.1 nM	Plasma and Serum	[374]
Amino-glycoside antibiotic	Amikacin	Gold	E1:50, E2:70, E3:-150	HyPurity Elite (150 mm × 4.6 mm, 5 μm)	Sodium-1-octane sulphonate, 1.5 g Anhydrous sodium sulfate, 20 g Tetrahydrofuran, 15 mL 0.2 M phosphate buffer (pH:3)	20	0.102 μM	Cerebrospinal fluid	[375]
Antiprotozoal	Artesunate Dihydro-artemisinin	Porous graphite	−400	Eclipse XDB-C18 (150 mm × 4.6 mm, 5 μm)	Acetonitrile:ESA (40:60, v/v)	0.17 h 0.72 h (MRT values)	3.04 nM 2.43 nM	Dog plasma	[376]

In another study, vitamin E was analyzed using LC-ED by Shintani [380]. ECD detection was selective for the compounds with tautomerism such as keto ⇌ enol or amine ⇌ imine with relatively low oxidation–reduction voltage. So in case of vitamin E analysis, ECD detection can be applied because phenolic OH ⇌ carbonyl tautomerism exists under the applied voltage. In addition, ECD detection was approximately ten times more sensitive than UV detection. Sensitivity and selectivity of ECD detection was almost identical to fluorescence detection.

Typical drugs determination examples by LC-ECD methods are listed in Table 9.12.

9.3.4 Biosensors in Pharmaceutical Analysis (See also Chap. 6)

Biosensor for analysis of drug use is one of the most popular topics in recent years. In this section, some examples of different biosensors for drug analysis and biological samples can be found as follows and in Table 9.13.

The interaction of ciprofloxacin with dsDNA was studied by Fotouhi et al. [424] using CV, fluorescence emission spectroscopy, and UV–vis spectroscopy. The presence of DNA resulted in a decrease in the current. A positive shift in the ciprofloxacin oxidation peak indicated the intercalative interaction. The corresponding heterogeneous rate constant (k_s) and the electron transfer coefficient (a) were calculated to free Cf and the bound Cf–DNA complex.

Asturias-Arribas et al. developed a tetrathiafulvalene (TTF) modified screen-printed biosensor based on the inhibition of the acetylcholinesterase activity for the determination of codeine [425]. Applying a potential of +250 mV, a 1 mM solution of acetylthiocholine in electrolyte solution pH 7.0 gave an oxidation signal due to the TTF catalytical oxidation of the enzymatically generated thiocholine. The electrochemical signal decreased by consecutive additions of codeine. It allowed linear calibration curves to be obtained between 20 and 200 μM with a reproducibility of 3.31 % ($n = 6$). Additionally, the developed biosensor was used for the determination of codeine in pharmaceutical commercial tablets and urine samples.

The interaction between codeine and morphine with dsDNA was assessed at pH 7.0 by Ensafi et al. [426]. DPV at a pencil graphite electrode showed that both molecules were electrochemically oxidized due to the presence of phenolic and amino groups in their structure. Detection limits of 0.137 and 0.15 μM were obtained for codeine and morphine, respectively. The biosensor was applied to the analysis of codeine and morphine in blood serum, urine samples, and pharmaceutical dosage forms.

Canbay and Akyilmaz fabricated MWCNT-nafion-cysteamine modified tyrosinase biosensor for determination of dopamine [427]. In this study, dopamine determination was realized by DPV in the range 0.4 and −0.15 V. In the optimization studies, some parameters such as pH, temperature, enzyme amount, and

Table 9.13 Biosensors for the electrochemical assay of drugs

Drug group	Drug name	Electrode type	Technique	LOD/LOQ	Applications	References
Local anesthetic	Cocaine	DNA/AuE	EIS	0.21 nM	Human urine, human serum	[381]
Antiprotozoal	Furazolidone	dsDNA/MWCNT/GCE	CV	0.11 μM	Serum	[382]
Vitamin	Folic acid	dsDNA/PGE	DPV	0.0106 μM	Tablet, Fortified wheat flour, Spinach	[383]
Gonadal hormone	Tamoxifen	HRP/PANI/PtE	CA	0.188 nM	Pharmaceutical dosage forms	[384]
Anticancer	Methotrexate	DNA/GCE	GCE	5 nM	Active compound	[385]
Antidepressant	Amitriptyline	Graphite (SiO_2/Al_2O_3/Nb_2O_5/DNA)	DPV	0.12 μM	Pharmaceutical dosage forms	[386]
Adrenergic agonist (Cardiac stimulant)	Dopamine	MWCNT/HRP/CP	SWV	2 μM	Pharmaceutical dosage forms	[387]
Vitamin	B_1	dsDNA/PMWCNTPE	DPV	1.46 μM	Serum, plasma, urine	[388]
Antimalarial	Artesunate	Gr/PA/HP	CV, EIS	0.031 nM	Parenteral artesunate, human urine, human serum, and human plasma	[389]
Sulphonamide antibiotic	Sulfadiazine	MWCNT-GCE/DNA	CV	0.12 μM	Human blood plasma sample	[390]
Digestive system and metabolism	Glutamate	GluOx/CMWCNT/AuNP/CHIT AuE	CV	1.6 μM	Serum	[391]
Fluoroquinolone antibiotic	Gemifloxacin	dsDNA/PGE	DPV	0.59 μM	Pharmaceutical dosage forms	[392]
Antiprotozoal	Quinacrine	DNA-Cu(II)/PAA/GCE	Amperometry	10 μM	Active compound	[393]
Vitamin	Biotin (B_8)	streptavidin modified magnetic microbeads	SWV	81 μM	Commercial dietary supplement, infant formula samples	[394]
Local anesthetic	Cocaine	Aptamer AuNPs/GO/SPCE	DPV	1 nM	Active compound	[395]
Analgesic and antipyretic	Acetaminophen	Banana tissue/polypyrrole CE	DPV	0.01 μM	Human plasma, milk samples	[396]
Adrenergic agonist (Cardiac stimulant)	Dopamine	PA-MNPs/GCE	Amperometry EIS	7.25 μM 14.1 μM	Active compound	[397]

(continued)

Table 9.13 (continued)

Drug group	Drug name	Electrode type	Technique	LOD/LOQ	Applications	References
Anticancer	Clofarabine	dsDNA/GCE	DPV	0.08 μM	Active compound	[398]
Vitamin	Riboflavin (B2)	DNA/PGE	DPV DPASV	0.90 μM 0.202 nM	Urine, pharmaceutical dosage forms	[399]
Antiepileptic	Gabapentin	DNA/AuE	CV	–	Active compound	[400]
Adrenergic agonists (Cardiac stimulant)	Dopamine Uric acid	DNA/PAMAM/ MWNT-Chit/AuE	DPV	0.03 μM 0.07 μM	Active compound	[401]
Anticancer	Leuprolide	DNA/PGE	DPAdSV	1.15 nM	Pharmaceutical dosage forms	[402]
Gonadal hormone	Fulvestrant	(dsDNA)/PGE	DPV	0.675 μM	Pharmaceutical dosage forms	[403]
Anticancer	Daunomycin Doxorubicin Idarubicine Mitoksantron	SPE/AuNPs/pTTBA/ dsDNA/CL/AuNPs	Amperometry	5.5 ± 0.3 fM 3.6 ± 0.2 fM 2.2 ± 0.1 fM 1.2 ± 0.05 fM	Spiked real urine samples	[404]
Vasoprotective	Rutin	DNA/MWNTs-COOH/ Fe_3O_4/GCE	DPV	7.5 nM	Active compound	[405]
Anticancer	Tamoxifen	dsDNA/CPE	ZCP	0.11 μM	Active compound	[406]
NSAIDs	Aspirin Salicylic acid	DNA/PPy/PtE	DPV	5.24 μM 0.086 μM	Active compound	[407]
Antibiotic	Pipemidic acid	GE/Tyr	FIA	18 nM	Pharmaceutical dosage forms	[408]
Anticancer	Mitomycin C	dsDNA immobilized PVF/AuE	DPV	269.2 nM	Active compound	[409]
Fluoroquinolone antibiotics	Sparfloxacin Moxifloxacin Gatifloxacin	DNA/GCE	DPV	0.08 μM 0.08 μM 0.05 μM	Pharmaceutical dosage forms	[410]
Anticancer	Irinotecan	dsDNA/PGE	DPV	–	Active compound	[411]
Antibiotic	Penicillin	MWCNTs)/hematein/ lactamase/GCE	CA	50 nM	Spiked milk	[412]
Anticancer	Adriamycin	(dsDNA)/Ag_{nano}/ PPAA/ MWCNTs-COOH	DPV	3.2 pM	Active compound	[413]

Antihypertensive / Diuretic	Amiloride	(dsDNA)/PGE	DPV	0.5 μM	Tablets, urine samples	[414]
Anticancer	Methotrexate	dsDNA/GCE	SWV	5 nM	Pharmaceutical dosage forms, human urine	[415]
Antiepileptic	Levetiracetam	HRP/SPCE	CA	17.5 μM	Pharmaceutical dosage forms	[416]
Antiviral	Efavirenz	dsDNA-treated PGE	DPAdSV	13.3 nM	Pharmaceutical dosage forms	[417]
Antiviral	Aloe-emodin	(dsDNA)/GCE	DPV	0.067 μM	Active compound	[418]
Anticancer	Taxol	Cysteamine/DNA/ SWNTs-film/ AuE	SWV	8.86 μM	Active compound	[419]
Vasoprotective	Rutin	DNA/MWCNTs-COOH/GCE	DPV	38.1 pM	Active compound	[420]
Anticancer	Taxol	DNA/SAM/AuE	EIS	12 nM	Human serum	[421]
Anticancer	Berberine	DNA/GCE	DPV	14.5 μM	Fish test DNA	[422]
Antiaging	Spermidine	DNA/CP/PtE	DPV	0.02 μM	Active compound	[423]

effect of MWCNT concentration were investigated. Afterwards some parameters such as linearity and reproducibility were determined. A standard deviation of ±0.04 and a coefficient of variation of 3.8 % were obtained for a 1 μM dopamine concentration ($n = 15$). Determination of dopamine was carried out in drug samples.

Huang et al. developed based on the Au@carbon dots-chitosan composite film (Au@CDs–CS/GCE) biosensor for determination of dopamine [428]. Under optimal conditions, selective detection of dopamine in a linear concentration range of 0.01 – 100.0 μM was obtained with the limit of 0.001 μM. At the same time, the Au@CDs-CS/GCE was also applied to the detection of DA content in DA's injection with satisfactory results, and the biosensor could keep its activity for at least 2 weeks.

An electrochemical biosensor based on the drop-casting of MWCNTs and microsomal cytochrome P4501A2 on graphite screen-printed electrode was realized by Baj-Rossi et al. [429]. The dynamic linear range for the amperometric detection of naproxen (NAP) had an upper limit of 300 μM with a limit of detection of 16 ± 1 μM. The MWCNT/msCYP1A2-SPE sensor was also calibrated for NAP detection in mouse serum that was previously extracted from mice, showing a slightly higher LOD (33 ± 18 μM).

Marco et al. developed a DNA modified MWCNT paste electrode for the determination of promethazine using a SWV method [430]. The best results were obtained by using SWV in 0.30 M phosphate buffer pH 7.0 and DNA immobilization onto SiAlNb surface using 0.8 mg mL^{-1} concentration. The sensitivity of the sensor was found to be 0.210 μA $μM^{-1}$.

Some typical drug assay using different electrochemical biosensors are listed in Table 9.13.

9.4 Conclusion

An analytical method dedicated to the analysis of a drug compound in a given matrix should be developed in a critical manner, considering different available instrumentations and by knowing their performances in terms of selectivity, sensitivity, ease of use, speed of analysis, etc. Electrochemistry is a well-established and continuously growing area with a number of possible applications in the pharmaceutical field. The scope of electroanalytical assays has largely benefited during the past decade from the development of new electrode material and innovative electrode modifications. The main advantage of electrochemical techniques is their high sensitivity with a wide concentration range up to the low nanomolar concentration level and their high versatility for a relatively moderate equipment cost.

Many of the applications described in this chapter are of limited relatively scope, but they are all of some significance in their own particular field and contribute to the general importance of electroanalytical assay in modern pharmaceutical analysis. This is a rapidly developing subject, and it can be anticipated that many new applications of electroanalytical methods may be offered in the near future.

References

1. Kissinger PT, Heineman WR (eds) (1996) Laboratory techniques in electroanalytical chemistry, 2nd edn. New York, Marcel Dekker
2. Smyth MR, Vos JG (eds) (1992) Analytical voltammetry, vol XXVII. Elsevier, Amsterdam
3. Wang J (ed) (1988) Electroanalytical techniques in clinical chemistry and laboratory medicine. VCH Publishers, New York
4. Wang J (1989) Voltammetry following nanoelectrolytic preconcentration. In: Bard AJ (ed) Electroanalytical chemistry, vol 16. Marcel Dekker, New York
5. Patriarche GJ, Chateau-Gosselin M, Vandenbalck JL, Zuman P (1979) Polarography and related electroanalytical techniques in pharmacy and pharmacology. In: Bard AJ (ed) Electroanalytical chemistry, vol 11. Marcel Dekker, New York
6. Vire J-C, Kauffmann J-M (1994) Curr Top Electrochem 3:493–515
7. Brooks MA, Tsai EW (1996) Electrochemistry in pharmaceutical analysis. In: Kissinger PT, Heineman WR (eds) Laboratory techniques in electroanalytical chemistry, 2nd edn. Marcel Dekker, New York
8. Wang J (ed) (2006) Analytical electrochemistry, 3rd edn. Wiley-VCH, New York
9. Zuman P, Perrin CL (1969) Organic polarography. Wiley, New York
10. Bard AJ, Faulkner LR (2001) Electrochemical methods, fundamentals and applications, 2nd edn. Wiley, New York
11. Kellner R, Mermet J-M, Otto M, Valcarcel M, Widmer HM (eds) (2004) Analytical chemistry, a modern approach to analytical science, 2nd edn. New York, Wiley
12. Brett CMA, Oliveira-Brett AM (eds) (1993) Electrochemistry, principles, methods and applications. Oxford University Press, New York
13. Bagotsky VS (ed) (2006) Fundamentals of electrochemistry, 2nd edn. Wiley, New York
14. Christian GD, Dasgupta P, Schug K (eds) (2013) Analytical chemistry, 7th edn. Wiley, New Jersey
15. Hart JP (1990) Electroanalysis of biologically important compounds. Ellis Harwood, London
16. Branina KH, Neyman E (1993) Electroanalytical stripping methods. In: Winefordner JD (ed) Chemical analysis: A series of monographs an analytical chemistry and its applications, vol 126. Wiley, New York
17. Ozkan SA (ed) (2012) Electroanalytical methods in pharmaceutical analysis and their validation, 1st edn. New York, HNB Pub
18. Ferguson BS, Hoggarth DA, Maliniak D, Ploense K, White RJ, Woodward N, Hsieh K, Bonham AJ, Eisenstein M, Kippin TE, Plaxco KW, Soh HT (2013) Sci Transl Med 5:213
19. Jahn S, Karst U (2012) J Chromatogr A 1259:16–49
20. Flanagan RJ, Perrett D, Whelpton R (eds) (2005) Electrochemical detection in HPLC analysis of drugs and poisons. Smith RM (series ed) RSC Pub, Cambridge
21. La Course WR (ed) (1997) Pulsed electrochemical detection in high performance liquid chromatography. Wiley, New York
22. Ozkan SA (2007) Chromatographia 66:S3–S13
23. Diculescu VC, Chiorcea Paquim AM, Oliveira Brett AM (2005) Sensors 5:377–393
24. Ozkan SA, Uslu B, Aboul-Enein HY (2003) Crit Rev Anal Chem 33:155–181
25. Ohannesian L, Streeter AJ (eds) (2002) Handbook of pharmaceutical analysis. Marcel Dekker, New York
26. Watson DG (2005) Pharmaceutical analysis, 2nd edn. Elsevier, New York
27. Gürler B, Özkorucuklu SP, Kir E (2013) J Pharm Biomed Anal 84:263–268
28. Elqudaby HM, Mohamed GG, Ali FA, Eid SM (2013) Arab J Chem 6:327–333
29. Ghalkhani M, Shahrokhian S (2013) Sens Actuators B 185:669–674
30. Zhang F, Gu S, Ding Y, Zhang Z, Li L (2013) Anal Chim Acta 770:53–61
31. Brahman PK, Dar RA, Pitre KS (2013) Sens Actuators B 176:307–314
32. Ensafi AA, Allafchian AR, Rezaei B (2013) Colloids Surf B 103:468–474
33. Devaraj M, Deivasigamani RK, Jeyadevan S (2013) Colloids Surf B 102:554–561

34. Ensafi AA, Allafchian AR (2013) Colloids Surf B 102:687–693
35. Zhang F, Gu S, Ding Y, Li L, Liu X (2013) Bioelectrochemistry 89:42–49
36. Zhang F, Li L, Luo L, Ding Y, Liu X (2013) J Appl Electrochem 43:263–269
37. Farias PAM, Castro AA, Cordoves AIP (2012) ECS Trans 43:267–287
38. Shetti NP, Malode SJ, Nandibewoor ST (2012) Bioelectrochemistry 88:76–83
39. Svorc L, Sochr J, Rievaj M, Tomcik P, Bustin D (2012) Bioelectrochemistry 88:36–41
40. Souza MRS, Aragao DM, Santos RHT, Garcia CAB, Alves JPH, Arguelho MLPM (2012) ECS Trans 43:289–296
41. Wen W, Zhao D-M, Zhang X-H, Xiong H-Y, Wang S-F, Chen W, Zhao Y-D (2012) Sens Actuators B 174:202–209
42. Ensafi AA, Allafchian AR, Rezaei B (2012) J Nanopart Res 14:1244–1255
43. Geto A, Amare M, Tessema M, Admassie S (2012) Anal Bioanal Chem 404:525–530
44. Peng J, Hou C, Hu X (2012) Sens Actuators B 169:81–87
45. Heli H, Pourbahman F, Sattarahmady N (2012) Anal Sci 28:503–510
46. Attia AK, El-Shal MA (2012) Anal Bioanal Electrochem 4:213–224
47. Brahman PK, Dar RA, Tiwari S, Pitre KS (2012) Rev Anal Chem 31:83–92
48. Radi A-E, El-Ghany NA, Wahdan T (2012) Monatsh Chem 143:697–702
49. Arvand M, Ansari R, Heydari L (2011) Mater Sci Eng C 31:1819–1825
50. Peng JY, Hou CT, Liu XX, Li HB, Hu XY (2011) Talanta 86:227–232
51. Jain R, Vikas (2011) Colloids Surf B 87:423–426
52. Castro AA, de Souza MVN, Rey NA, Farias PAM (2011) J Braz Chem Soc 22:1662–1668
53. Wan H, Zhao F, Zeng B (2011) Colloids Surf B 86:247–250
54. Hu Y, Li J, Zhang Z, Zhang H, Luo L, Yao S (2011) Anal Chim Acta 698:61–68
55. Attia AK, Saber RA (2011) Anal Bioanal Electrochem 3:291–301
56. Cordoves AIP, Farias PAM (2011) Curr Pharm Anal 7:71–78
57. Habib IHI, Rizk MS, El-Aryan TR (2011) Pharm Chem J 44:705–710
58. Ghalkhani M, Fernandes IPG, Oliveira SCB, Shahrokhian S, Oliveira-Brett AM (2011) Bioelectrochemistry 80:175–181
59. Fotouhi L, Nemati M, Heravi MM (2011) J Appl Electrochem 41:137–142
60. Wan H, Zhao F, Wu W, Zeng B (2011) Colloids Surf B 82:427–431
61. Castro AA, Aucélio RQ, Rey NA, Miguel EM, Farias PAM (2011) Comb Chem High Throughput Screening 14:22–27
62. Sheng Q, Liu R, Zhang S, Zheng J (2014) Biosens Bioelectron 51:191–194
63. Fotouhi L, Bahmani F (2013) Electroanalysis 25:757–764
64. Mirmoghtadaie L, Ensafi A, Kadivar M, Norouzi P (2013) Mater Sci Eng C 33:1753–1758
65. Radhapyari K, Kotoky P, Khan R (2013) Mater Sci Eng C 33:583–587
66. Rafique B, Khalid AM, Akhtar K, Jabbar A (2013) Biosens Bioelectron 44:21–26
67. Marco JP, Borges KB, Tarley CRT, Ribeiro ES, Pereira AC (2013) J Electroanal Chem 704: 159–168
68. Ribeiro FAS, Tarley CRT, Borges KB, Pereira AC (2013) Sens Actuators B 185:743–754
69. Brahman PK, Dar RA, Pitre KS (2013) Sens Actuators B 177:807–812
70. Radhapyari K, Kotoky P, Das MR, Khan R (2013) Talanta 111:47–53
71. Fotouhi L, Hashkavayi AB, Heravi MM (2013) Int J Biol Macromol 53:101–106
72. Batra B, Pundir CS (2013) Biosens Bioelectron 47:496–501
73. Bozal B, Dogan-Topal B, Uslu B, Ozkan SA, Aboul-Enein HY (2012) Curr Anal Chem 8: 528–534
74. Gu T, Hasebe Y (2012) Biosens Bioelectron 33:222–227
75. Kergaravat SV, Gomez GA, Fabiano SN, Chavez TIL, Pividori MI, Hernandez SR (2012) Talanta 97:484–490
76. Jiang B, Wang M, Chen Y, Xie J, Xiang Y (2012) Biosens Bioelectron 32:305–308
77. Rounaghi G, Kakhki RM (2012) Bull Mater Sci 35:811–816
78. Chandra S, Arora K, Bahadur D (2012) Mater Sci Eng B 177:1531–1537

79. Satana HE, Pontinha ADR, Diculescu VC, Oliveira-Brett AM (2012) Bioelectrochemistry 87:3–8
80. Ensafi AA, Bafrooei EH, Amini M (2012) Biosens Bioelectron 31:376–381
81. Jalali F, Dorraji PS (2012) J Pharm Biomed Anal 70:598–601
82. Liu X, Peng Y, Qu X, Ai S, Han R, Zhu X (2011) J Electroanal Chem 654:72–78
83. Dogan-Topal B, Ozkan SA (2011) Talanta 83:780–788
84. Dogan-Topal B, Ozkan SA (2011) Electrochim Acta 56:4433–4438
85. Chandra P, Zaidi SA, Noh H-B, Shim Y-B (2011) Biosens Bioelectron 28:326–332
86. Bian CL, Zeng QX, Yang LJ, Xiong HY, Zhang XH, Wang SF (2011) Sens Actuators B 156: 615–620
87. X-x G, Z-j S, J-j S, J-f S (2011) Biosens Bioelectron 26:4001–4005
88. Yousef Elahi M, Bathaie SZ, Kazemi SH, Mousavi MF (2011) Anal Biochem 411:176–184
89. Bertolino FA, De Vito IE, Messina GA, Fernández H, Raba J (2011) J Electroanal Chem 651: 204–210
90. Kuralay F, Erdem A, Abacı S, Özyörük H, Yıldız A (2010) J Appl Electrochem 40: 2039–2050
91. Radi A-E, Wahdan T, Anwar Z, Mostafa H (2010) Electroanalysis 22:2665–2671
92. Topkaya SN, Aydinlik S, Aladag N, Ozsoz M, Ozkan-Ariksoysal D (2010) Comb Chem High Throughput Screening 13:582–589
93. Chen B, Ma M, Su X (2010) Anal Chim Acta 674:89–95
94. Zhang Y, Zhang K, Ma H (2009) Anal Biochem 387:13–19
95. Mirmomtaz E, Ensafi AA, Soleimanian-Zad S (2009) Electrochim Acta 54:1141–1146
96. Wang F, Wu Y, Liu J, Ye B (2009) Electrochim Acta 54:1408–1413
97. Alonso-Lomillo MA, Domínguez-Renedo O, Hernández-Martín A, Arcos-Martínez MJ (2009) Anal Biochem 395:86–90
98. Dogan-Topal B, Uslu B, Ozkan SA (2009) Biosens Bioelectron 24:2358–2364
99. Lin XH, Wan HY, Zhang YF, Chen JH (2008) Talanta 74:944–950
100. Cheng WX, Peng DY, Lu CH, Fang CW (2008) Russ J Electrochem 44:1052–1057
101. Niu S, Zhao M, Hu L, Zhang S (2008) Sens Actuators B 135:200–205
102. Mehdinia A, Kazemi SH, Bathaie SZ, Alizadeh A, Shamsipur M, Mousavi MF (2008) Anal Biochem 375:331–338
103. Tian X, Song Y, Dong H, Ye B (2008) Bioelectrochemistry 73:18–22
104. Ghanbari K, Bathaie SZ, Mousavi MF (2008) Biosens Bioelectron 23:1825–1831
105. Ashrafi AM, Kurbanoglu S, Vytřas K, Uslu B, Ozkan SA (2014) J Electroanal Chem 712: 178–184
106. Ashrafi AM, Gumustas M, Vytřas K, Nematollahi D, Uslu B, Mikysek T, Jirásko R, Ozkan SA (2013) Electrochim Acta 109:381–388
107. Dogan-Topal B, Bozal-Palabiyik B, Uslu B, Ozkan SA (2013) Sens Actuators B 177:841–847
108. da Silva H, Pacheco JG, Magalhães JM, Viswanathan S, Delerue-Matos C (2014) Biosens Bioelectron 52:56–61
109. Agrawal B, Chandra P, Goyal RN, Shim YB (2013) Biosens Bioelectron 47:307–312
110. Ensafi AA, Karimi-Maleh H (2012) Drug Test Analysis 4:970–977
111. Teradal NL, Kalanur SS, Prashanth SN, Seetharamappa J (2012) J Appl Electrochem 42: 917–923
112. Wang S, Huang Z, Liu M, Ding H (2012) Sens Lett 10:117–121
113. Wang F, Gao Y, Gao L, Xing T (2011) J Chin Chem Soc 58:450–456
114. Hosseiny Davarani SS, Kalate Bojdi M, Mehdinia A (2011) Chem Pharm Bull 59:1209–1213
115. Sataraddi SR, Nandibewoor ST (2011) Der Pharma Chemica 3:253–265
116. Chandra P, Noh H-B, Won M-S, Shim Y-B (2011) Biosens Bioelectron 26:4442–4449
117. Erdem A, Karadeniz H, Caliskan A (2011) Analyst 136:1041–1045
118. Šelešovská R, Bandžuchová L, Navrátil T (2011) Electroanalysis 23:177–187
119. Zhang D, Zhang Y, He L (2013) Electroanalysis 25:2136–2144

120. Bozal-Palabiyik B, Dogan-Topal B, Uslu B, Can A, Ozkan SA (2013) J Solid State Electrochem 17:2815–2822
121. Kurbanoglu S, Dogan-Topal B, Uslu B, Can A, Ozkan SA (2013) Electroanalysis 25: 1473–1482
122. Karadas N, Sanli S, Akmese B, Dogan-Topal B, Can A, Ozkan SA (2013) Talanta 115: 911–919
123. Ye L, Xiang M, Zhang Y, Luo L, Gao Y, Yu J, Cha J (2013) Int J Electrochem Sci 8: 12726–12734
124. Onay-Besikci A, Ozkan SA (2012) Drugs versus chemicals. In: Ozkan SA (ed) Electroanalytical methods in pharmaceutical analysis and their validation, 1st edn. HNB Pub, New York
125. Amiri-Aref M, Raoof JB, Ojani R (2013) Colloids Surf B 109:287–293
126. Jain R, Dhanjai, Sharma S (2013) Colloids Surf A 436:178–184
127. Arvand M, Palizkar B (2013) Mater Sci Eng C 33:4876–4883
128. Maashhadizadeh MH, Refahati R, Amereh E (2013) Anal Bioanal Electrochem 5:270–282
129. Dogan-Topal B (2013) J Solid State Electrochem 17:1059–1066
130. Gupta VK, Jain AK, Shoora SK (2013) Electrochim Acta 93:248–253
131. Wang Q, Das MR, Li M, Boukherroub R, Szunerits S (2013) Bioelectrochemistry 93:15–22
132. Švorc L, Sochr J, Svítková J, Rievaj M, Bustin D (2013) Electrochim Acta 87:503–510
133. Mashhadizadeh MH, Afshar E (2012) Electroanalysis 24:2193–2202
134. Alipour E, Gasemlou S (2012) Anal Methods 4:2962–2969
135. Leite FRF, Maroneze CM, de Oliveira AB, Santos WTPD, Damos FS, Luz RDCS (2012) Bioelectrochemistry 86:22–29
136. Abbar J, Nandibewoor S (2012) Crit Rev Anal Chem 42:272–281
137. Esfandiari Baghbamidi S, Beitollahi H, Karimi-Maleh H, Soltani-Nejad S, Soltani-Nejad V, Roodsaz S (2012) J Anal Method Chem 2012:305872. doi:10.1155/2012/305872
138. Mazloum-Ardakani M, Taleat Z, Khoshroo A, Beitollahi H, Dehghani H (2012) Biosens Bioelectron 35:75–81
139. Saberi R-S, Shahrokhian S (2012) Bioelectrochemistry 84:38–43
140. Lin WY, Pan ML, Wang HY, Su YO, Huang PW (2012) Med Chem Res 21:4389–4394
141. Abbar JC, Nandibewoor ST (2012) Ind Eng Chem Res 51:111–118
142. Heli H, Faramarzi F, Sattarahmady N (2012) J Solid State Electrochem 16:45–52
143. Navaee A, Salimi A, Teymourian H (2012) Biosens Bioelectron 31:205–211
144. Mersal GAM (2012) Food Anal Methods 5:520–529
145. Pruneanu S, Pogacean F, Biris AR, Ardelean S, Canpean V, Blanita G, Dervishi E, Biris AS (2011) J Phys Chem C 115:23387–23394
146. Chung I-C, Chang C-C, Chiu H-S, Jiang S-F, Lee M-H, Chung C-L, Liu B-D, Huang C-Y, Lin HY (2011) J Nanosci Nanotechnol 11:10633–10638
147. Wang Y (2011) Colloids Surf B 88:614–621
148. Demircigil BT, Dogan-Topal B, Ozkan SA (2011) Collect Czech Chem Commun 76: 1717–1736
149. Atta NF, Galal A, Azab SM (2011) Analyst 136:4682–4691
150. Mazloum-Ardakani M, Ganjipour B, Beitollahi H, Amini MK, Mirkhalaf F, Naeimi H, Nejati-Barzoki M (2011) Electrochim Acta 56:9113–9120
151. Nigović B, Spajić J (2011) Talanta 86:393–399
152. Langmaier J, Garay F, Kivlehan F, Chaum E, Lindner E (2011) Anal Chim Acta 704:63–67
153. Gholivand MB, Khodadadian M (2011) Talanta 85:1680–1688
154. Yang G, Chen Y, Li L, Yang Y (2011) Clin Chim Acta 412:1544–1549
155. Pérez-Ortiz M, Bollo S, Zapata-Urzúa C, Yáñez C, Álvarez-Lueje A (2011) Anal Lett 44: 1683–1698
156. Manjunatha JG, Swamy BEK, Mamatha GP, Gilbert O, Srinivas MT, Sherigara BS (2011) Der Pharma Chemica 3:236–249
157. Sanghavi BJ, Srivastava AK (2011) Electrochim Acta 56:4188–4196

158. Parvin MH (2011) Electrochem Commun 13:366–369
159. Karimi MA, Hatefi-Mehrjardi A, Ardakani MM, Ardakani RB, Mashhadizadeh MH, Sargazi S (2011) Russ J Electrochem 47:34–41
160. Shahrokhian S, Kamalzadeh Z, Hamzehloei A (2013) Bioelectrochemistry 90:36–43
161. Mashhadizadeh MH, Afshar E (2013) Electrochim Acta 87:816–823
162. Ağın F, Karadaş N, Uslu B, Özkan SA (2013) Maced J Chem Chem Eng 32:41–55
163. Raj MA, John SA (2013) Anal Chim Acta 771:14–20
164. Ansari M, Kazemi S, Khalilzadeh MA, Karimi-Maleh H, Zanousi MBP (2013) Int J Electrochem Sci 8:1938–1948
165. Macíková P, Skopalová J, Cankař P, Papoušková B, Straková R, Jirovský D, Maier V (2013) Electroanalysis 25:205–212
166. Atta NF, Galal A, Azab SM (2012) Electroanalysis 24:1431–1440
167. Mersal GAM (2012) J Solid State Electrochem 16:2031–2039
168. Ahmar H, Fakhari AR (2012) Anal Methods 4:812–818
169. Ensafi AA, Karimi-Maleh H (2011) Drug Test Anal 3:325–330
170. Ensafi AA, Dadkhah M, Karimi-Maleh H (2011) Colloids Surf B 84:148–154
171. Mazloum-Ardakani M, Beitollahi H, Amini MK, Mirkhalaf F, Mirjalili B-F (2011) Biosens Bioelectron 26:2102–2106
172. Sattarahmady N, Heli H, Dehdari Vais R (2013) Biosens Bioelectron 48:197–202
173. Prasad BB, Prasad A, Tiwari MP, Madhuri R (2013) Biosens Bioelectron 45:114–122
174. Yan J, Liu S, Zhang Z, He G, Zhou P, Liang H, Tian L, Zhou X, Jiang H (2013) Colloids Surf B 111:392–397
175. Thomas T, Mascarenhas RJ, Swamy BEK, Martis P, Mekhalif Z, Sherigara BS (2013) Colloids Surf B 110:458–465
176. Zeng Y, Zhou Y, Kong L, Zhou T, Shi G (2013) Biosens Bioelectron 45:25–33
177. Abbar JC, Nandibewoor ST (2013) Colloids Surf B 106:158–164
178. Kor K, Zarei K (2013) Electroanalysis 25:1497–1504
179. Jain R (2013) Dhanjai. Colloids Surf B 105:278–283
180. Javanbakht M, Fathollahi F, Divsar F, Ganjali MR, Norouzi P (2013) Sens Actuators B 182: 362–367
181. Habibi D, Faraji AR, Gil A (2013) Sens Actuators B 182:80–86
182. Ardila JA, Oliveira GG, Medeiros RA, Fatibello-Filho O (2013) J Electroanal Chem 690: 32–37
183. Karimi-Maleh H, Khalilzadeh MA, Ranjbarha Z, Beitollahi H, Ensafi AA, Zareyee D (2012) Anal Methods 4:2088–2094
184. Xu L, Guo Q, Yu H, Huang J, You T (2012) Talanta 97:462–467
185. Shrivastas R, Satsangee SP, Jain R (2012) ECS Trans 50(54):23–36
186. Sun W, Wang Y, Zhang Y, Ju X, Li G, Sun Z (2012) Anal Chim Acta 751:59–65
187. Freitas KHG, Fatibello-Filho O, de Mattos IL (2012) Braz J Pharm Sci 48:639–649
188. Shakibaian V, Parvin MH (2012) J Electroanal Chem 683:119–124
189. Deepa MB, Mamatha GP, Arthoba Naik Y, Sherigara BS, Manjappa S, Vijaya B (2012) J Chem Pharm Res 4:2803–2816
190. Karadas N, Sanli S, Gumustas M, Ozkan SA (2012) J Pharm Biomed Anal 66:116–125
191. Bui M-PN, Li CA, Seong GH (2012) Biochip J 6:149–156
192. Tsai H-Y, Lin Z-H, Chang H-T (2012) Biosens Bioelectron 35:479–483
193. Galal A, Atta NF, El-Ads EH (2012) Talanta 93:264–273
194. Ensafi AA, Arabzadeh A (2012) J Anal Chem 67:486–496
195. Jain R, Sharma R (2012) J Appl Electrochem 42:341–348
196. Stoiljkovic ZZ, Avramov Ivic ML, Petrovic SD, Mijin DZ, Stevanovic SI, Lacnjevac UC, Marinkovic AD (2012) Int J Electrochem Sci 7:2288–2303
197. Abbar JC, Malode SJ, Nandibewoor ST (2012) Bioelectrochemistry 83:1–7
198. Malode SJ, Abbar JC, Shetti NP, Nandibewoor ST (2012) Electrochim Acta 60:95–101
199. Zhang W, Chai Y, Yuan R, Chen S, Han J, Yuan D (2012) Anal Chim Acta 756:7–12

200. Hadi M, Rouhollahi A (2012) Anal Chim Acta 721:55–60
201. Shadjou N, Hasanzadeh M, Saghatforoush L, Mehdizadeh R, Jouyban A (2011) Electrochim Acta 58:336–347
202. Nigović B, Marušić M, Jurić S (2011) J Electroanal Chem 663:72–78
203. Park J, Takmakov P, Wightman RM (2011) J Neurochem 119:932–944
204. Matos IDO, Alves WA (2011) ACS Appl Mater Interfaces 3:4437–4443
205. Li Y, Liu X, Liu X, Mai N, Li Y, Wei W, Cai Q (2011) Colloids Surf B 88:402–406
206. Huang K-J, Jing Q-S, Wu Z-W, Wang L, Wei C-Y (2011) Colloids Surf B 88:310–314
207. Ensafi AA, Karimi-Maleh H, Mallakpour S, Rezaei B (2011) Colloids Surf B 87:480–488
208. Noroozifar M, Khorasani-Motlagh M, Akbari R, Parizi MB (2011) Biosens Bioelectron 28: 56–63
209. Chuekachang S, Phanichphant S, Chaiyasit S (2011) Mol Cryst Liq Cryst 538:292–297
210. Zhang Y, Yuan R, Chai Y, Li W, Zhong X, Zhong H (2011) Biosens Bioelectron 26: 3977–3980
211. Koehne JE, Marsh M, Boakye A, Douglas B, Kim IY, Chang SY, Jang DP, Bennet KE, Kimble C, Andrews R, Meyyappan M, Lee KH (2011) Analyst 136:1802–1805
212. Behpour M, Ghoreishi SM, Honarmand E, Salavati-Niasari M (2011) J Electroanal Chem 653:75–80
213. Jain R, Yadav RK (2011) Drug Test Anal 3:171–175
214. Xiao C, Chu X, Yang Y, Li X, Zhang X, Chen J (2011) Biosens Bioelectron 26:2934–2939
215. Atta NF, Galal A, Ahmed RA (2011) Bioelectrochemistry 80:132–141
216. dos Santos SX, Cavalheiro ETG (2011) Anal Lett 44:850–862
217. Kul D, Uslu B, Ozkan SA (2011) Anal Lett 44:1341–1357
218. Kaur B, Pandiyan T, Satpati B, Srivastava R (2013) Colloids Surf B 111:97–106
219. Khoobi A, Ghoreishi SM, Masoum S, Behpour M (2013) Bioelectrochemistry 94:100–107
220. Ardila JA, Sartori ER, Rocha-Filho RC, Fatibello-Filho O (2013) Talanta 103:201–206
221. Bali Prasad B, Jauhari D, Prasad Tiwari M (2013) Biosens Bioelectron 50:19–27
222. Gholivand MB, Mohammadi-Behzad L (2013) Anal Biochem 438:53–60
223. Molero L, Faundez M, del Valle MA, del Río R, Armijo F (2013) Electrochim Acta 88: 871–876
224. Bagheri A, Hosseini H (2012) Bioelectrochemistry 88:164–170
225. Sharma DK, Mourya GL, Jhankal KK, Jones LA, Bhargava SK (2012) Der Pharmacia Lettre 4:1599–1606
226. Brahman PK, Dar RA, Tiwari S, Pitre KS (2012) Colloids Surf A 396:8–15
227. Pan M, Fang G, Duan Z, Kong L, Wang S (2012) Biosens Bioelectron 31:11–16
228. Dogan-Topal B, Kul D, Ozkan SA, Uslu B (2011) J Appl Electrochem 41:1253–1260
229. Jain R, Vikas, Radhapyari K (2011) Drug Test Anal 3:743–747
230. Radi AEM, Eissa SH (2011) Collect Czech Chem Commun 76:13–25
231. Rafiee B, Fakhari AR (2013) Biosens Bioelectron 46:130–135
232. de Lima CA, Spinelli A (2013) Electrochim Acta 107:542–548
233. Arvand M, Gholizadeh TM (2013) Colloids Surf B 103:84–93
234. Khaskheli AR, Fischer J, Barek J, Vyskočil V, Sirajuddin, Bhanger MI (2013) Electrochim Acta 101:238–242
235. Thomas T, Mascarenhas RJ, Cotta F, Guha KS, Swamy BEK, Martis P, Mekhalif Z (2013) Colloids Surf B 101:91–96
236. Dalmasso PR, Pedano ML, Rivas GA (2012) Sens Actuators B 173:732–736
237. Gowda JI, Nandibewoor ST (2012) Ind Eng Chem Res 51:15936–15941
238. Amiri M, Alimoradi M, Nekoueian K, Bezaatpour A (2012) Ind Eng Chem Res 51: 14384–14389
239. Babu RS, Prabhu P, Anuja S, Narayanan SS (2012) J Chem Pharm Res 4:3592–3600
240. Xu F, Ru H-Y, Sun L-X, Zou Y-J, Jiao C-L, Wang T-Y, Zhang J-M, Zheng Q, Zhou H-Y (2012) Biosens Bioelectron 38:27–30
241. Arvand M, Gholizadeh TM, Zanjanchi MA (2012) Mater Sci Eng C 32:1682–1689

242. Devadas B, Rajkumar M, Chen S-M, Saraswathi R (2012) Int J Electrochem Sci 7:3339–3349
243. Sadeghi S, Motaharian A, Moghaddam AZ (2012) Sens Actuators B 168:336–344
244. Raoof JB, Ojani R, Amiri-Aref M, Chekin F (2012) Russ J Electrochem 48:450–456
245. Ghavami R, Navaee A (2012) Microchim Acta 176:493–499
246. Hasanzadeh M, Shadjou N, Saghatforoush L, Dolatabadi JEN (2012) Colloids Surf B 92: 91–97
247. Gopu G, Muralidharan B, Vedhi C, Manisankar P (2012) Ionics 18:231–239
248. Faria EO, Lopes Junior ACV, Souto DEP, Leite FRF, Damos FS, de Cassia SLR, dos Santos AS, Franco DL, DosSantos WTP (2012) Electroanalysis 24:1141–1146
249. Sanghavi BJ, Srivastava AK (2011) Anal Chim Acta 706:246–254
250. Hou X, Shen G, Meng L, Zhu L, Guo M (2011) Russ J Electrochem 47:1262–1267
251. Kauffmann JM, van Antwerpen P, Sarakbi A, Feier B, Tarik S, Aydogmus Z (2011) Electroanalysis 23:2643–2650
252. Atta NF, Galal A, Abu-Attia FM, Azab SM (2011) J Mater Chem 21:13015–13024
253. Fan Y, Liu J-H, Lu H-T, Zhang Q (2011) Colloids Surf B 85:289–292
254. Vediappan K, Lee CW (2011) Curr Appl Phys 11:995–1000
255. Amiri M, Pakdel Z, Bezaatpour A, Shahrokhian S (2011) Bioelectrochemistry 81:81–85
256. Lu T-L, Tsai Y-C (2011) Sens Actuators B 153:439–444
257. Habibi B, Jahanbakhshi M, Pournaghi-Azar MH (2011) Anal Biochem 411:167–175
258. Samadi-Maybodi A, Nejad-Darzi SKH, Ilkhani H (2011) Anal Bioanal Electrochem 3: 134–145
259. Patil RH, Hegde RN, Nandibewoor ST (2011) Colloids Surf B 83:133–138
260. Norouzi P, Ghaheri N (2011) Anal Bioanal Electrochem 3:87–101
261. Yildiz G, Aydoğmuş Z, Kauffmann JM (2013) Electroanalysis 25:1796–1802
262. Ghadimi H, Tehrani RMA, Ali ASM, Mohamed N, Ab Ghani S (2013) Anal Chim Acta 765: 70–76
263. Karadas N, Bozal-Palabiyik B, Uslu B, Ozkan SA (2013) Sens Actuators B 186:486–494
264. Sun W, Hu J (2013) J Anal Chem 68:694–699
265. Piech R, Paczosa-Bator B (2013) Cent Eur J Chem 11:736–741
266. Saxena S, Shrivastava R, Satsangee SP (2012) Maced J Chem Chem Eng 31:195–203
267. Kreft GL, de Braga OC, Spinelli A (2012) Electrochim Acta 83:125–132
268. Wu G-H, Wu Y-F, Liu X-W, Rong M-C, Chen X-M, Chen X (2012) Anal Chim Acta 745: 33–37
269. Malode SJ, Shetti NP, Nandibewoor ST (2012) Colloids Surf B 97:1–6
270. Zi L, Li J, Mao Y, Yang R, Qu L (2012) Electrochim Acta 78:434–439
271. Li Y, Wu S, Luo P, Liu J, Song G, Zhang K, Ye B (2012) Anal Sci 28:497–502
272. Ziyatdinova GK, Giniyatova ER, Budnikov HC (2012) J Anal Chem 67:467–473
273. Amare M, Admassie S (2012) Bull Chem Soc Ethiop 26:73–84
274. Hu Y, Zhang Z, Li J, Zhang H, Luo L, Yao S (2012) Biosens Bioelectron 31:190–196
275. Farghali RA, Ahmed RA (2012) Int J Electrochem Sci 7:13008–13019
276. Sopha H, Hocevar SB, Pihlar B, Ogorevc B (2012) Electrochim Acta 60:274–277
277. Raoof JB, Baghayeri M, Ojani R (2012) Colloids Surf B 95:121–128
278. Keeley GP, McEvoy N, Nolan H, Kumar S, Rezvani E, Holzinger M, Cosnier S, Duesberg GS (2012) Anal Methods 4:2048–2053
279. Svorc L, Sochr J, Tomcik P, Rievaj M, Bustin D (2012) Electrochim Acta 68:227–234
280. Revin SB, John SA (2012) Analyst 137:209–215
281. Balasoiu S-C, Stefan-van Staden R-I, van Staden JF, Ion R-M, Radu G-L, Aboul-Enein HY (2011) Electrochim Acta 58:290–295
282. Kannan P, John SA (2011) Biosens Bioelectron 30:276–281
283. Suresh Babu R, Prabhu P, Sriman Narayanan S (2011) Colloids Surf B 88:755–763
284. Karimi-Maleh H, Keyvanfard M, Alizad K, Fouladgar M, Beitollahi H, Mokhtari A, Gholami-Orimi F (2011) Int J Electrochem Sci 6:6141–6150
285. Heli H, Yadegari H, Karimian K (2011) J Exp Nanosci 6:488–508

286. Kuralay F, Vural T, Bayram C, Denkbas EB, Abaci S (2011) Colloids Surf B 87:18–22
287. Liu X, Luo L, Ding Y, Ye D (2011) Bioelectrochemistry 82:38–45
288. Mbouguen JCK, Kenfack IT, Walcarius A, Ngameni E (2011) Talanta 85:754–762
289. Kalachar HCB, Arthoba Naik Y, Basavanna S, Viswanatha R, Venkatesha TG, Sheela T (2011) J Chem Pharm Res 3:530–539
290. Baś B, Jakubowska M, Górski L (2011) Talanta 84:1032–1037
291. Goyal RN, Rana ARS, Aziz MA, Oyama M (2011) Anal Chim Acta 693:35–40
292. Enache TA, Oliveira-Brett AM (2011) Bioelectrochemistry 81:46–52
293. El-Desoky HS, Ghoneim MM (2011) Talanta 84:223–234
294. da Silva IS, Araújo MFA, Ferreira HA, Varela JJG Jr, Tanaka SMCN, Tanaka AA, Angnes L (2011) Talanta 83:1701–1706
295. Lertanantawong B, O'Mullane AP, Zhang J, Surareungchai W, Somasundrum M, Bond AM (2009) Anal Chem 80:6515–6525
296. Silva FAS, da Silva MGA, Lima PR, Meneghetti MR, Kubota LT, Goulart MOF (2013) Biosens Bioelectron 50:202–209
297. Agin F, Karadas N, Uslu B, Ozkan SA (2013) Curr Pharm Anal 9:299–309
298. Jain A, Jain A, Jain A (2013) J Young Pharm 5:70–72
299. Altinöz S, Uyar B (2013) Anal Methods 5:5709–5716
300. Hoang VD, Huyen DT, Phuc PH (2013) J Anal Method Chem 2013, 367914
301. Hoang VD, Yen NT (2013) Trop J Pharm Res 12:783–790
302. Alarfaj NA (2013) J Anal Chem 68:335–340
303. Çelebier M, Süslü I, Altinöz S (2013) Anal Methods 5:1301–1306
304. Mandil H, Sakur AA, Alulu S (2012) Asian J Chem 24:2980–2984
305. Singh N (2012) Indian J Hematol Blood Transfus 28:97–104
306. Ghoneim MM, Hassanein AM, Salahuddin NA, El-Desoky HS, Elfiky MN (2012) J Braz Chem Soc 23:1594–1605
307. Ramadan AA, Mandil H (2012) Int J Pharm Pharm Sci 4:255–261
308. Carvalho JM, Da Silva AR, Da Cunha ALMC, Aucélio RQ, Alberti ALM, Leandro KC (2012) Quim Nova 35:988–992
309. Al-Ghamdi AF, Hefnawy MM (2012) Arab J Chem 5:383–389
310. Al-Ghamdi AF, Hefnawy MM, Al-Majed AA, Belal FF (2012) Chem Cent J 6, 15
311. Rege PV, Sathe PA, Salvi VS (2011) Int J Pharm Res 3:9–12
312. Jugade R, Jeyaseelan C, Joshi AP (2011) Int J Pharm Sci Rev Res 11:38–41
313. Gandhi S, Mehta V, Rajput S (2011) Indian J Pharm Sci 73:583–586
314. Rohini Kumar PC, Balaji K, Pushpalatha M, Sridevi C, Suresh Reddy C (2011) Der Pharmacia Lettre 3:311–319
315. Choure R, Vaidja N, Pitre KS (2011) Rev Anal Chem 30:53–57
316. Ghoneim MM, El-Desoky HS, Abdel-Galeil MM (2011) Bioelectrochemistry 81:65–73
317. Taşkoparan Yilmaz Ü, Somer G (2011) Turk J Chem 35:201–213
318. Bernard V, Fojt L, Skorpikova J, Mornstein V (2011) Indian J Biochem Biophys 48:59–62
319. Ghoneim MM, El-Desoky HS, Abdel-Galeil MM (2011) Bioelectrochemistry 80:162–168
320. El-Desoky HS, Ghoneim MM, Habazy AD (2011) J Braz Chem Soc 22:239–247
321. Nageswara Reddy C, Reddy Prasad P, Sreedhar NY (2011) Int J PharmTech Res 3: 1125–1131
322. Temerk YM, Kamal MM, Ibrahim MS, Ibrahim HSM, Schuhmann W (2011) Electroanalysis 23:1638–1644
323. Oliveira TMBF, Ribeiro FWP, Soares JES, De Lima-Neto P, Correia AN (2011) Anal Biochem 413:148–156
324. Ahmadi F, Ghasemi S, Rahimi-Nasrabadi M (2011) Collect Czech Chem Commun 76: 371–382
325. Yola ML, Özaltın N (2011) Rev Chim 62:420–426
326. Zayed SIM (2011) Acta Chim Slov 58:75–80
327. Ramadan AA, Mandil H, Hafez B (2011) Asian J Chem 23:403–406

328. Jain R, Vikas, Rather JA (2011) Colloids Surf B 82:333–339
329. Ramadan AA, Mandil H (2010) Anal Biochem 404:1–7
330. Youssef RM, Maher HM (2010) Drug Test Anal 2:392–396
331. El-Enany N, El-Sherbiny D, Belal F (2010) Int J Biomed Sci 6:128–134
332. Gupta VK, Jain R, Jadon N, Radhapyari K (2010) J Colloid Interface Sci 350:330–335
333. Jain R, Gupta VK, Jadon N, Radhapyari K (2010) J Electroanal Chem 648:20–27
334. El-Desoky HS, Ghoneim MM, Abdel-Galeil MM (2010) J Braz Chem Soc 21:669–679
335. Jain R, Gupta VK, Jadon N, Radhapyari K (2010) Anal Biochem 407:79–88
336. Mandil H, Sakur AA, Alulu S (2010) Asian J Chem 22:2129–2135
337. Leandro KC, Moreira JC, Farias PAM (2010) Anal Lett 43:1951–1957
338. El-Desoky HS, Ghoneim MM, Abdel-Galeil MM (2010) Acta Chim Slov 57:332–340
339. Jain R, Yadav RK, Dwivedi A (2010) Colloids Surf A 359:25–30
340. Süslü I, Özaltın N, Altınöz S (2010) J AOAC Int 93:562–568
341. Gholivand MB, Khodadadian M, Majnooni MB, Bahrami G (2010) Anal Lett 43:269–279
342. Jeyaseelan C, Joshi AP (2009) Int J ChemTech Res 1:596–601
343. Jain R, Mishra R, Dwivedi A (2009) J Sci Ind Res India 68:945–950
344. El-Enany N, El-Brashy A, Belal F, El-Bahay N (2009) Port Electrochim Acta 27:113–125
345. Gartner P, Zschimmer K, Knoth H (2009) Pharmazie 64:371–375
346. Jain R, Mishra R, Dwivedi A (2009) J Sci Ind Res India 68:540–547
347. Jain R, Mishra R, Dwivedi A (2009) Colloids Surf A 337:74–79
348. Sheribah ZA, El-Brashy AM, El-Gamal RM (2009) J AOAC Int 92:419–427
349. Rajput SJ, Raj HA (2009) Indian J Pharm Sci 71:50–52
350. Biçer E, Çetinkaya P (2009) Croat Chem Acta 82:573–582
351. Skrzypek S, Nosal-Wiercinska A, Ciesielski W (2009) Collect Czech Chem Commun 74: 1455–1466
352. Ghoneim EM, El-Attar MA, Ghoneim MM (2009) Chem Anal (Warsaw) 54:389–402
353. Bozal B, Doğan-Topal B, Uslu B, Özkan SA, Aboul-Enein HY (2009) Anal Lett 42: 2322–2338
354. Ramadan AA, Mandil H, Genco T (2009) Asian J Chem 21:7387–7397
355. Alghamdi AH, Alghamdi AF, Alomar MA (2009) Port Electrochim Acta 27:645–655
356. Turan Ş, Durmuş Z, Kiliç E (2009) Curr Pharm Anal 5:416–423
357. Al-Ghamdi AF (2009) J Saudi Chem Soc 13:235–241
358. Süslü I, Özaltın N, Altınöz S (2009) J Appl Electrochem 39:1535–1543
359. Ensafi AA, Khayamian T, Taei M (2009) Talanta 78:942–948
360. El-Desoky HS, Beltagi AM, Ghoneim MM (2009) J AOAC Int 92:806–812
361. Zayed SIM, Issa YM (2009) Bioelectrochemistry 75:9–12
362. Ghoneim EM, El-Attar MA, Ghoneim MM (2009) J AOAC Int 92:597–603
363. Solangi AR, Mallah A, Khuhawar MY, Bhanger MI (2010) J Iran Chem Soc 6:71–76
364. Somer G, Doğan M (2008) Bioelectrochemistry 74:96–100
365. Ensafi AA, Hajian R (2008) Anal Sci 24:1449–1454
366. Korany MA, Hewala II, Abdel-Hay KM (2008) J AOAC Int 91:1051–1058
367. Alvarez-Lueje A, Brain-Isasi S, Nunez-Vergara LJ, Squella JA (2008) Talanta 75:691–696
368. El-Enany N, Belal F, Rizk M (2008) J Biochem Bioph Methods 70:889–896
369. Ghoneim MM, El-Desoky HS, El-Ries MA, Abd-Elaziz AM (2008) Chem Pap-Chem Zvesti 62:127–134
370. Haggag R, Belal S, Shaalan R (2008) Sci Pharm 76:33–48
371. Jain R, Dwivedi A, Mishra R (2008) J Colloid Interface Sci 318:296–301
372. El-Sherbiny D, El-Enany N, Belal F (2008) Anal Lett 41:806–821
373. Dogan B, Uslu B, Ozkan SA, Zuman P (2008) Anal Chem 80:209–216
374. Ghoneim MM, El-Attar MA (2008) Chem Anal (Warsaw) 53:689–702
375. Nigovic B, Komorsky-Lovric S, Devcic D (2008) Croat Chem Acta 81:453–459
376. Süslü I, Altınöz S (2008) Pharmazie 63:428–433
377. Habib IHI, Weshahy SA, Toubar S, El-Alamin MMA (2008) Pharm Chem J 42:438–442

378. Habib IHI, Weshahy SA, Toubar S, El-Alamin MMA (2008) Port Electrochim Acta 26: 315–324
379. Fernández-Torres R, Villar Navarro M, Bello López MA, Callejón Mochón M, Jiménez Sánchez JC (2008) Talanta 77:241–248
380. Neves MMPS, Nouws HPA, Delerue-Matos C (2008) Anal Lett 41:2794–2804
381. Alghamdi AH (2008) Chem Pap-Chem Zvesti 62:339–344
382. Ensafi AA, Hajian R (2008) J Braz Chem Soc 19:405–412
383. Habib IHI, Weshahy SA, Toubar SS, El-Alamin MMA (2008) Pharmazie 63:337–341
384. Al-Ghamdi AH, Al-Ghamdi AF, Al-Omar MA (2008) Anal Lett 41:90–103
385. Ramadan AA, Mandil H, Hafez B (2013) Int J Pharm Pharm Sci 5:434–440
386. Sun W, Zhang Y, Ju X, Zhan T (2013) J Indian Chem Soc 90:891–895
387. Ahuja S, Scypinski S (eds) (2001) Handbook of modern pharmaceutical analysis, vol 3. Academic, London
388. Wang L-H, Liu H-H (2013) Curr Anal Chem 9:143–149
389. Zaater MF, Tahboub YR, Ghanem E (2012) J Chromatogr Sci 50:763–768
390. Yosypchuk O, Barek J, Vyskocil V (2012) Anal Bioanal Chem 404:693–699
391. Squellerio I, Caruso D, Porro B, Veglia F, Tremoli E, Cavalca V (2012) J Pharm Biomed Anal 71:111–118
392. Liu L, Li S, Chen Z (2012) J Pharm Biomed Anal 61:252–255
393. Liu L, Chen Z (2012) Anal Chim Acta 737:99–104
394. Bartosova Z, Riman D, Jakubec P, Halouzka V, Hrbac J, Jirovsky D (2012) Sci World J 2012, 295802
395. Liu L, Chen Y, Zhang Y, Wang F, Chen Z (2011) Biomed Chromatogr 25:938–942
396. Li Y, Wang F, Chen Z (2011) J Sep Sci 34:514–519
397. Khan A, Khan MI, Iqbal Z, Shah Y, Ahmad L, Nazir S, Watson DG, Khan JA, Nasir F, Khan A (2011) Talanta 84:789–801
398. Khan A, Iqbal Z, Watson DG, Khan A, Khan I, Muhammad N, Muhammad S, Nasib HA, Iqbal N, Faiz-ur-Rehman, Kashif M (2011) J Chromatogr B 879:1725–1731
399. Khamanga SM, Walker RB (2011) Talanta 83:1037–1049
400. Hojo K, Hakamata H, Kusu F (2011) J Chromatogr B 879:751–755
401. Bugamelli F, Marcheselli C, Barba E, Raggi MA (2011) J Pharm Biomed Anal 54:562–567
402. Bartošová Z, Jirovský D, Horna A (2011) J Chromatogr A 1218:7996–8001
403. Kato D, Komoriya M, Nakamoto K, Kurita R, Hirono S, Niwa O (2011) Anal Sci 27:703–707
404. Khan A, Khan MI, Iqbal Z, Ahmad L, Shah Y, Watson DG (2010) J Chromatogr B 878: 2782–2788
405. Hubbard KE, Wells A, Owens TS, Tagen M, Fraga CH, Stewart CF (2010) Biomed Chromatogr 24:626–631
406. Chan SW, Lin G, Yamamoto K, Yew DTW, Rudd JA (2010) J Pharm Biomed Anal 53: 705–709
407. Caceres MIR, Cabanillas AG, Diaz TG, Canas MAM (2010) J Chromatogr B 878:398–402
408. Park HJ, Mah E, Bruno RS (2010) Anal Biochem 407:151–159
409. Ito N, Hakamata H, Kusu F (2010) Anal Methods 2:174–179
410. Pashkova E, Pirogov A, Bendryshev A, Ivanaynen E, Shpigun O (2009) J Pharm Biomed Anal 50:671–674
411. Locatelli I, Mrhar A, Grabnar I (2009) J Pharm Biomed Anal 50:905–910
412. Li X, Franke AA (2009) J Chromatogr B 877:853–856
413. Kotani A, Miyaguchi Y, Kohama M, Ohtsuka T, Shiratori T, Kusu F (2009) Anal Sci 25: 1007–1011
414. Chiu MH, Yang HH, Liu CH, Zen JM (2009) J Chromatogr B 877:991–994
415. Andrade LS, de Moraes MC, Rocha-Filho RC, Fatibello-Filho O, Cass QB (2009) Anal Chim Acta 654:127–132
416. Zhang X, Zhang D-q (2009) Liaoning Zhongyiyao Daxue Xuebao 11:255–257

417. Marszałł ML, Lebiedzińska A, Czarnowski W, Makarowski R, Kłos M, Szefer P (2009) J Chromatogr B 877:3151–3158
418. Brajanoski G, Hoogmartens J, Allegaert K, Adams E (2008) J Chromatogr B 867:149–152
419. Gu Y, Li Q, Melendez V, Weina P (2008) J Chromatogr B 867:213–218
420. Jia P, Wang S, Meng X, Lan W, Luo J, Liao S, Xiao C, Zheng X, Li L, Liu Q, Zheng J, Zhou Y, Zheng X (2013) Talanta 107:103–110
421. Zhou Y, Yan H, Xie Q, Huang S, Liu J, Li Z, Ma M, Yao S (2013) Analyst 138:7246–7253
422. Birbeck JA, Mathews TA (2013) Anal Chem 85:7398–7404
423. Shintani H (2013) Pharmaceut Anal Acta 4:207. doi:10.4172/2153-2435.1000207
424. Fotouhi L, Atoofi Z, Heravi MM (2013) Talanta 103:194–200
425. Asturias-Arribas L, Alonso-Lomillo MA, Domínguez-Renedo O, Arcos-Martínez MJ (2013) Talanta 111:8–12
426. Ensafi AA, Heydari-Bafrooei E, Rezaei B (2013) Biosens Bioelectron 41:627–633
427. Canbay E, Akyilmaz E (2014) Anal Biochem 444:8–15
428. Huang Q, Zhang H, Hu S, Li F, Weng W, Chen J, Wang Q, Bao X (2014) Biosens Bioelectron 52:277–280
429. Baj-Rossi C, Jost TR, Cavallini A, Grassi F, De Micheli G, Carrara S (2014) Biosens Bioelectron 53:283–287
430. Marco JP, Borges KB, Tarley CRT, Ribeiro ES, Pereira AC (2013) Sens Actuators B 177: 251–259

About the Authors

Ahmad Sarakbi received his Master's degree in Sciences in 2003 from Al-Baath University, Homs, Syria and his Ph.D. in Biomedical and Pharmaceutical Sciences in 2014 from the Université libre de Bruxelles. He is currently performing a post doc at the Université libre de Bruxelles at the Faculty of Pharmacy. His research interest includes the development of electrochemical sensors for liquid chromatography.

Ana Maria Oliveira Brett is professor in the Chemistry Department at Coimbra University, Portugal. Her research areas are in bioelectrochemistry, the study of electron transfer reactions of antioxidants to understand the free-radical-induced damaging aspects of the chemistry of disease processes, bioelectroanalysis, and the development of enzymes and DNA biosensors. She is a member of several scientific societies and is presently the President of the International Bioelectrochemical Society (BES). She is Fellow of IUPAC and Royal Society of Chemistry (CChem, FRSC). Her research activity is documented by more than 200 papers and

S.A. Ozkan et al., *Electroanalysis in Biomedical and Pharmaceutical Sciences*, Monographs in Electrochemistry, DOI 10.1007/978-3-662-47138-8

chapters in multiauthor books, coauthor of 2 undergraduate/graduate textbooks "*Electrochemistry. Principles, methods and applications*," 1993 and "*Electroanalysis*," 1998, both Oxford University Press.

Bengi Uslu received Pharmacist degree in 1990. She received her M.Sc. and Ph.D. degrees in the Faculty of Pharmacy in 1993 and 1999, respectively, from Ankara University. She was visiting Prof. Emil Palecek at the Institute of Biophysics, Lab of Biophysical Chemistry, and Molecular Oncology in Czech Republic with The Scientific and Technological Research Council of Turkey's fellowship in 2005. Her research interest focuses on electrochemical drug analysis, nanosensor, and biosensors. She worked there on protein electrochemistry. She had been an Associate Professor at Ankara University from 2003 to 2009. She received Ankara University Scientific Promotion Award in 2008 and Ankara University Scientific Support Award in 2014. Since December 2009, she has been a Professor at the same department. She is author or coauthor of more than 100 published papers.

Christopher Brett is a full professor of chemistry at the University of Coimbra, Portugal. His research interests include new nanostructured electrode materials and modified electrode surfaces, electrochemical sensors and biosensors, electroactive polymers, corrosion and its inhibition, and applications of electrochemistry in the environmental, food, and pharmaceutical areas, on which he has published more than 200 peer-reviewed research articles, as well as book chapters and coauthoring two textbooks. He is currently a member of the editorial board of the journals Electrochimica Acta, Electroanalysis, and Analytical Letters. He was President of the International Society of Electrochemistry (ISE) and is currently a member of the Bureau of the International Union of Pure and Applied Chemistry (IUPAC).

Cobra Parsajoo received her Master degree in analytical sciences in 2009 from the Institute for Advanced Studies in Basic Sciences (IASBS)—Zanjan, Iran. She is currently Ph.D. student in Biomedical and Pharmaceutical Sciences at the Université libre de Bruxelles. Her interest is the development of flow systems with an enzyme immobilized sensor for inhibition studies.

Eric Rozet is Postdoctoral Researcher at the Analytical Chemistry Laboratory of the University of Liege. He has more than 9 years of experience in nonclinical statistics: robust optimization, validation, transfer, and uncertainty assessment. Eric has a B.Sc. degree in Bioengineering and a Ph. D. degree in Pharmaceutical Sciences.

Jean-Michel Kauffmann was born in Arlon, Belgium, in 1954. He became a pharmacist from the University Libre de Bruxelles (ULB), in 1977, and earned a Ph.D. degree in pharmaceutical sciences in 1983 from the same University. He became full Professor in 1991 and Head of the laboratory of Instrumental Analysis and Bioelectrochemistry and of the Laboratory of Toxicology. He was Vice-Dean of Pharmaceutical Institute (2003–2007), Dean of Faculty of Pharmacy of ULB (2007–2011), and Vice-Dean of Faculty of Pharmacy ULB (2011–2013). He was president of the Belgian Society of

Pharmaceutical Sciences (2000–2006) and a member of the Flemish Science Foundation (FWO) Committee (1999–2005), the Belgian National Science Foundation (FNRS) Committee (2006–2009), Academia Iberoamericana de Farmacia, Sevilla Spain, since 1997, National Academy of Pharmacy Paris, France, since 2004, Belgian Royal Academy of Medicine, Brussels, Belgium, since 2013, and Doctor Honoris Causa at the Faculty of Medicine & Pharmacy University of Cluj Napoca Romania (2004).

His research interests include development and characterization of electrochemical (bio)sensors and electrochemical sensors for HPLC analysis of peptides and small molecules, development of portable electrochemical sensors for trace metal analysis, study of the electrochemical behavior of pharmacologically active compounds as a mean for the early screening of drug metabolization (drug stability): CV, LC-EC-MS, and development of miniaturized electrochemical sensors for on-site drug therapy monitoring.

He has published 157 papers and 23 chapters in books in the field of analytical chemistry with special emphasis on electroanalysis, biosensors, and modified electrodes for the investigation of compounds of pharmaceutical interest. He has expertise in the development of electroanalytical methods and in characterization of electrochemical sensors and biosensors with immobilized enzymes and antibodies. He is the Editor-in-Chief of Talanta since 1995 and a member of the Editorial board of: Electroanalysis, Analytical Letters, Talanta, Current Separations, Chimica Acta Turcica, J. Pharm. Belgique, FABAD, J. Pharm. Sciences of Ankara, The Open Anal.Chem. Journal, Anal.Pharm.Française, Arabian J. Chem., Farmacia (Romanian review).

Petr Zuman enrolled at Charles University in Prague in 1945 and graduated in early 1948. After this, he joined the research group of Jaroslav Heyrovský (Nobel prize in chemistry 1959) and was awarded RNDr (equivalent to a Ph.D.) in 1950. He was one of the first researchers in the newly created Central Polarographic Institute in Prague and within a few years he was appointed Head of the Organic Polarography Division. In 1960, Petr was awarded the Doctor of Science degree by the Czechoslovak Academy of Sciences. Following the 1967 political upheaval in Czechoslovakia, Zuman obtained a three-and-one-half-year appointment as a research fellow at the University of Birmingham in England. In 1970, he accepted a position in the Chemistry Department at Clarkson University in Potsdam, New York, where he remained until his retirement. Postretirement Zuman has been a Distungusihed Emeritus Research Professor at Clarkson University and continues to contribute to the chemistry department.

He was interested in reductions of aldehydes, ketones, quinones, nitro-, and nitrosocompounds, as well as in compounds with azomethine (C=N) bonds) and their reactions with various nucleophiles, followed by reductions of single C−OH and C-NR2 bonds activated by adjacent C=O or C=N and C−X bonds. Homogeneous kinetics and equilibria involving additions of H_2O, OH^-, RS^-, CN^-, and primary amines have been followed.

The compounds of pharmaceutical and biological interest were studied barbiturates; thiobarbiturates; uracils and thiouracils; various steroids; vitamins like B2, B6, and B12, as well as ascorbic acid; reactions of 2,3-simercaptopropanol (BAL used for elimination of heavy metal ions from organism); pesticides; herbicides; compounds like Mitomycin C, Abacivir, and sildenafil citrate (Viagra); antibiotics like penicillin, cefetamet, cefamycin, cephalosporins, and cefepime; and bile salts (and their reactions with heavy metal ions). Interactions of polymer lignin with various pesticides were as well investigated as well as its alkaline cleavage yielding useful products. Most recently, the mechanism of reactions of amines with orthophthalaldehyde were investigated, and it was detected that the reaction involves seven interlinked equilibria. The reaction is widely used for determination of amino acids. Absence of understanding of the mechanism leads to empirical choice of reaction conditions. More than 25 different reaction conditions have been reported in the literature, in which recommended concentrations differ in up to two orders of magnitude.

Another active area are studies of electroreduction and electrooxidation of some organic compounds, such as aromatic nitrocompounds, various pesticides including maleic hydrazide, 1,3,5- and 1,2,4-triazines, selenous acid, mitomycin C, and phenols. Structure–reactivity relationships are investigated, for example, the transfer of electronic effects through an N–N single bond, ring formation of some 2-amino-1,4-benzoquinones, or interactions between two reducible groups in a molecule. Polarographic reduction of pesticides has been used to study their adsorption on lignin to determine their bioavailability in applications in forest nurseries. Studies of alkaline cleavage of lignin at room temperature will form a basis for the use of lignin (which is a renewable raw material) for future industrial applications.

He has won numerous awards, including the prestigious 1975 Benedetti–Pichler award given annually by the American Microchemical Society, as well as numerous visiting professorships at institutions around the world. He has been a prolific author with well over 400+ papers and 15 books published, many considered classics in electrochemical analysis. Petr has been a member of the editorial board of the Microchemical Journal, has published on a regular basis, and has critically reviewed many manuscripts and books. He is fluent in five languages—Czech, German, English, Polish, and Serbo-Croat.

Philippe Hubert is Professor of Analytical Chemistry and head of the Department of Pharmaceutical Sciences at the University of Liege. He has published more than 200 peer-reviewed articles. His research focuses on separation sciences, vibrational spectroscopy, method validation, and chemometrics aspects including quality by design.

Sibel A. Ozkan is presently working as a Full Professor of Analytical Chemistry at Ankara University, Faculty of Pharmacy, from where she received her graduate degree in 1985. She obtained her M.Sc. (1988) and Ph.D. (1994) degrees in the same university, Analytical Chemistry Department. She has become full professor in 2003. She has been involved in several analytical chemistry projects related to electroanalytical chemistry and separation techniques on drug active compounds and DNA–Drug interactions. Her research interests are centered on fundamental aspects in the areas, analysis of pharmaceuticals with using separation techniques especially on liquid chromatography, method development and their validation, electrochemical techniques, novel electrode materials, nanostructured materials, surface-modified electrodes, analysis of pharmaceuticals from their dosage forms, and biological samples.

She received Ankara University Scientific Support Award in 2003 and in 2008 Academy of Pharmacy Science Award of Turkish Pharmaceutical Association.

She is the Editorial Board Member of the many journals like Talanta, Journal of Pharmaceutical and Biomedical Analysis, Chromatographia, Current Drug Therapy, The Open Analytical Chemistry Journal, Global Journal Of Analytical Chemistry, The Open Chemical And Biomedical Methods Journal, Journal Of Analytical Methods In Chemistry, Journal Of Analytical And Pharmaceutical Chemistry, The Open Chemical And Biomedical Methods Journal, Journal of Turkish Pharmaceutical Sciences, etc.

She was the Vice Editor of the Special Issue of *Chromatographia* Vol. 66, 2007, "*Application of Separation Techniques in Turkey*" with Dr. Henk Lingeman (Editor of Chromatographia). She was Guest Editor of the Special Issue of *Combinatorial Chemistry* and *High Throughput Screening "Applications of Analytical Methods: Prospects for High Throughput Screening of Pharmaceutically Active Compounds*," Volume 13, issues 6; 7; 8, 2010 and for the Special Issue of *The Open Chemical And Biomedical Methods Journal* 2010 (volume 3). She was also Guest Editor for the Special Issue of *Chromatographia* (SPRINGER), "New Developments for the Sensitive and Validated Drug Analysis Using Chromatographic Methods," published in 2012.

She has published more than 170 original and review papers and scientific book namely "Electroanalytical methods in pharmaceutical analysis and their validation" published in 2012.

Stéphanie Patris received her Master in Pharmaceutical Sciences degree in 2003 and her Ph.D. in Biomedical and Pharmaceutical Sciences in 2014 both from the Université libre de Bruxelles, Belgium. She is now assistant professor at the Faculty of Pharmacy at the Université libre de Bruxelles, Belgium. Her research interest includes development of biosensors and affinity sensors for point of care testing.

About the Editor

Fritz Scholz is Professor at the University of Greifswald, Germany. Following studies of chemistry at Humboldt University, Berlin, he obtained a Dr. rer. nat. and a Dr. sc. nat. (habilitation) from that University. In 1987 and 1989, he worked with Alan Bond in Australia. His main interest is in electrochemistry and electroanalysis. He has published more than 300 scientific papers, and he is editor and coauthor of the book "Electroanalytical Methods" (Springer, 2002, 2005, 2010, and Russian Edition: BINOM, 2006), coauthor of the book "Electrochemistry of Immobilized Particles and Droplets" (Springer 2005), coeditor of the "Electrochemical Dictionary" (Springer, 2008; 2nd ed. 2012), and coeditor of volumes 7a and 7b of the "Encyclopedia of Electrochemistry" (Wiley-VCH 2006). In 1997, he has founded the *Journal of Solid State Electrochemistry* (Springer) and serves as Editor-in-Chief since that time. In 2014, he has founded the journal *ChemTexts—The Textbook Journal* (Springer). He is the editor of the series "Monographs in Electrochemistry" (Springer) in which modern topics of electrochemistry are presented. Scholz introduced the technique "Voltammetry of Immobilized Microparticles" for studying the electrochemistry of solid compounds and materials; he introduced three-phase electrodes to determine the Gibbs energies of ion transfer between immiscible liquids, and currently he is studying the interaction of free oxygen radicals with metal surfaces, as well as the interaction of liposomes with the surface of mercury electrodes in order to assess membrane properties. In the last years, his research also comprises the interaction of electrode surfaces with free oxygen radicals.

S.A. Ozkan et al., *Electroanalysis in Biomedical and Pharmaceutical Sciences*, Monographs in Electrochemistry, DOI 10.1007/978-3-662-47138-8

Index

S.A. Ozkan et al., *Electroanalysis in Biomedical and Pharmaceutical Sciences*, Monographs in Electrochemistry, DOI 10.1007/978-3-662-47138-8

MIX
Papier aus verantwortungsvollen Quellen
Paper from responsible sources
FSC® C105338
FSC
www.fsc.org

If you have any concerns about our products,
you can contact us on
ProductSafety@springernature.com

In case Publisher is established outside the EU,
the EU authorized representative is:
Springer Nature Customer Service Center GmbH
Europaplatz 3, 69115 Heidelberg, Germany

Printed by Libri Plureos GmbH
in Hamburg, Germany